1750

LINEAR ALGEBRA
AND ITS
APPLICATIONS

LINEAR ALGEBRA
AND ITS
APPLICATIONS

GILBERT STRANG
Massachusetts Institute of Technology

ACADEMIC PRESS   New York   San Francisco   London

A Subsidiary of Harcourt Brace Jovanovich, Publishers

ACADEMIC PRESS, INC.
111 Fifth Avenue, New York, New York 10003

United Kingdom Edition published by
ACADEMIC PRESS, INC. (LONDON) LTD.
24/28 Oval Road, London NW1

Library of Congress Cataloging in Publication Data

Strang, Gilbert.
    Linear algebra and its applications

    Bibliography: p.
    Includes index.
    1.   Algebras, Linear.   I.   Title.
QA184.S8   512$'$.5   75-26637
ISBN 0−12−673650-2

AMS (MOS) 1970 Subject Classifications: 15-01, 52-01, 65-01

CONTENTS

4 DETERMINANTS

5 EIGENVALUES AND EIGENVECTORS

6 POSITIVE DEFINITE MATRICES

7 COMPUTATIONS WITH MATRICES

8 LINEAR PROGRAMMING AND GAME THEORY

PREFACE

I believe that the teaching of linear algebra has become too abstract. This is a sweeping judgment, and perhaps it is too sweeping to be true. But I feel certain that a text can explain the essentials of linear algebra, and develop the ability to reason mathematically, without ignoring the fact that *this subject is as useful and central and applicable as calculus.* It has a simplicity which is too valuable to be sacrificed.

Of course there are good reasons for the present state of courses in linear algebra: The subject is an excellent introduction to the precision of a mathematical argument, and to the construction of proofs. These virtues I recognize and accept (and hope to preserve); I enjoyed teaching in exactly this way. Nevertheless, once I began to experiment with alternatives at M.I.T., another virtue became equally important: Linear algebra allows and even encourages a very satisfying combination of both elements of mathematics—abstraction and application.

As it is, too many students struggle with the abstraction and never get to see the application. And too many others, especially those who are outside mathematics departments, never take the course. Even our most successful students tend to become adept at abstraction, but inept at any calculation—solving linear equations by Cramer's rule, for example, or understanding eigenvalues only as roots of the characteristic equation. There is a growing desire to make our teaching more useful than that, and more open.

We hope to treat linear algebra in a way which makes sense to a wide variety of students at all levels. This does not imply that we have written a cookbook; the subject deserves better than that. It does imply less concentration on rigor for its own sake, and more on understanding—*we try to explain rather than to deduce.* Some definitions are formal, but others are allowed to come to the surface in the middle of a discussion. In the same way, some proofs are intended to be orderly and precise, but not all. In every case the underlying theory has to be there; it is the core of the subject, but it can be motivated and reinforced by examples.

One specific difficulty in constructing the course is always present, and is hard to postpone: How should it start? Most students come to the first class already knowing something about linear equations. Nevertheless, we are convinced that linear algebra must begin with the fundamental problem of n equations in n unknowns, and that it must teach the simplest and most useful method of solution—Gaussian elimination (not determinants!). Fortunately, even though this method is simple, there are a number of insights that are central to its understanding and new to almost every student. The most important is the equivalence between elimination and matrix factorization; the coefficient matrix is transformed into a product of triangular matrices. This provides a perfect introduction to matrix notation and matrix multiplication.

The other difficulty is to find the right speed. If matrix calculations are already familiar, then *Chapter 1 must not be too slow*; the next chapter is the one which demands hard work. Its goal is a genuine understanding, deeper than elimination can give, of the equation $Ax = b$. I believe that the introduction of four fundamental subspaces—the column space of A; the row space; and their orthogonal complements, the two nullspaces—is an effective way to generate examples of linear dependence and independence, and to illustrate the ideas of basis and dimension and rank. The orthogonality is also a natural extension to n dimensions of the familiar geometry of three-dimensional space. And of course those four subspaces are the key to $Ax = b$.

Chapters 1–5 are really the heart of a course in linear algebra. They contain a large number of applications to physics, engineering, probability and statistics, economics, and biology. (There is also the geometry of a methane molecule, and even an outline of factor analysis in psychology, which is the one application that my colleagues at M.I.T. refuse to teach!) At the same time, you will recognize that this text can certainly not explain every possible application of matrices. It is simply a first course in linear algebra. Our goal is not to develop all the applications, but to prepare for them—and that preparation can only come by understanding the theory.

This theory is well established. After the vector spaces of Chapter 2, we study projections and inner products in Chapter 3, determinants in Chapter 4, and eigenvalues in Chapter 5. I hope that engineers and others will look especially at Chapter 5, where we concentrate on the uses of diagonalization (including the spectral theorem) and save the Jordan form for an appendix. Each chapter is followed by a set of review exercises, and is so organized that its last section is optional; this applies also to Section 3.4 on pseudoinverses. In a one-semester or a one-quarter course, the instructor must decide whether the positive definite matrices of Chapter 6 or the linear programs of Chapter 8 are the more essential to his class; I believe that Sections 8.1 and 8.4 will allow a brief but worthwhile introduction to linear programming and game theory.

The book also contains the essentials of three quite different courses. One is on numerical linear algebra, which involves all of Chapter 1, the essential facts of Chapters 2 to 6, and then Chapter 7 on computations and Section 8.2 on the simplex method. Another is "linear algebra for statistics," which must treat Chapters 3 and 6 much more completely. And the third possibility is to regard inequalities as coequal with equations, as economists do, and go as quickly as possible from $Ax = b$ to linear programming and duality.

We should like to ask one favor of the mathematician who simply wants to teach basic linear algebra. That is the true purpose of the book, and we hope he will not be put off by the "operation counts," and the other remarks about numerical computation, which arise especially in Chapter 1. From a practical viewpoint these comments are obviously important. Also from a theoretical viewpoint they have a serious purpose—to reinforce a detailed grasp of the elimination sequence, by actually counting the steps. I normally ask the class to make this count during the first or second lecture, with completely unpredictable results. But there is no need to discuss this or any other computer-oriented topic in class; any text ought to supplement as well as summarize the lectures.

In short, a book is needed that will permit the applications to be taught successfully, in combination with the underlying mathematics. That is the book I have tried to write.

For help in writing it, I take this special opportunity to give thanks to Tom Slobko for his encouragement, to Ursula for typing everything with such gentle grace, and to my family who are precious above all. Beyond this there is an earlier debt, which I can never fully repay. It is to my parents, and I now dedicate the book to them, hoping that they will understand how much they gave to it: Thank you both.

GILBERT STRANG

$$\begin{bmatrix} 1 \end{bmatrix}$$

GAUSSIAN
ELIMINATION

The central problem of linear algebra is the solution of simultaneous linear equations. The most important case, and the simplest, is when the number of unknowns equals the number of equations. Therefore we begin with this problem: *n equations in n unknowns*.

Two ways of solving simultaneous equations are proposed, almost in a sort of competition, from high school texts on. The first is the method of ***elimination***: Multiples of the first equation in the system are subtracted from the other equations, in such a way as to remove the first unknown from those equations. This leaves a smaller system, of $n - 1$ equations in $n - 1$ unknowns. The process is repeated over and over until there remains only one equation and one unknown, which can be solved immediately. Then it is not hard to go backward, and find all the other unknowns in reverse order; we shall work out an example in a moment. A second and more sophisticated way introduces the idea of ***determinants***. There is an exact formula, called Cramer's rule, which gives the solution (the correct values of the unknowns) as a ratio of two n by n determinants. It is not always obvious from the examples that are worked in a textbook ($n = 3$ or $n = 4$ is about the upper limit on the patience of a reasonable human being) which way is better.

In fact, the more sophisticated formula involving determinants is a disaster, and elimination is the algorithm that is constantly used to solve large systems of simultaneous equations. Our first goal is to understand this algorithm. It is generally called *Gaussian elimination*.

The algorithm is deceptively simple, and in some form it may already be familiar to the reader. But there are four aspects that lie deeper than the simple mechanics of elimination, and which—together with the algorithm itself—we want to explain in this chapter. They are:

(1)   The interpretation of the elimination method as a factorization of the coefficient matrix. We shall introduce *matrix notation* for the system of simultaneous equations, writing the n unknowns as a vector x and the n equations in the matrix shorthand $Ax = b$. Then *elimination amounts to factoring A into a product LU of a lower triangular matrix L and an upper triangular matrix U.* This is a basic and very useful observation.

Of course, we have to introduce matrices and vectors in a systematic way, as well as the rules for their multiplication. We also define the transpose A^{T} and the inverse A^{-1} of a matrix A.

(2)   In most cases the elimination method works without any difficulties or modifications. In some exceptional cases it breaks down—either because the equations were originally written in the wrong order, which is easily fixed by exchanging them, or else because the equations $Ax = b$ fail to have a unique solution. In the latter case there may be no solution, or infinitely many. We want to understand how, at the time of breakdown, the elimination process identifies each of these possibilities.

(3)   It is essential to have a rough count of the *number of arithmetic operations* required to solve a system by elimination. In many practical problems the decision of how many unknowns to introduce—balancing extra accuracy in a mathematical model against extra expense in computing—is governed by this operation count.

(4)   We also want to see, intuitively, how sensitive to *roundoff error* the solution x might be. Some problems are sensitive; others are not. Once the source of difficulty becomes clear, it is easy to guess how to try to control it. Without control, a computer could carry out millions of operations, rounding each result to a fixed number of digits, and produce a totally useless "solution."

The final result of this chapter will be an elimination algorithm which is about as efficient as possible. It is essentially the algorithm that is in constant use in a tremendous variety of applications. And at the same time, understanding it in terms of matrices—the coefficient matrix, the matrices that carry out an elimination step or an exchange of rows, and the final triangular factors L and U—is an essential foundation for the theory.

1.2 ■ AN EXAMPLE OF GAUSSIAN ELIMINATION

The way to understand this subject is by example. We begin in three dimensions with the system

$$
\begin{aligned}
2u + v + w &= 1 \\
4u + v \phantom{{}+ w} &= -2 \\
-2u + 2v + w &= 7.
\end{aligned}
\tag{1}
$$

The problem is to find the unknown values of u, v, and w, and we shall apply Gaussian elimination. (Gauss is recognized as the greatest of all mathematicians, but certainly not because of this invention, which probably took him ten minutes. Ironically, however, it is the most frequently used of all the ideas that bear his name.) The method starts by *subtracting multiples of the first equation from the others, so as to eliminate u from the last two equations.* This requires that we

(a)   subtract 2 times the first equation from the second;

(b)   subtract -1 times the first equation from the third.

The result is an equivalent system of equations

$$2u + v + w = 1$$
$$-1v - 2w = -4 \tag{2}$$
$$3v + 2w = 8.$$

The coefficient 2, which multiplied the first unknown u in the first equation, is known as the *pivot* in this first elimination step.

At the second stage of elimination, we ignore the first equation. The other two equations involve only the two unknowns v and w, and the same elimination procedure can be applied to them. *The pivot for this stage is* -1, and a multiple of this second equation will be subtracted from the remaining equations (in this case there is only the third one remaining) so as to eliminate the second unknown v. This means that we

(c)   subtract -3 times the second equation from the third.

The elimination process is now complete, at least in the "forward" direction, and leaves the simplified system

$$2u + v + w = 1$$
$$-1v - 2w = -4 \tag{3}$$
$$-4w = -4.$$

There is an obvious order in which to solve this system. The last equation gives $w = 1$; substituting into the second equation, we find $v = 2$; then the first equation gives $u = -1$. This simple process is called ***back-substitution***.

It is easy to understand how the elimination idea can be extended to n equations in n unknowns, no matter how large the system may be. At the first stage, we use multiples of the first equation to annihilate all coefficients below the first pivot. Next, the second column is cleared out below the second pivot; and so on. Finally, the last equation contains only the last unknown. Back-substitution yields the answer in the opposite order, beginning with the last unknown, then solving for the next to last, and eventually for the first.

EXERCISE 1.2.1   Apply elimination and back-substitution to solve

$$
\begin{aligned}
2u - 3v \quad\quad &= 3 \\
4u - 5v + \; w &= 7 \\
2u - \; v - 3w &= 5.
\end{aligned}
$$

What are the pivots? List the three operations in which a multiple of one row is subtracted from another.

EXERCISE 1.2.2   Solve the system

$$
\begin{aligned}
2u - \; v \quad\quad\quad\quad &= 0 \\
-u + 2v - \; w \quad\quad &= 0 \\
- \; v + 2w - \; z &= 0 \\
- \; w + 2z &= 5.
\end{aligned}
$$

We want to ask two questions. They may seem a little premature—after all, we have barely got the algorithm working—but their answers will shed more light on the method itself. The first question is whether this elimination procedure always leads to the solution. *Under what circumstances could the process break down?* The answer is: If none of the pivots are zero, there is only one solution to the problem and it is found by forward elimination and back-substitution. But if any of the pivots happen to be zero, the elimination technique cannot proceed.

If the first pivot were zero, for example, the elimination of u from the other equations would be impossible. The same is true at every intermediate stage. Notice that an intermediate pivot may become zero during the elimination process (as in Exercise 1.2.3 below) even though in the original system the coefficient in that place was not zero. Roughly speaking, *we do not know whether the pivots are nonzero until we try*, by actually going through the elimination process.

In most cases this problem of a zero pivot can be cured, and elimination can proceed to find the unique solution to the problem. In other cases, a breakdown is unavoidable since the equations have either no solution or infinitely many. We postpone to a later section the analysis of breakdown.

The second question is very practical, in fact it is financial. *How many separate arithmetical operations does elimination require for a system of n equations in n unknowns?* If n is large, a computer is going to take our place in carrying out the elimination (you may have such a program available, or be able to write one) but since all the steps are known in advance, we should be able to predict how long the computer will take. For the moment we ignore the right-hand sides of the equations, and count only the operations on the left. These operations are of two kinds. One is a division by the pivot in order to find out what multiple (say l) of the pivotal equation is to be subtracted from an equation below it. Then when we actually do this subtraction of one equation from another, we continually meet a "multiply–subtract" combination; the terms in the pivotal equation are multiplied by l, and then subtracted from the equation beneath it.

Suppose we agree to call each division, and each multiplication–subtraction, a single operation. At the beginning, when the first equation has length n, *it takes n operations for every zero we achieve in the first column*—one to find the multiple l, and the others to find the new entries along the row. There are $n - 1$ rows underneath the first one, and therefore $n - 1$ zeros to be produced below the first pivot, so *the first stage of elimination needs $n(n - 1) = n^2 - n$ operations*. After that stage, the first column is set. Now notice that later stages are faster because the equations are becoming progressively shorter; at the second stage we are working with only $n - 1$ equations in $n - 1$ unknowns. When the elimination is down to k equations, only $k(k - 1) = k^2 - k$ operations are needed to clear out the column below the pivot—by the same reasoning that applied to the first stage, when k equaled n. Altogether, therefore, *the total number of arithmetical operations on the left side of the equations is*

$$P = (n^2 - n) + \cdots + (k^2 - k) + \cdots + (1^2 - 1).$$

(Notice that there was no work to do at the last stage, $1^2 - 1 = 0$, when we are down to one equation in one unknown.) This sum P is known to equal

$$P = \sum_1^n k^2 - \sum_1^n k = \frac{1}{3} n \left(n + \frac{1}{2} \right) (n + 1) - \frac{1}{2} n (n + 1) = \frac{n^3 - n}{3}.$$

(Here calculus is actually useful as a check. The integral of x^2 from 0 to n is $n^3/3$, and the integral of x is $n^2/2$. These are exactly the leading terms in the two sums.) If n is at all large, *a very good estimate for the number of operations is $P \approx n^3/3$.*

Back-substitution is considerably faster. The last unknown is found from one operation (a division by the last pivot), the second to last unknown requires two (a multiplication–subtraction and then a division), and so on. The kth step involves only k operations. Therefore, back-substitution requires altogether

$$Q = \sum_{k=1}^n k = \frac{1}{2} n (n + 1) \approx \frac{n^2}{2} \quad \text{operations.}$$

A few years ago, almost every mathematician would have guessed that these numbers were essentially optimal, in other words that a general system of order n could not be solved with much fewer than $n^3/3$ multiplications. (There were even theorems to demonstrate it, but they did not allow for all possible methods.) Astonishingly, that guess has been proved wrong, and *there now exists a method that requires only $Cn^{\log_2 7}$ operations*! Fortunately for elimination, the constant C is by comparison so large, and so many more additions are required, and the computer programming is so awkward, that the new method is largely of theoretical interest. It seems to be completely unknown whether the exponent can be made any smaller.

EXERCISE 1.2.3   Apply elimination to the system

$$u + v + w = -2$$
$$3u + 3v - w = 6$$
$$u - v + w = -1.$$

When a zero pivot arises, exchange that equation for the one below it, and proceed. What coefficient of v in the third equation, in place of the present -1, would make it impossible to proceed—and force the elimination method to break down?

EXERCISE 1.2.4   For a system of two equations like

$$au + bv = 0$$
$$cu + dv = 1$$

list explicitly the $P = 2$ individual operations that are applied to the left side.

EXERCISE 1.2.5   With reasonable assumptions on computer speed and cost, how large a system can be solved for $1, and for $1000? Use $n^3/3$ as the operation count, and you might pay $1000 an hour for a computer that could average a million operations a second.

EXERCISE 1.2.6   (very optional)   Normally the multiplication of two complex numbers

$$(a + ib)(c + id) = (ac - bd) + i(bc + ad)$$

involves the four separate multiplications ac, bd, bc, ad. Ignoring i, can you compute the quantities $ac - bd$ and $bc + ad$ with only three multiplications? (You may do additions, such as forming $a + b$ before multiplying, without any penalty.)

EXERCISE 1.2.7   Use elimination to solve

$$u + v + w = 6$$
$$u + 2v + 2w = 11$$
$$2u + 3v - 4w = 3.$$

1.3 ■ MATRIX NOTATION AND MATRIX MULTIPLICATION

So far, with our 3 by 3 example, we have been able to write out all the equations in full. We could even list in detail each of the elimination steps, subtracting a multiple of one row from another, which put the system of equations into a simpler form. For a large system, however, this way of keeping track of the elimination would be hopeless; a much more concise record is needed. We shall now introduce matrix notation to describe the original system of equations, and matrix multiplication to describe the operations that make it simpler.

Notice that in our example

$$2u + v + w = 1$$
$$4u + v \qquad = -2$$
$$-2u + 2v + w = 7$$

three different types of quantities appear. There are the unknowns u, v, w; there are the right sides 1, -2, 7; and finally, there is a set of nine numerical coefficients on the left side (one of which happens to be zero). For the column of numbers on the right side—the *inhomogeneous terms* in the equations—we introduce the vector notation

$$b = \begin{bmatrix} 1 \\ -2 \\ 7 \end{bmatrix}.$$

This is a three-dimensional column vector. To represent it geometrically, we can take its three components as the coordinates of a point in three-dimensional space. Then every point in the space is matched with a three-dimensional vector (which we may visualize as an arrow, or a directed line segment, which starts at the origin and ends at the point).

The basic operations are the addition of two such vectors and the multiplication of a vector by a scalar. Geometrically, $2b$ is a vector in the same direction as b but twice as long; $-2b$ goes in the opposite direction; and $b + c$ is found by placing the starting point of the vector c at the end point of b. Algebraically, this just means that vector operations are carried out *component by component*:

$$2b = 2 \begin{bmatrix} 1 \\ -2 \\ 7 \end{bmatrix} = \begin{bmatrix} 2 \\ -4 \\ 14 \end{bmatrix}, \qquad -2b = \begin{bmatrix} -2 \\ 4 \\ -14 \end{bmatrix},$$

$$b + c = \begin{bmatrix} 1 \\ -2 \\ 7 \end{bmatrix} + \begin{bmatrix} 1 \\ -4 \\ -4 \end{bmatrix} = \begin{bmatrix} 2 \\ -6 \\ 3 \end{bmatrix}.$$

Two vectors can be added only if they have the same dimension, that is, the same number of components.

The three unknowns in the equation are also represented by a vector:

$$\text{the unknown is} \quad x = \begin{bmatrix} u \\ v \\ w \end{bmatrix}; \quad \text{the solution is} \quad x = \begin{bmatrix} -1 \\ 2 \\ 1 \end{bmatrix}.$$

Again these are three-dimensional column vectors. For the array of nine coefficients, we introduce a "matrix" with three rows and three columns. This is the *coefficient matrix*

$$A = \begin{bmatrix} 2 & 1 & 1 \\ 4 & 1 & 0 \\ -2 & 2 & 1 \end{bmatrix}.$$

Notice that, because the number of equations in our example agrees with the number of unknowns, A is a *square matrix* (of order three). More generally, we might have n equations in n unknowns—with a square coefficient matrix of order n. Or still more generally, we might have m equations and n unknowns. In this case the coefficient matrix will be rectangular, with m rows and n columns—in other words, it will be an "m by n matrix."

Matrices are added to each other, or multiplied by numerical constants, exactly as vectors are—one component at a time. In fact we may regard vectors as special cases of matrices; they are matrices with only one column. As before, two matrices can be added only if they have the same shape:

$$\begin{bmatrix} 2 & 1 \\ 3 & 0 \\ 0 & 4 \end{bmatrix} + \begin{bmatrix} 1 & 2 \\ -3 & 1 \\ 1 & 2 \end{bmatrix} = \begin{bmatrix} 3 & 3 \\ 0 & 1 \\ 1 & 6 \end{bmatrix}, \qquad 2\begin{bmatrix} 2 & 1 \\ 3 & 0 \\ 0 & 4 \end{bmatrix} = \begin{bmatrix} 4 & 2 \\ 6 & 0 \\ 0 & 8 \end{bmatrix}.$$

Multiplication of a Matrix and a Vector

Now we put this notation to use. We propose to rewrite the system (1) of three equations in three unknowns in the simplified matrix form $Ax = b$. Written out in full, this form is

$$\begin{bmatrix} 2 & 1 & 1 \\ 4 & 1 & 0 \\ -2 & 2 & 1 \end{bmatrix} \begin{bmatrix} u \\ v \\ w \end{bmatrix} = \begin{bmatrix} 1 \\ -2 \\ 7 \end{bmatrix}.$$

The right side is clear enough; it is the column vector of inhomogeneous terms. The left side consists of the vector x, premultiplied by the matrix A. Obviously, this multiplication will be defined *exactly so as to reproduce the original system* (1). Therefore, the first component of the product Ax must come from "multiplying" the first row of A into the column vector x:

$$\begin{bmatrix} 2 & 1 & 1 \end{bmatrix} \begin{bmatrix} u \\ v \\ w \end{bmatrix} = \begin{bmatrix} 2u + v + w \end{bmatrix}. \tag{4}$$

This equals the first component of b; $2u + v + w = 1$ is the first equation in our system. The second component of the product Ax is determined by the second row of A—it is $4u + v$—and the third component $-2u + 2v + w$ comes from the third row. Thus the matrix equation $Ax = b$ is precisely equivalent to the three simultaneous equations with which we started.

The operation in Eq. (4) is fundamental to all matrix multiplications. It starts with a row vector and a column vector of matching lengths, and it produces a single number. This single quantity is called the *inner product* or *scalar product* or *dot product* of the two vectors. In other words, the product of a 1 by n matrix, which is a *row vector*, and an n by 1 matrix, alias a *column vector*, is a 1 by 1 matrix:

$$\begin{bmatrix} a_1 & a_2 & \cdots & a_n \end{bmatrix} \begin{bmatrix} b_1 \\ b_2 \\ \vdots \\ b_n \end{bmatrix} = \begin{bmatrix} a_1b_1 + a_2b_2 + \cdots + a_nb_n \end{bmatrix}.$$

EXAMPLE

$$\begin{bmatrix} 2 & 4 & 1 \end{bmatrix} \begin{bmatrix} 3 \\ -1 \\ 0 \end{bmatrix} = \begin{bmatrix} 2 \end{bmatrix}.$$

The rule for multiplying a vector by a matrix extends immediately from our problem of order three to the general case of an n by n matrix. The vector must have n components, to match the length of the rows of A. Using letters rather than specific values for the components of A and x, the product is

$$Ax = \begin{bmatrix} a_{11} & a_{12} & \cdots & a_{1n} \\ a_{21} & a_{22} & \cdots & a_{2n} \\ \vdots & \vdots & \cdots & \vdots \\ a_{n1} & a_{n2} & \cdots & a_{nn} \end{bmatrix} \begin{bmatrix} x_1 \\ x_2 \\ \vdots \\ x_n \end{bmatrix} = \begin{bmatrix} a_{11}x_1 + \cdots + a_{1n}x_n \\ a_{21}x_1 + \cdots + a_{2n}x_n \\ \vdots + \cdots + \vdots \\ a_{n1}x_1 + \cdots + a_{nn}x_n \end{bmatrix}.$$

Notice the notation for the entries a_{ij} of a matrix. The first subscript gives the row number, and the second subscript is the column number. Thus, the entry in row i and column j is a_{ij}. Notice also that we could use a summation notation

$$\sum_{j=1}^{n} a_{ij}x_j$$

to describe the ith component of the product Ax. Such summations are simpler to work with than writing out everything in full, but they are not as good as matrix notation itself.

EXAMPLE

$$A = \begin{bmatrix} 2 & 3 \\ 4 & 0 \end{bmatrix}, \qquad x = \begin{bmatrix} 1 \\ 5 \end{bmatrix}.$$

$$Ax = \begin{bmatrix} (2)(1) + (3)(5) \\ (4)(1) + (0)(5) \end{bmatrix} = \begin{bmatrix} 17 \\ 4 \end{bmatrix}. \tag{5}$$

In this multiplication, Ax *is actually a combination of the two columns of A, each column weighted by the corresponding entry of x.* In other words, Ax can be found all at once, instead of an entry at a time, in the following way:

$$Ax = \begin{bmatrix} 2 \\ 4 \end{bmatrix}(1) + \begin{bmatrix} 3 \\ 0 \end{bmatrix}(5) = \begin{bmatrix} 17 \\ 4 \end{bmatrix}. \tag{6}$$

This rule will be used over and over throughout the book, and therefore we repeat it for emphasis. The product Ax can be found from the individual entries, as in (5), or *it can be computed by using whole columns at once*:

$$\begin{bmatrix} 2 & 1 & 1 \\ 4 & 1 & 0 \\ -2 & 2 & 1 \end{bmatrix}\begin{bmatrix} -1 \\ 2 \\ 1 \end{bmatrix} = (-1)\begin{bmatrix} 2 \\ 4 \\ -2 \end{bmatrix} + 2\begin{bmatrix} 1 \\ 1 \\ 2 \end{bmatrix} + \begin{bmatrix} 1 \\ 0 \\ 1 \end{bmatrix}.$$

The product Ax is a combination of the columns of A. In this example, the combination turns out to be equal to

$$b = \begin{bmatrix} 1 \\ -2 \\ 7 \end{bmatrix}, \quad \text{so that} \quad x = \begin{bmatrix} -1 \\ 2 \\ 1 \end{bmatrix}$$

is confirmed as the solution of $Ax = b$.

EXERCISE 1.3.1   Compute, first entry by entry and then column by column, the product

$$Ax = \begin{bmatrix} 4 & 0 & 1 \\ 0 & 1 & 0 \\ 4 & 0 & 1 \end{bmatrix}\begin{bmatrix} 3 \\ 4 \\ 5 \end{bmatrix}.$$

EXERCISE 1.3.2   Compute the products

$$Ax = \begin{bmatrix} 1 & 2 \\ 3 & -3 \\ 0 & 4 \\ 0 & 1 \end{bmatrix}\begin{bmatrix} 1 \\ -1 \end{bmatrix} \quad \text{and} \quad \begin{bmatrix} -4 & 1 & 3 \end{bmatrix}\begin{bmatrix} -4 \\ 1 \\ 3 \end{bmatrix}.$$

If an m by n matrix A multiplies an n-dimensional vector x, what is the size and shape of Ax?

EXERCISE 1.3.3   Suppose that, comparing the beginning and end of 1977,

(a)   of those who started the year in California, 80 percent stayed in and 20 percent moved out;

(b)   of those who started the year outside California, 90 percent stayed out and 10 percent moved in.

Following these laws, express the following questions in "matrix form" and answer them.

(i) If there are 30 million inside and 200 million outside at the beginning of the year, what is the situation at the end of the year?

(ii) If there are 30 million inside and 200 million outside at the end of the year, what was the situation at the beginning of the year?

(iii) What special distribution at the beginning of the year would come out unchanged at the end of the year? In other words, how should the number u originally inside be related to the number v originally outside, in order to end the year with the same u and v?

EXERCISE 1.3.4 Draw a pair of perpendicular axes and mark off the points $x = 2$, $y = 1$ and $x = 0$, $y = 3$. Put in the vectors from the origin to those two points, and also the one to their sum $\begin{bmatrix} 2 \\ 1 \end{bmatrix} + \begin{bmatrix} 0 \\ 3 \end{bmatrix}$. Complete the parallelogram.

May we add two remarks to Exercise 1.3.3? The first is to emphasize the *linearity* of the problem. If the numbers were doubled to 60 and 400 million, all the answers would be doubled. Or if we solved the problem also for 40 million and 250 million, then the solutions for 70 million and 450 million would be exactly the sum of the two separate solutions.

The second remark applies to part (ii) of the exercise, where you were to compute the number u of people originally inside California and the number v originally outside. You must have needed two equations, and probably you set them up separately—first an equation like $.8u + .1v = 30$, to ensure that 30 million would end the year inside, and then an equation to produce the 200 million outside. Our remark is this: The "column at a time" idea would be to think of these two equations simultaneously. The goal is to produce the vector $b = \begin{bmatrix} 30 \\ 200 \end{bmatrix}$, giving the number inside and the number outside. The u people originally inside will end the year distributed like $\begin{bmatrix} .8u \\ .2u \end{bmatrix}$, and the v originally outside end up as $\begin{bmatrix} .1v \\ .9v \end{bmatrix}$. These must combine to produce b:

$$u \begin{bmatrix} .8 \\ .2 \end{bmatrix} + v \begin{bmatrix} .1 \\ .9 \end{bmatrix} = \begin{bmatrix} 30 \\ 200 \end{bmatrix}.$$

So, from this "column at a time" point of view, we are computing the weights u and v which should multiply the two columns on the left in order to produce the desired column b on the right. In part (iii), the desired column is $\begin{bmatrix} u \\ v \end{bmatrix}$ itself! Then the year ends as it started.

The Matrix Form of One Elimination Step

So far we have a convenient shorthand $Ax = b$ for the original system of equations. What about the operations that are carried out during elimination? In our example, the first step was to subtract 2 times the first equation from the second. On the right side of the equation, this means that 2 times the first component of b was subtracted from the second component, and *we claim that this same result is achieved if we multiply the vector b by the following special* *matrix:*

$$E = \begin{bmatrix} 1 & 0 & 0 \\ -2 & 1 & 0 \\ 0 & 0 & 1 \end{bmatrix}.$$

This is verified just by obeying the rule for multiplying a matrix and a vector:

$$Eb = \begin{bmatrix} 1 & 0 & 0 \\ -2 & 1 & 0 \\ 0 & 0 & 1 \end{bmatrix} \begin{bmatrix} 1 \\ -2 \\ 7 \end{bmatrix} = \begin{bmatrix} 1 \\ -4 \\ 7 \end{bmatrix}.$$

The first and third components, 1 and 7, stayed the same (because of the form we chose for the first and third rows of E). The new second component is the correct value -4; it appeared in Eq. (2), after the first elimination step.

To maintain equality we must apply the same operation to both sides of $Ax = b$. In other words, we must premultiply the vector Ax by the matrix E. Again this subtracts 2 times the first component from the second, leaving the first and third components unchanged. After this step the new and simpler system (equivalent to the old) is just $E(Ax) = Eb$. It is simpler because of the zero that was created below the first pivot; it is equivalent because we can recover the original system just by adding 2 times the first equation back to the second. So the two systems have exactly the same solution x. This is the justification for Gaussian elimination; the solution does not change as the steps are carried out, but it becomes progressively easier to find.

Matrix Multiplication

Now we ask, What is the coefficient matrix in the new system $E(Ax) = Eb$? The single step of Gaussian elimination, which is represented by the matrix E, carries the original coefficient matrix

$$A = \begin{bmatrix} 2 & 1 & 1 \\ 4 & 1 & 0 \\ -2 & 2 & 1 \end{bmatrix}$$

into the new matrix

$$EA = \begin{bmatrix} 2 & 1 & 1 \\ 0 & -1 & -2 \\ -2 & 2 & 1 \end{bmatrix}$$

Again, the first and third rows are unchanged, while 2 times the first row has been subtracted from the second. *The new coefficient matrix combines the effect of first multiplying x by the original coefficient matrix A, and then multiplying Ax by E.* This new matrix is called the product of the two old ones, and it is denoted by EA. Notice that we maintain the order in which the matrices appear. To repeat:

1A The product EA is a single matrix that gives the same result as first multiplying by A and then by E: For every vector x,

$$(EA)x = E(Ax). \tag{7}$$

This rule will produce, in Fig. 1.1 and in Eq. (8), the laws of matrix multiplication.

Suppose we are given any pair of matrices E and A, possibly rectangular; how do we use this rule to compute their product EA? First, the shapes of E and A have to be right for the product to make sense. If they are square matrices, as in our example, then they must be of the same size. If E and A are rectangular, they must *not* have the same shape; *the number of columns in E has to equal the number of rows in A*. In other words, if E is l by m, and A is m by n—E has m columns and A has m rows—then the multiplications are all possible. They start with an n-vector x, produce an m-vector Ax, and then finally a vector EAx with l components. (We omit parentheses in EAx because the rule (7) was chosen exactly so as to make them unnecessary.) The product EA has l rows, like E, and n columns like A : l by m times m by n leaves l by n.

We now have to compute what EA actually is. The usual way is to describe each individual entry, for example the entry in row i and column j: *This entry is the inner product of the ith row of E and the jth column of A*. The last row and column are indicated in Fig. 1.1, and their product gives the last entry in EA:

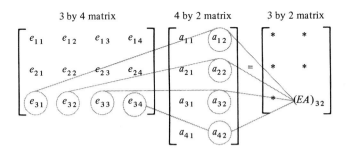

Fig. 1.1. An illustration of matrix multiplication.

$$(EA)_{32} = \begin{bmatrix} e_{31} & e_{32} & e_{33} & e_{34} \end{bmatrix} \begin{bmatrix} a_{12} \\ a_{22} \\ a_{32} \\ a_{42} \end{bmatrix} = e_{31}a_{12} + e_{32}a_{22} + e_{33}a_{32} + e_{34}a_{42}. \tag{8}$$

EXAMPLE 1

$$\begin{bmatrix} 2 & 3 \\ 4 & 0 \end{bmatrix} \begin{bmatrix} 1 & 2 & 0 \\ 5 & -1 & 0 \end{bmatrix} = \begin{bmatrix} 17 & 1 & 0 \\ 4 & 8 & 0 \end{bmatrix}.$$

The entry 17, for instance, is $(2)(1) + (3)(5)$, the inner product of the first row and first column. The entry 8 is $(4)(2) + (0)(-1)$, from the second row and second column. The third column is zero in A, and therefore it is zero in EA.

EXAMPLE 2

$$\begin{bmatrix} 0 & 1 \\ 1 & 0 \end{bmatrix} \begin{bmatrix} 2 & 3 \\ 7 & 8 \end{bmatrix} = \begin{bmatrix} 7 & 8 \\ 2 & 3 \end{bmatrix}.$$

This matrix E produces a row exchange in A.

EXERCISE 1.3.5 Find the product of

$$E = \begin{bmatrix} 1 & 0 & 0 \\ -2 & 1 & 0 \\ 1 & 0 & 1 \end{bmatrix} \quad \text{and} \quad A = \begin{bmatrix} 2 & 1 & 1 \\ 4 & 1 & 0 \\ -2 & 2 & 1 \end{bmatrix}.$$

EXERCISE 1.3.6 Find the product of

$$E = \begin{bmatrix} 1 & 0 & 0 \\ -2 & 1 & 0 \\ -5 & 3 & 1 \end{bmatrix} \quad \text{and} \quad A = \begin{bmatrix} 2 & 1 & 1 \\ 4 & 1 & 0 \\ -2 & 2 & 1 \end{bmatrix}.$$

EXERCISE 1.3.7 Multiply a 1 by 2 matrix and a 2 by 1 matrix, $E = \begin{bmatrix} 1 & -4 \end{bmatrix}$ and $A = \begin{bmatrix} 4 \\ 1 \end{bmatrix}$. If E is l by m and A is m by n, why is it that lmn separate multiplications are involved in EA?

EXERCISE 1.3.8 Multiply a 3 by 2 matrix and a 2 by 1 matrix:

$$E = \begin{bmatrix} 8 & -3 \\ -5 & 2 \\ 1 & 0 \end{bmatrix}, \quad A = \begin{bmatrix} 2 \\ 1 \end{bmatrix}.$$

Notice that the last exercise was just like multiplying a matrix by a vector; *the rule for matrix multiplication EA agrees with the original rule for computing Ax.*

Furthermore, just as for Ax, the product EA can be computed using whole columns at once. For instance, the first column of EA is exactly E times the first column of A. Look again at our example

$$\begin{bmatrix} 2 & 3 \\ 4 & 0 \end{bmatrix} \begin{bmatrix} 1 & 2 & 0 \\ 5 & -1 & 0 \end{bmatrix} = \begin{bmatrix} 17 & 1 & 0 \\ 4 & 8 & 0 \end{bmatrix}.$$

Taking the product column by column,

$$\begin{bmatrix} 2 & 3 \\ 4 & 0 \end{bmatrix} \begin{bmatrix} 1 \\ 5 \end{bmatrix} = \begin{bmatrix} 17 \\ 4 \end{bmatrix}, \quad \begin{bmatrix} 2 & 3 \\ 4 & 0 \end{bmatrix} \begin{bmatrix} 2 \\ -1 \end{bmatrix} = \begin{bmatrix} 1 \\ 8 \end{bmatrix}, \quad \begin{bmatrix} 2 & 3 \\ 4 & 0 \end{bmatrix} \begin{bmatrix} 0 \\ 0 \end{bmatrix} = \begin{bmatrix} 0 \\ 0 \end{bmatrix}.$$

This illustrates an important point: *The jth column of EA depends only on the jth column of A, not on any of the other columns.* Furthermore, this column of EA is a combination of all the columns of E, weighted by the numbers a_{ij} in that jth column of A. The rule for multiplication of two matrices actually becomes an extension of the earlier rule for multiplying a matrix and a column vector, with only a simple change: Instead of one column, there are now several columns side by side.

This observation is useful in verifying one of the most important properties of matrix multiplication. Suppose we are given three matrices A, B, and C, possibly rectangular, and suppose their shapes permit them to be multiplied in that order: The number of columns in A and B match, respectively, the number of rows in B and C. Then the property is this:

1B   Matrix multiplication is associative: $(AB)C = A(BC)$.

If C happens to be just a vector, in other words a matrix with just one column, then this property is identical with (7): It was the whole basis for our rule of matrix multiplication. And if C has several columns $c_1, \ldots, c_n$, we have only to apply this rule several times. On the one hand, the first column of $(AB)C$ is $(AB)c_1$; on the other hand, the first column of BC is Bc_1, and then the first column of $A(BC)$ is $A(Bc_1)$. We know that $(AB)c_1 = A(Bc_1)$, by the rule (7), and the same applies to every other column.

EXERCISE 1.3.9   Check the associative law for

$$A = \begin{bmatrix} 1 & 0 & 0 \\ 0 & 1 & 0 \\ 1 & 0 & 1 \end{bmatrix}, \quad B = \begin{bmatrix} 1 & 0 & 0 \\ -2 & 1 & 0 \\ 0 & 0 & 1 \end{bmatrix}, \quad C = \begin{bmatrix} 2 & 1 & 1 \\ 4 & 1 & 0 \\ -2 & 2 & 1 \end{bmatrix}.$$

We want to get on with the connection between matrix multiplication and Gaussian elimination, but there are two more properties to mention first—one property that matrix multiplication has, and another which it *does not have*. The property that it does possess is:

1C   Matrix operations are distributive:

$$A(B + C) = AB + AC \quad \text{and} \quad (B + C)D = BD + CD.$$

Of course the shapes of these matrices must be properly matched—B and C have the same shape, so they can be added, and A and D are the right size for premultiplication and postmultiplication. The proof of this law is too boring for words.

The property that fails to hold is a little more interesting:

1D   Matrix multiplication is not commutative: Usually $FE \neq EF$.

EXAMPLE   Suppose E is the matrix introduced earlier, whose effect was to subtract twice the first equation from the second:

$$E = \begin{bmatrix} 1 & 0 & 0 \\ -2 & 1 & 0 \\ 0 & 0 & 1 \end{bmatrix}.$$

Suppose F is the matrix we would meet at the last step of elimination, when -3 times the second equation is subtracted from the third (or 3 times the second equation is added to the third):

$$F = \begin{bmatrix} 1 & 0 & 0 \\ 0 & 1 & 0 \\ 0 & 3 & 1 \end{bmatrix}.$$

Then we compare

$$FE = \begin{bmatrix} 1 & 0 & 0 \\ -2 & 1 & 0 \\ -6 & 3 & 1 \end{bmatrix} \quad \text{with} \quad EF = \begin{bmatrix} 1 & 0 & 0 \\ -2 & 1 & 0 \\ 0 & 3 & 1 \end{bmatrix}.$$

Obviously, the order makes a difference. In the first case, where we apply E to the system of equations and then F, the second equation is altered by twice the first row *before* it is used to change the third equation; consequently the entry -6 appears in the (3, 1) position. This is the order actually met in elimination. If, on the contrary, F is applied first, the entry -6 does not appear: In this order the third equation feels no effect from the first. So $FE \neq EF$.

The Identity Matrix

There is one important n by n matrix that does commute with any n by n matrix. In fact it leaves the other matrix unchanged; it acts like multiplication by 1. The matrix with this property is called the *identity matrix*; it has 1's along the main diagonal, and 0's everywhere else:

$$I = \begin{bmatrix} 1 & 0 & 0 & 0 \\ 0 & 1 & 0 & 0 \\ 0 & 0 & 1 & 0 \\ 0 & 0 & 0 & 1 \end{bmatrix}, \quad \text{when} \quad n = 4.$$

It is easy to verify that $IA = AI = A$ for any matrix A of the same order.

EXERCISE 1.3.10 Show that the only matrices $A = \begin{bmatrix} a & b \\ c & d \end{bmatrix}$ that commute with both

$$B = \begin{bmatrix} 1 & 0 \\ 0 & 0 \end{bmatrix} \quad \text{and} \quad C = \begin{bmatrix} 0 & 1 \\ 0 & 0 \end{bmatrix}$$

are multiples of the identity: $a = d$ and $b = c = 0$. Verify that these A commute with every 2 by 2 matrix, and are therefore the only matrices to do so.

EXERCISE 1.3.11 Find examples of 2 by 2 matrices such that

(a) $A^2 = -I$, A having only real entries;
(b) $B^2 = 0$, although $B \neq 0$;
(c) $CD = -DC$, not allowing the case $CD = 0$;
(d) $EF = 0$, although no entries of E or F are zero.

EXERCISE 1.3.12 True or false; give a specific counterexample when false.

(a) If the first and third columns of B are the same, so are the first and third columns of AB.
(b) If the first and third rows of B are the same, so are the first and third rows of AB.
(c) If the first and third rows of A are the same, so are the first and third rows of AB.
(d) $(AB)^2 = A^2B^2$.

EXERCISE 1.3.13 The first row of AB is a linear combination of all the rows of B. What are the weights in this combination, and what is the first row of AB, if

$$A = \begin{bmatrix} 2 & 1 & 4 \\ 0 & -1 & 1 \end{bmatrix}, \quad B = \begin{bmatrix} 1 & 1 \\ 0 & 1 \\ 1 & 0 \end{bmatrix} ?$$

GAUSSIAN ELIMINATION = TRIANGULAR FACTORIZATION ■ 1.4

We want to look again at Gaussian elimination and see what it means in terms of matrices. The starting point was the system $Ax = b$:

$$Ax = \begin{bmatrix} 2 & 1 & 1 \\ 4 & 1 & 0 \\ -2 & 2 & 1 \end{bmatrix} \begin{bmatrix} u \\ v \\ w \end{bmatrix} = \begin{bmatrix} 1 \\ -2 \\ 7 \end{bmatrix}.$$

Then there were three elimination steps:

(i)   Subtract 2 times the first equation from the second;
(ii)   Subtract -1 times the first equation from the third;
(iii)   Subtract -3 times the second equation from the third.

The result was an equivalent but simpler system, with a new coefficient matrix which we denote by U:

$$Ux = \begin{bmatrix} 2 & 1 & 1 \\ 0 & -1 & -2 \\ 0 & 0 & -4 \end{bmatrix} \begin{bmatrix} u \\ v \\ w \end{bmatrix} = \begin{bmatrix} 1 \\ -4 \\ -4 \end{bmatrix}. \tag{9}$$

The coefficient matrix is now **upper triangular**—all the entries below the main diagonal are zero.

The right side, which is a new vector c, was derived from the original vector b by the same steps that took A into U. Thus, Gaussian elimination amounted to:

Start with A and b;
Apply steps (i), (ii), (iii), in that order;
End with U and c.

The last stage is the solution of $Ux = c$ by back-substitution, but for the moment we are not concerned with that; we concentrate on the relation between A and U.

The matrix E that accomplishes step (i), subtracting twice the first row from the second, was already introduced in the previous section. It will be denoted now by E_{21}, indicating by the subscripts that it changes row 2 by a multiple of row 1, and also that it produces a zero in the $(2, 1)$ entry of the coefficient matrix. Recall that, because the multiple of the first row was 2, the matrix that subtracted this multiple was

$$E_{21} = \begin{bmatrix} 1 & 0 & 0 \\ -2 & 1 & 0 \\ 0 & 0 & 1 \end{bmatrix}.$$

Similarly, the elimination steps (ii) and (iii) can be described in matrix terms; we multiply by the matrices

$$E_{31} = \begin{bmatrix} 1 & 0 & 0 \\ 0 & 1 & 0 \\ 1 & 0 & 1 \end{bmatrix} \quad \text{and} \quad E_{32} = \begin{bmatrix} 1 & 0 & 0 \\ 0 & 1 & 0 \\ 0 & 3 & 1 \end{bmatrix}.$$

These are called *elementary matrices*, and it is easy to see how they work in general. To subtract a multiple l_{ij} of equation j from equation i, we form E_{ij} in this way: Start from the identity matrix I, and replace the zero in row i, column j, by $-l_{ij}$. Notice that all these elementary matrices are **lower triangular**, and they all have 1's on the main diagonal.

The three matrix operations that take A into U are therefore

$$E_{32}E_{31}E_{21}A \;=\; U. \tag{10}$$

Similarly, since the same operations apply to the inhomogeneous terms,

$$E_{32}E_{31}E_{21}b \;=\; c. \tag{11}$$

We could, if we wanted to, multiply the E's together to find a single matrix that takes A and b into U and c:

$$E_{32}E_{31}E_{21} \;=\; \begin{bmatrix} 1 & 0 & 0 \\ -2 & 1 & 0 \\ -5 & 3 & 1 \end{bmatrix}. \tag{12}$$

Notice that it is again lower triangular; it appeared in Exercise 1.3.6. The associative law is very useful here: The E's can be multiplied together, and then multiplied by A, with the same result as starting with A and using each E in turn. Notice also the entry -5, in the bottom left corner. This was not one of the multipliers in the elimination process; steps (i) and (iii) combined to give a -6 and step (ii) contributed $+1$.

Now comes the important question: How would we get back from U to A? *How can we undo the steps of Gaussian elimination?*

A single step, say step (i), is not hard to undo. Instead of subtracting, we *add* twice the first row to the second. (Not twice the second row to the first!) In this way, the elementary matrix E_{21} is *inverted* by another elementary matrix, with $+2$ in the same location which previously contained -2:

$$E_{21}{}^{-1} \;=\; \begin{bmatrix} 1 & 0 & 0 \\ 2 & 1 & 0 \\ 0 & 0 & 1 \end{bmatrix}.$$

The product of E_{21} and $E_{21}{}^{-1}$, taken in either order, is the identity; one operation cancels the other. In matrix terms, each matrix is the *inverse of the other*:

$$E_{21}{}^{-1}E_{21} = I, \qquad E_{21}E_{21}{}^{-1} = I. \tag{13}$$

Similarly, the second and third elementary matrices can be inverted by adding back what was subtracted in steps (ii) and (iii):

$$E_{31}{}^{-1} = \begin{bmatrix} 1 & 0 & 0 \\ 0 & 1 & 0 \\ -1 & 0 & 1 \end{bmatrix}, \qquad E_{32}{}^{-1} = \begin{bmatrix} 1 & 0 & 0 \\ 0 & 1 & 0 \\ 0 & -3 & 1 \end{bmatrix}.$$

We now have the matrices that undo each of the separate elimination steps (i), (ii), and (iii). The next problem is to undo the whole process at once, and see what matrix takes U back to A. Notice that, *since step (iii) was the last* *one to be taken in going from A to U, it must be the first one to be inverted when we go in the reverse direction.* This is a general rule: Inverses come in the opposite order from the original sequence of operations. (It will be repeated when inverses are discussed more systematically in the next section.) The second step to be inverted is (ii), and the last is step (i). The result is, starting from U, to bring back the matrix A:

$$A = E_{21}^{-1}E_{31}^{-1}E_{32}^{-1}U. \tag{14}$$

The reader can mentally substitute $U = E_{32}E_{31}E_{21}A$ into this equation, and see that the order of the matrices is right; all the products like $E_{32}^{-1}E_{32}$ are simply the identity.

Now we can compute the matrix L that takes U back to A. It has to be the product of the three matrices that reverse the individual steps, and it is already present in Eq. (14):

$$L = E_{21}^{-1}E_{31}^{-1}E_{32}^{-1}, \quad \text{so that} \quad A = LU. \tag{15}$$

This matrix L and the matrix $E_{32}E_{31}E_{21}$, which took A into U, are inverses of of one another:

$$L = (E_{32}E_{31}E_{21})^{-1} \quad \text{or} \quad L^{-1} = E_{32}E_{31}E_{21}. \tag{16}$$

We now have the key formulas, but the most important fact about L can be seen only by multiplying it out:

$$L = \begin{bmatrix} 1 & 0 & 0 \\ 2 & 1 & 0 \\ 0 & 0 & 1 \end{bmatrix} \begin{bmatrix} 1 & 0 & 0 \\ 0 & 1 & 0 \\ -1 & 0 & 1 \end{bmatrix} \begin{bmatrix} 1 & 0 & 0 \\ 0 & 1 & 0 \\ 0 & -3 & 1 \end{bmatrix} = \begin{bmatrix} 1 & 0 & 0 \\ 2 & 1 & 0 \\ -1 & -3 & 1 \end{bmatrix}. \tag{17}$$

Certainly L is lower triangular, with 1's on the main diagonal. The special thing is that *the entries below the diagonal are exactly the multipliers* 2, -1, *and* -3 *used in the three elimination steps.* Normally we expect, when three matrices are multiplied together, that there will be no way to read off the answer directly from the individual entries in the three matrices. But in the present case, the matrices come in just the right order so that their product can be written down immediately. If a computer stores each multiplier l_{ij}—the quantity that multiplies row j when it is subtracted from row i to produce a zero in the (i, j) entry—then these multipliers not only form a complete record of Gaussian elimination, *they fit right into the matrix L that takes U back to A.* Remember also that this must be the same matrix that takes c back to b: $Lc = b$.

We emphasize that this did not happen for the matrix product $E_{32}E_{31}E_{21}$, computed in (12). Multiplied in this order, there was a cancellation of terms and a resulting entry of -5 in the (3, 1) place.

EXERCISE 1.4.1   Multiply the matrix in (12) by L to verify that this matrix is actually L^{-1}, the inverse of L. It is only L, not its inverse, that can be written down at sight.

We should try to understand why, for a system of any size, the order is right for the matrix product L; the multipliers fit in without any cancellation. Consider what the order would be in the case of 4 by 4 matrices:

$$L = E_{21}{}^{-1}E_{31}{}^{-1}E_{41}{}^{-1}E_{32}{}^{-1}E_{42}{}^{-1}E_{43}{}^{-1}I.$$

The identity matrix is put on the far right mostly for convenience; no harm is done, and it allows us to think of the operations as being applied to the rows of the matrix I. The first operation is $E_{43}{}^{-1}$, which multiplies the third row $[0, 0, 1, 0]$ by l_{43} and adds it to the fourth row; that puts l_{43} in the right place. Then $E_{42}{}^{-1}$ and $E_{32}{}^{-1}$ add back multiples of the second row $[0, 1, 0, 0]$ to the third and fourth row—which puts in l_{42} and l_{32} without disturbing l_{43}. And finally the first column of L is filled in by the three operations $E_{41}{}^{-1}$, $E_{31}{}^{-1}$, and $E_{21}{}^{-1}$.

It is exactly these same operations, in this order, that would recover the original system $Ax = b$ from the upper triangular system $Ux = c$. They add back the multiples that were subtracted off to produce U. Since their combined effect is to multiply by L, the matrix form of Gaussian elimination is easy to summarize:

1E As long as no pivots are zero, the matrix A can be written as a product LU of a lower triangular matrix L and an upper triangular matrix U. The entries of L on the main diagonal are 1's; below the diagonal, they are the multipliers l_{ij} of row j which are subtracted from row i during elimination. U is the coefficient matrix which appears after elimination and before back-substitution; its diagonal entries are the pivots.

EXAMPLE

$$A = \begin{bmatrix} 1 & 2 \\ 3 & 4 \end{bmatrix} \quad \text{goes into} \quad U = \begin{bmatrix} 1 & 2 \\ 0 & -2 \end{bmatrix}, \quad \text{with} \quad L = \begin{bmatrix} 1 & 0 \\ 3 & 1 \end{bmatrix}.$$

EXERCISE 1.4.2 Apply elimination to produce the factors L and U for

$$A = \begin{bmatrix} 2 & 1 \\ 8 & 7 \end{bmatrix} \quad \text{and} \quad A = \begin{bmatrix} 1 & 0 \\ 8 & 1 \end{bmatrix}.$$

EXERCISE 1.4.3 Factor A into LU, and write down the upper triangular system $Ux' = c$ which appears after elimination, for

$$Ax = \begin{bmatrix} 2 & 3 & 3 \\ 0 & 5 & 7 \\ 6 & 9 & 8 \end{bmatrix} \begin{bmatrix} u \\ v \\ w \end{bmatrix} = \begin{bmatrix} 2 \\ 2 \\ 5 \end{bmatrix}.$$

The original system of equations, which was $Ax = b$ or $LUx = b$, is transformed by elimination into $Ux = c$. In matrix terms it is multiplied by L^{-1}, and therefore $c = L^{-1}b$. (But note that we did not explicitly form, and in actual computation *never should form*, the matrix L^{-1}. The elimination was done step by step, before L^{-1} was ever heard of.) Finally, the upper triangular system $Ux = c$ is solved by back-substitution; the last equation gives $x_n = c_n/u_{nn}$, then the next to last equation yields x_{n-1}, and so on. In matrix terms this solution is

$$x = U^{-1}c, \quad \text{or} \quad x = U^{-1}L^{-1}b, \quad \text{or} \quad x = A^{-1}b. \quad (18)$$

Again the matrix U^{-1} is not formed explicitly, and never should be. It would be a waste of time since even if we knew this matrix completely, it would still take the same $n^2/2$ operations to multiply c by U^{-1} as are needed in back-substitution. A similar remark applies to almost every application of A^{-1}; it is the solution $x = A^{-1}b$ which we want, and not all the entries in the inverse.

This does not mean that if we had a new vector on the right side, say b', we should repeat the whole elimination process. It often occurs in design problems that the same matrix A appears with a number of different inhomogeneous terms b, and it is essential to avoid repeating the $n^3/3$ operations of elimination. Instead, the computer simply looks at the record it has kept of the elimination steps and applies them to the new right-hand side b'; it need not apply them all over again to A because the result will only be the same U.

To summarize: Once the factors L and U have been computed, *the solution x' for any new right side b' can be found in only n^2 operations—$n^2/2$ in the forward direction to find c', and $n^2/2$ in back-substitution to find x'.*

EXAMPLE The matrix A in our original equation had the following LU factorization, copying the U from Eq. (9) and the L from (17):

$$A = \begin{bmatrix} 2 & 1 & 1 \\ 4 & 1 & 0 \\ -2 & 2 & 1 \end{bmatrix} = \begin{bmatrix} 1 & 0 & 0 \\ 2 & 1 & 0 \\ -1 & -3 & 1 \end{bmatrix} \begin{bmatrix} 2 & 1 & 1 \\ 0 & -1 & -2 \\ 0 & 0 & -4 \end{bmatrix} = LU.$$

Suppose we were given the same A but a new right-hand side b', for example

$$2u' + v' + w' = 8$$
$$4u' + v' = 11 \quad (19)$$
$$-2u' + 2v' + w' = 3.$$

Then, knowing L and U, we first solve $Lc' = b'$:

$$\begin{bmatrix} 1 & 0 & 0 \\ 2 & 1 & 0 \\ -1 & -3 & 1 \end{bmatrix} \begin{bmatrix} c_1' \\ c_2' \\ c_3' \end{bmatrix} = \begin{bmatrix} 8 \\ 11 \\ 3 \end{bmatrix}, \quad \text{or} \quad c' = \begin{bmatrix} 8 \\ -5 \\ -4 \end{bmatrix}.$$

This is exactly what we would get on the right side after elimination in (19); but having once done the elimination and having recorded L, this was much faster. Now the usual back-substitution in $Ux' = c'$ produces the solution

$$\begin{bmatrix} 2 & 1 & 1 \\ 0 & -1 & -2 \\ 0 & 0 & -4 \end{bmatrix} \begin{bmatrix} u' \\ v' \\ w' \end{bmatrix} = \begin{bmatrix} 8 \\ -5 \\ -4 \end{bmatrix}, \quad \text{or} \quad x' = \begin{bmatrix} 2 \\ 3 \\ 1 \end{bmatrix}.$$

Once the LU factorization is known, the problem is reduced to two triangular systems—in other words, to a forward-substitution and a back-substitution.

We want to add two more remarks on these triangular factors:

(1)   The LU form is "unsymmetric" in one respect: U has the pivots along its main diagonal, where L always has 1's. This is easy to correct, by factoring from U a diagonal matrix D made up entirely of the pivots $d_1, d_2, \ldots, d_n$:

$$U = \begin{bmatrix} d_1 & & & \\ & d_2 & & \\ & & \ddots & \\ & & & d_n \end{bmatrix} \begin{bmatrix} 1 & u_{12}/d_1 & u_{13}/d_1 & \cdot \\ & 1 & u_{23}/d_2 & \cdot \\ & & \ddots & \cdot \\ & & & 1 \end{bmatrix}.$$

Now the triangular decomposition of A is written as $A = LDU$: L is lower triangular with 1's on the diagonal, U is upper triangular with 1's on the diagonal, and D is the diagonal matrix of pivots. (It is conventional, though sometimes confusing, to go on denoting this new upper triangular matrix by the same letter U.) The $A = LDU$ decomposition treats L and U evenly.

Our example, with the same L and U that were written out just above, becomes

$$A = \begin{bmatrix} 1 & 0 & 0 \\ 2 & 1 & 0 \\ -1 & -3 & 1 \end{bmatrix} \begin{bmatrix} 2 & 1 & 1 \\ 0 & -1 & -2 \\ 0 & 0 & -4 \end{bmatrix} = \begin{bmatrix} 1 & 0 & 0 \\ 2 & 1 & 0 \\ -1 & -3 & 1 \end{bmatrix} \begin{bmatrix} 2 & & \\ & -1 & \\ & & -4 \end{bmatrix} \begin{bmatrix} 1 & \frac{1}{2} & \frac{1}{2} \\ 0 & 1 & 2 \\ 0 & 0 & 1 \end{bmatrix}$$

$$= LDU.$$

(2)   We may have given the impression, in describing each step of the elimination process, that there was no freedom to do the calculations in a different order. That is wrong; there is *some* freedom, and there is a "Crout algorithm" which arranges the calculations in a slightly different way. But there is certainly not complete freedom since row operations in a random order could easily destroy at one step the zeros that were created at a previous step. And also, *there is no freedom in the final L, D, and U.* That is our main point:

1F If $A = L_1 D_1 U_1$ and $A = L_2 D_2 U_2$, where the L's are lower triangular with unit diagonal, the U's are upper triangular with unit diagonal, and the D's are diagonal matrices with no zeros on the diagonal, then $L_1 = L_2$, $D_1 = D_2$, $U_1 = U_2$. The LDU factorization is uniquely determined by A.

Proof We are given that $L_1 D_1 U_1 = L_2 D_2 U_2$. We must use the fact that L_1^{-1} has the same properties (lower triangular, unit diagonal) as L_1; both are just products of elementary matrices. Similarly there exists a U_2^{-1}, upper triangular with unit diagonal, such that $U_2 U_2^{-1} = I$. And obviously any diagonal matrix like D_1 has an inverse that is also diagonal:

$$\begin{bmatrix} d_1 & & & \\ & d_2 & & \\ & & \ddots & \\ & & & d_n \end{bmatrix}^{-1} = \begin{bmatrix} 1/d_1 & & & \\ & 1/d_2 & & \\ & & \ddots & \\ & & & 1/d_n \end{bmatrix}.$$

Therefore, premultiplying by L_1^{-1} and D_1^{-1}, and postmultiplying by U_2^{-1}, our equation becomes

$$U_1 U_2^{-1} = D_1^{-1} L_1^{-1} L_2 D_2.$$

The left side is a product of two upper triangular matrices with unit diagonal. Such a product must be another matrix of the same kind. On the other hand, the right side is a *lower* triangular matrix. This forces both sides to be just the identity matrix—the only matrix that is at the same time upper triangular with unit diagonal, and also lower triangular. Thus $U_1 U_2^{-1} = I$, and after multiplication by U_2, we have $U_1 = U_2$.

Similarly $L_1 = L_2$, and finally $D_1 = D_2$.

EXERCISE 1.4.4 Assuming no zero pivots, find the *LDU* factorization of a general 2 by 2 matrix $A = \begin{bmatrix} a & b \\ c & d \end{bmatrix}$.

EXERCISE 1.4.5 Find the factors L, D, and U for

$$A = \begin{bmatrix} 2 & -1 & 0 \\ -1 & 2 & -1 \\ 0 & -1 & 2 \end{bmatrix}, \qquad b = \begin{bmatrix} 6 \\ 0 \\ -6 \end{bmatrix}.$$

What is the intermediate vector c during elimination, and what is the solution to $Ax = b$?

EXERCISE 1.4.6 For two systems of order $n = 150$, sharing the same coefficient matrix A, why is the solution of the second system 50 times cheaper than the first?

EXERCISE 1.4.7 Solve $Ax = b$, knowing the factors

$$L = \begin{bmatrix} 1 & 0 & 0 \\ -1 & 1 & 0 \\ 0 & -1 & 1 \end{bmatrix}, \qquad U = \begin{bmatrix} 1 & -1 & 0 \\ 0 & 1 & -1 \\ 0 & 0 & 1 \end{bmatrix}, \qquad b = \begin{bmatrix} 2 \\ -3 \\ 4 \end{bmatrix}.$$

Forward elimination is the same as $Lc = b$, and back-substitution is $Ux = c$.

ROW EXCHANGES, INVERSES, AND ROUNDOFF ERRORS ■ 1.5

We now have to face a problem which has so far been avoided, the possible appearance of a zero pivot. This problem might occur at the very beginning of elimination; the top left entry a_{11} might be zero, and therefore useless as a pivot. We take the simplest example:

$$\begin{bmatrix} 0 & 2 \\ 3 & 4 \end{bmatrix} \begin{bmatrix} u \\ v \end{bmatrix} = \begin{bmatrix} b_1 \\ b_2 \end{bmatrix}. \tag{20}$$

The difficulty is clear; no multiple of the first equation can be used to annihilate the coefficient 3.

The remedy is equally clear: Exchange the two equations, so that the coefficient 3 moves to the top left and becomes the pivot. In this simple case the matrix would then be upper triangular already; it would itself be U, and its lower triangular factor L would be the identity matrix I. The system

$$3u + 4v = b_2$$

$$2v = b_1$$

can be solved immediately by back-substitution.

The reordering can be described in matrix terms, once we identify the matrix P that exchanges the two rows. This ***permutation matrix*** is

$$P = \begin{bmatrix} 0 & 1 \\ 1 & 0 \end{bmatrix},$$

and its product with A does exchange the rows:

$$PA = \begin{bmatrix} 0 & 1 \\ 1 & 0 \end{bmatrix} \begin{bmatrix} 0 & 2 \\ 3 & 4 \end{bmatrix} = \begin{bmatrix} 3 & 4 \\ 0 & 2 \end{bmatrix}.$$

Of course P has the same effect on b, exchanging b_1 and b_2.

Consider a more general case. Suppose that several steps have been completed, and then a zero pivot appears. (On p. 167, we shall be able to give a precise formula for the pivots, but elimination will still be the best way to evaluate this formula and discover whether or not the pivot is actually zero.) We may suppose, for example, that the first $k - 1$ pivots were nonzero, and elimination has produced zeros below these pivots. Now what happens if the kth pivot is zero?

A zero pivot raises two possibilities: The trouble may be easy to fix, or it may be serious. Either there is a later equation that can be exchanged with the kth in order to provide a nonzero pivot, or else none of these other equations can help. In other words, we look at the kth column below the zero pivot; if some entry is nonzero, say in row l, then an exchange of rows k and l allows elimination to proceed. If the column is entirely zero, from the pivot on down, then the matrix is ***singular*** (and so was the original matrix A with which the elimination began). Elimination has pinpointed this singularity; if we look at the tail end of the system, there are too few unknowns left to match the number of remaining equations. We should find the unknowns k through n appearing in equations k through n, but because of the zeros in column k, that unknown is not to be found. This is the situation to be discussed in the next chapter; either the system has infinitely many solutions, or (more likely) no solution at all.

Notice that if all the pivots are nonzero except the last, we are certainly in this singular case; below the nth pivot, there is nowhere to look.

Example of a successful sequence with pivoting (the nonsingular case):

$$
\begin{bmatrix} 1 & 3 & 2 \\ 2 & 6 & 9 \\ 2 & 8 & 8 \end{bmatrix} \rightarrow \begin{bmatrix} 1 & 3 & 2 \\ 0 & 0 & 5 \\ 0 & 2 & 4 \end{bmatrix} \rightarrow \begin{bmatrix} 1 & 3 & 2 \\ 0 & 2 & 4 \\ 0 & 0 & 5 \end{bmatrix}.
$$

Example of an unsuccessful sequence (the singular case):

$$
\begin{bmatrix} 1 & 3 & 2 \\ 2 & 6 & 9 \\ 3 & 9 & 8 \end{bmatrix} \rightarrow \begin{bmatrix} 1 & 3 & 2 \\ 0 & 0 & 5 \\ 0 & 0 & 2 \end{bmatrix}.
$$

EXERCISE 1.5.1   Solve, by pivoting when necessary,

$$
\begin{aligned}
u + 4v + 2w &= -2 \\
-2u - 8v + 3w &= 32 \\
v + w &= 1.
\end{aligned}
$$

EXERCISE 1.5.2   Carry out elimination on the singular system

$$
\begin{aligned}
u + v &= b_1 \\
3u + 3v &= b_2
\end{aligned}
$$

and determine for which vectors b a solution exists.

In this section we stay with the nonsingular case, which means that a nonzero pivot can be found by an exchange of rows. (Of course, a later pivot might again be zero, and require still another row exchange; this might even happen several times, though it is very improbable. And we note that a computer need not necessarily take time to shift whole rows; each exchange could be just a matter of bookkeeping.) Our question is this: *What becomes of the LU (or LDU) factorization, when there are exchanges of rows?*

Certainly we arrive at an upper triangular matrix U, just as before. That is the object of elimination. But now the process involves not only matrices E_{ij} which subtract multiples of one row from another, but also permutation matrices P_{kl}. The matrix P_{kl}, when it premultiplies another matrix, exchanges rows k and l of that matrix. (The form of P_{kl} can be found by a neat trick. Let the matrix it multiplies be the identity I; then $P_{kl}I = P_{kl}$ is the identity matrix with rows k and l exchanged.) As an example,

$$P_{24} = \begin{bmatrix} 1 & 0 & 0 & 0 \\ 0 & 0 & 0 & 1 \\ 0 & 0 & 1 & 0 \\ 0 & 1 & 0 & 0 \end{bmatrix}.$$

It is equally easy to compute the inverse matrix P_{kl}^{-1}, which undoes the permutation P_{kl}; we have simply to exchange the two rows again. Therefore $P_{kl}^{-1} = P_{kl}$.

Now the matrix that leads from U back to A could be computed as before; it is a product of the E_{ij}^{-1} and also the P_{kl}^{-1} which invert each individual elimination step. Because of the permutations, however, this product would not be a lower triangular L; if zero pivots arise and row exchanges are required, *the original matrix A could not be factored into triangular L and U with nonzero diagonals.* That factorization is lost.

Nevertheless, there is a simple observation which almost allows us to recapture $A = LU$. Suppose we look at the list of row exchanges that are needed during elimination, and *carry them out before elimination starts.* In other words, we replace the original A by a permuted matrix PA; the matrix P is still a permutation matrix, with a single 1 in every row and column. In fact, P is a product of the elementary permutation matrices P_{kl}, and combines all the row exchanges at once.

The observation is this: *Elimination can be carried out on the matrix PA, which has its rows put into the right order ahead of time.* The pivots are exactly the same ones that were obtained when row exchanges were done as needed during elimination. In other words, the matrix PA admits an LU (or LDU) factorization of the standard kind, with no zero pivots. There is only one minor effect of doing all the permutations first: Suppose that a multiple of row 1 is subtracted from row 2, and later rows 2 and 3 have to be exchanged to yield a nonzero pivot. Then for PA, with rows 2 and 3 exchanged ahead of time, the multiple of row 1 will be subtracted from row 3.

EXAMPLE   The basic elimination sequence is

$$A = \begin{bmatrix} 1 & 2 & 3 \\ 2 & 4 & 2 \\ 0 & 1 & 1 \end{bmatrix} \xrightarrow{E_{21}} \begin{bmatrix} 1 & 2 & 3 \\ 0 & 0 & -4 \\ 0 & 1 & 1 \end{bmatrix} \xrightarrow{P_{23}} \begin{bmatrix} 1 & 2 & 3 \\ 0 & 1 & 1 \\ 0 & 0 & -4 \end{bmatrix} = U.$$

If the row exchange is done first, then $P = P_{23}$, and

$$PA = \begin{bmatrix} 1 & 2 & 3 \\ 0 & 1 & 1 \\ 2 & 4 & 2 \end{bmatrix} = \begin{bmatrix} 1 & 0 & 0 \\ 0 & 1 & 0 \\ 2 & 0 & 1 \end{bmatrix}\begin{bmatrix} 1 & 2 & 3 \\ 0 & 1 & 1 \\ 0 & 0 & -4 \end{bmatrix} = LU.$$

Twice the first row of PA is subtracted from its *third* row—E_{21} has been switched to E_{31}—and that step is inverted by L.

The entire theory of Gaussian elimination, which was the main goal of this chapter, can be summarized as follows:

> **1G**   In the nonsingular case, there is a permutation matrix P that reorders the rows of A so that PA admits a factorization with nonzero pivots: $PA = LU$ (or $PA = LDU$, whichever is preferred). In this case there is a unique solution to $Ax = b$, and it is found by elimination with row exchanges. In the singular case, no reordering can avoid a zero pivot.

EXERCISE 1.5.3   Find the $PA = LDU$ factorization of

$$A = \begin{bmatrix} 0 & 1 \\ 2 & 3 \end{bmatrix}.$$

EXERCISE 1.5.4   What permutation matrix P will allow a factorization of PA into LDU if

$$A = \begin{bmatrix} 1 & 1 & 1 \\ 1 & 1 & 2 \\ 1 & 2 & 5 \end{bmatrix}?$$

EXERCISE 1.5.5   Decide whether the following systems are singular or nonsingular, and whether they have a solution:

$$\begin{aligned} v - w &= 2 \\ u - v \quad &= 2 \\ u \quad - w &= 2 \end{aligned} \quad \text{and} \quad \begin{aligned} v - w &= 0 \\ u - v \quad &= 0 \\ u \quad - w &= 0. \end{aligned}$$

Inverses

The underlying theory for the singular case will not be complete until Section 2.4. Even the nonsingular case, which we have identified by the appearance of nonzero pivots, badly needs to be discussed in a manner independent of elimination. But we have already begun to use the notations E_{ij}^{-1}, U^{-1}, and A^{-1}, and it seems proper to state some of the basic rules for inverses.

The first is that a "left-inverse" B and a "right-inverse" C, if both exist, must be identical.

> **1H**   If $BA = I$ and $AC = I$, then $B = C$.

Proof   Since multiplication is associative, $B(AC) = (BA)C$, or $BI = IC$, or $B = C$. For rectangular matrices, one inverse can exist without the other—

but not for square matrices. Nonsingular n by n matrices A are those that possess a two-sided inverse:

$$AA^{-1} = A^{-1}A = I.$$

This property was verified for the elementary matrices E_{ij} and their inverses. Then L^{-1} was built up from the rule for inverses of products:

1I If A and B are invertible, then so is AB; and its inverse is found from A^{-1} and B^{-1} in reverse order:

$$(AB)^{-1} = B^{-1}A^{-1}. \tag{21}$$

Proof Again the associative law removes parentheses, and

$$(AB)(B^{-1}A^{-1}) = ABB^{-1}A^{-1} = AIA^{-1} = AA^{-1} = I.$$
$$(B^{-1}A^{-1})(AB) = B^{-1}A^{-1}AB = B^{-1}IB = B^{-1}B = I.$$

Extending this rule to a product of three matrices, the inverse of $A = LDU$ will be the triple product $A^{-1} = U^{-1}D^{-1}L^{-1}$. It is true, as we said earlier, that only very rarely should such an inverse be computed. If the required answer is $A^{-1}b$, then it is a waste of time to find A^{-1} separately. But it is interesting to know how much time it would take, and how it might be done. One good way is to consider the equation $AA^{-1} = I$ a column at a time: If x_j is the jth column of A^{-1}, and e_j is the jth column of the identity matrix, then clearly $Ax_j = e_j$. This produces n systems of equations, with the n different right-hand sides $e_1, \ldots, e_n$. Since the coefficient matrix A is the same for every system, we apply elimination only once to A. In fact, the simplest plan is to follow the **Gauss–Jordan method**: Carry out elimination on all n equations simultaneously.

EXAMPLE

$$\begin{bmatrix} A & e_1 & e_2 & e_3 \end{bmatrix} = \begin{bmatrix} 2 & 1 & 1 & 1 & 0 & 0 \\ 4 & 1 & 0 & 0 & 1 & 0 \\ -2 & 2 & 1 & 0 & 0 & 1 \end{bmatrix} \rightarrow \begin{bmatrix} 2 & 1 & 1 & 1 & 0 & 0 \\ 0 & -1 & -2 & -2 & 1 & 0 \\ 0 & 3 & 2 & 1 & 0 & 1 \end{bmatrix}$$

$$\rightarrow \begin{bmatrix} 2 & 1 & 1 & 1 & 0 & 0 \\ 0 & -1 & -2 & -2 & 1 & 0 \\ 0 & 0 & -4 & -5 & 3 & 1 \end{bmatrix}.$$

This completes the forward elimination. In the last matrix, the first three columns give the familiar upper triangular U. The other three columns, which are the three right-hand sides after they have been prepared for back-substitution, are the same L^{-1} we have had all along. This is the effect of applying the elementary operations E_{21}, E_{31}, and E_{32} to the identity matrix: $L^{-1} = E_{32}E_{31}E_{21}$. Now when we go on to back-substitution, the solutions x_1, x_2, x_3 to the three systems will be the columns of A^{-1}. The Gauss–Jordan method finds these columns simultaneously, by continuing with the elimination—except that now we subtract multiples of one row *from the rows above it*. Beginning by creating a zero above the pivot in the second column,

$$[A \quad I] \to \begin{bmatrix} 2 & 1 & 1 & 1 & 0 & 0 \\ 0 & -1 & -2 & -2 & 1 & 0 \\ 0 & 0 & -4 & -5 & 3 & 1 \end{bmatrix} \to \begin{bmatrix} 2 & 0 & -1 & -1 & 1 & 0 \\ 0 & -1 & -2 & -2 & 1 & 0 \\ 0 & 0 & -4 & -5 & 3 & 1 \end{bmatrix}$$

$$\to \begin{bmatrix} 2 & 0 & 0 & \frac{1}{4} & \frac{1}{4} & -\frac{1}{4} \\ 0 & -1 & 0 & \frac{1}{2} & -\frac{1}{2} & -\frac{1}{2} \\ 0 & 0 & -4 & -5 & 3 & 1 \end{bmatrix} \to \begin{bmatrix} 1 & 0 & 0 & \frac{1}{8} & \frac{1}{8} & -\frac{1}{8} \\ 0 & 1 & 0 & -\frac{1}{2} & \frac{1}{2} & \frac{1}{2} \\ 0 & 0 & 1 & \frac{5}{4} & -\frac{3}{4} & -\frac{1}{4} \end{bmatrix}$$

$$= [I \quad A^{-1}].$$

At the last stage, we divided through by the pivots. Now the coefficient matrix in the left half is the identity, and the three systems that we are solving have been simplified to

$$Ix_1 = \begin{bmatrix} \frac{1}{8} \\ -\frac{1}{2} \\ \frac{5}{4} \end{bmatrix}, \qquad Ix_2 = \begin{bmatrix} \frac{1}{8} \\ \frac{1}{2} \\ -\frac{3}{4} \end{bmatrix}, \qquad Ix_3 = \begin{bmatrix} -\frac{1}{8} \\ \frac{1}{2} \\ -\frac{1}{4} \end{bmatrix}. \tag{22}$$

These are the three columns of A^{-1}. The Gauss–Jordan method applied a total elimination to A, producing zeros above the pivots as well as below, and simultaneously applied the same operations to the columns of the identity matrix: A is changed to I in the left half of our 3 by 6 matrix, while the same operations on the right half must carry I into A^{-1}.

Remark   Purely out of curiosity, we might count the number of operations required to find A^{-1}. The normal count for each new right-hand side is n^2, half in the forward direction and half in back-substitution. With n different right-hand sides this makes n^3, and after including the $n^3/3$ operations on A itself, the total would be $4n^3/3$.

This result is a little too high, however, because of the special form of the right-hand sides e_j. In the first half of our calculations (the forward elimination) the only changes to make in e_j occur below the 1 in the jth place. It is true that e_1 requires the usual $n^2/2$ steps, but for e_j the count is effectively changed to $(n - j)^2/2$. (Nothing happens to e_j until elimination has reached the jth row of A, and then the remaining part has only $n - j$ components.) In other words, the process that takes I into L^{-1} works exclusively with lower triangular matrices, and the zeros above the diagonal mean an operation count of only

$$\frac{n^2}{2} + \cdots + \frac{(n-j)^2}{2} + \cdots + \frac{1}{2} \approx \frac{n^3}{6}. \tag{23}$$

This is to be combined with the $n^3/3$ operations that are applied once and for all to A, and the $n(n^2/2)$ back-substitution steps which finally produce the x_j—whether done separately, or simultaneously as in the Gauss–Jordan method. *The final operation count for computing A^{-1} is*

$$\frac{n^3}{6} + \frac{n^3}{3} + n\left(\frac{n^2}{2}\right) = n^3.$$

This count is remarkably low. In fact, since matrix multiplication already takes n^3 steps, it requires as many operations to compute A^2 as it does to compute A^{-1}. That fact seems almost unbelievable. (And computing A^3 requires twice as many, as far as we can see.) Nevertheless, if A^{-1} is not needed, it should not be computed.

EXERCISE 1.5.6   What is the inverse of $AB^{-1}C$?

EXERCISE 1.5.7   Find the inverses of

$$A = \begin{bmatrix} 0 & 1 \\ 1 & 0 \end{bmatrix} \quad \text{and} \quad B = \begin{bmatrix} \cos\theta & -\sin\theta \\ \sin\theta & \cos\theta \end{bmatrix}.$$

EXERCISE 1.5.8   Show that $\begin{bmatrix} 1 & 1 \\ 3 & 3 \end{bmatrix}$ has no inverse, by trying to solve

$$\begin{bmatrix} 1 & 1 \\ 3 & 3 \end{bmatrix}\begin{bmatrix} a & b \\ c & d \end{bmatrix} = \begin{bmatrix} 1 & 0 \\ 0 & 1 \end{bmatrix}.$$

EXERCISE 1.5.9   Describe the effect of multiplying a matrix A by E, and write down E^{-1} if

$$E = \begin{bmatrix} 1 & 0 & 8 \\ 0 & 1 & 0 \\ 0 & 0 & 1 \end{bmatrix}.$$

EXERCISE 1.5.10   Use the Gauss–Jordan method, solving $Ax_j = e_j$ simultaneously, to find the inverses of

$$A = \begin{bmatrix} a & b \\ c & d \end{bmatrix} \quad \text{and} \quad A = \begin{bmatrix} 2 & -1 & 0 \\ -1 & 2 & -1 \\ 0 & -1 & 2 \end{bmatrix}.$$

Roundoff Error

In theory the nonsingular case is completed. Row exchanges may be necessary to avoid zero pivots; they do achieve that result if the matrix is nonsingular, and the solution of $Ax = b$ can be computed by back-substitution. In practice, however, other row exchanges may be equally necessary—or the computed solution can easily become worthless.

Remember that for a system of moderate size, say 100 by 100, elimination involves a third of a million operations. With each operation we must expect a roundoff error. In normal floating-point computations, say to the base 10, every number n is represented by a decimal m and an exponent c: $n = m10^c$, with $\frac{1}{10} \leq |m| < 1$. The number of digits in m is the word-length of the computer. Without undertaking any kind of error analysis for floating-point computation, we can still point out the obvious: If floating-point numbers are added, and their exponents c differ say by two, then the last two digits in the smaller number will be more or less lost:

$$.345 + .00123 \rightarrow .346.$$

The question is, to what extent do all these individual roundoff errors contribute to the final error in the solution?

This is not an easy problem. It was attacked by John von Neumann, who was the leading mathematician at the time when computers suddenly made a million operations possible. In fact the combination of Gauss and von Neumann gives the simple elimination algorithm a remarkably distinguished history, although even von Neumann got only a very complicated estimate of the roundoff error; it was Wilkinson who found the right way to answer the question, and his books are now classics.

Two simple examples, borrowed from the texts by Noble and by Forsythe and Moler, will illustrate three of the most important points about roundoff error: The examples are

$$A = \begin{bmatrix} 1. & 1. \\ 1. & 1.0001 \end{bmatrix} \quad \text{and} \quad A' = \begin{bmatrix} .0001 & 1. \\ 1. & 1. \end{bmatrix}.$$

The first point is:

1J Some matrices are extremely sensitive to small changes, and others are not. The matrix A is ill-conditioned (that is, sensitive); A' is well-conditioned.

If we change very slightly the entries of A', or of the inhomogeneous vector b, the solution of $A'x = b$ is only slightly changed. On the other hand, consider two very close right-hand sides for the matrix A:

$$u + \qquad v = 2 \qquad\qquad u + \qquad v = 2$$

$$u + 1.0001v = 2.0001; \qquad u + 1.0001v = 2.0002.$$

The solution to the first is $u = 1$, $v = 1$; the solution to the second is $u = 0$, $v = 2$. *A change in the fifth digit of b was amplified to a change in the first digit of the solution, and no numerical method can avoid this sensitivity to small perturbations.* The ill-conditioning can be shifted from one place to another in the computations, but it cannot be removed; the true solution is very sensitive, and the computed solution cannot be less so.

The second point is:

1K Even a well-conditioned matrix can be ruined by a poor algorithm.

We regret to say that for the matrix A', a straightforward Gaussian elimination is among the poor algorithms. Suppose .0001 is accepted as the first pivot, and 10,000 times the first row is subtracted from the second. The lower right entry becomes -9999, but roundoff to three places gives 10,000. Every trace has disappeared of the entry 1 which was originally there.

Consider the specific example

$$.0001u + v = 1$$
$$u + v = 2. \tag{24}$$

After elimination the second equation *should* read

$$-9999v = -9998, \quad \text{or} \quad v = .99990.$$

Instead, roundoff will produce $-10,000v = -10,000$, or $v = 1$. So far the destruction of the second equation is not reflected in a poor solution; v is actually correct to three figures. As back-substitution continues, however, the first equation with the correct v should become

$$.0001u + .9999 = 1, \quad \text{or} \quad u = 1.$$

Instead, accepting for v a value that is wrong only in the fourth place, this equation becomes

$$.0001u + 1 = 1, \quad \text{or} \quad u = 0.$$

The computed u is completely mistaken. Even though A' is well-conditioned, a straightforward elimination is violently unstable. The factors L, D, and U, whether exact or approximate, are completely out of scale with the original matrix:

$$A' = \begin{bmatrix} 1 & 0 \\ 10,000 & 1 \end{bmatrix} \begin{bmatrix} .0001 & 0 \\ 0 & -9999 \end{bmatrix} \begin{bmatrix} 1 & 10,000 \\ 0 & 1 \end{bmatrix}.$$

The small pivot .0001 brought instability.

Qualitatively, A *is nearly singular while A' is not.* If we change the last entry in A into a 1—or make a corresponding little change somewhere else in the matrix—then it becomes singular. Not so for A'. Near-singularity is almost the same thing as ill-conditioning; we shall try to make this distinction clear in the second section of Chapter 7. But the reason why the well-conditioned A' was spoiled by the elimination algorithm, and the remedy, are already visible. This is our third point:

1L   Just a a zero pivot forced a theoretical change in the elimination algorithm, so a very small pivot forces a practical change. Unless it has special assurances to the contrary, a computer must compare each pivot with all the other possible pivots in the same column. Choosing the largest of these candidates, and exchanging the corresponding rows so as to make this largest value the pivot, is called *partial pivoting*.

In the matrix A', the possible pivot .0001 would be compared with the entry below it, which equals 1, and a row exchange would take place immediately. In matrix terms, this is just multiplication by a permutation matrix as before. The new matrix $A'' = PA'$ has the factorization

$$A'' = \begin{bmatrix} 1. & 1. \\ .0001 & 1. \end{bmatrix} = \begin{bmatrix} 1 & 0 \\ .0001 & 1 \end{bmatrix} \begin{bmatrix} 1 & 0 \\ 0 & .9999 \end{bmatrix} \begin{bmatrix} 1 & 1 \\ 0 & 1 \end{bmatrix}$$

The two pivots are now 1 and .9999, and they are perfectly in scale; previously they were .0001 and -9999.

Partial pivoting is distinguished from the still more conservative strategy of *complete pivoting*, which looks not only in the kth column but also in all later columns for the largest possible pivot. With complete pivoting, not only a row but also a column exchange is needed to move this largest value into the pivot. (In other words, there is a renumbering of the unknowns, or a *post*multiplication by a permutation matrix.) The difficulty with being so conservative is the expense; searching through all the remaining columns for the largest pivot is extremely time-consuming, and partial pivoting is normally quite adequate.

We have finally arrived at the basic algorithm of numerical linear algebra: *elimination with partial pivoting*. Some further refinements, such as watching to see whether a whole row or column needs to be rescaled, are still possible. But essentially the reader now knows what a computer does with a system of linear equations. Compared with the "theoretical" description—*find A^{-1}, and multiply by b*—our description has consumed a lot of the reader's time (and patience). I wish there were an easier way to explain how x is actually found, but I do not think there is.

EXERCISE 1.5.11   (recommended)   Compute the inverse of the 3 by 3 Hilbert matrix

$$A = \begin{bmatrix} 1 & \frac{1}{2} & \frac{1}{3} \\ \frac{1}{2} & \frac{1}{3} & \frac{1}{4} \\ \frac{1}{3} & \frac{1}{4} & \frac{1}{5} \end{bmatrix}$$

in two ways using the ordinary Gauss–Jordan elimination sequence: (i) by exact computation, and (ii) by rounding off each number to three figures. Note: This is a case where pivoting does not help; A is ill-conditioned and incurable.

EXERCISE 1.5.12   Compare the pivots in direct elimination to those with partial pivoting for the matrix

$$A = \begin{bmatrix} .001 & 0 \\ 1 & 1000 \end{bmatrix}.$$

(This is actually an example that needs rescaling before elimination.)

EXERCISE 1.5.13   Explain why with partial pivoting all the multipliers l_{ij} in L satisfy $|l_{ij}| \leq 1$. Deduce that if the original entries of A satisfy $|a_{ij}| \leq 1$, then after the first stage of elimination (producing zeros in the first column) all entries are bounded by 2; after k stages they are bounded by 2^k. Can you construct a 3 by 3 example A, with all $|a_{ij}| \leq 1$ and $|l_{ij}| \leq 1$, whose last pivot is 4?

BAND MATRICES, SYMMETRIC MATRICES, AND APPLICATIONS ■ 1.6

In this section we have two goals. The first is to explain one way in which large systems of linear equations can arise in practice. So far this book has not mentioned any applications, and the truth is that to describe a large and completely realistic problem in structural engineering or economics would lead us far afield. But there is one natural and important application that does not require a lot of preparation.

The other goal is to illustrate, by this same application, the special properties that coefficient matrices frequently have. It is unusual to meet large matrices that look as if they were constructed at random. Almost always there is a pattern, visible even at first sight—frequently a pattern of symmetry, and of very many zero entries. In the latter case, since a sparse matrix contains far fewer than n^2 pieces of information, the computations ought to work out much more simply than for a full matrix. We shall look particularly at the special properties of *band matrices*, whose nonzero entries are all concentrated near the main diagonal, to see how these properties are reflected in the elimination process.

Our example comes from changing a continuous problem into a discrete one. The continuous problem will have infinitely many unknowns (it asks for $u(x)$ at every x), and it cannot be solved exactly on a computer. Therefore it has to be approximated by a discrete problem—the more unknowns we keep, the better will be the accuracy and the greater the expense. As a simple but still very typical continuous problem, our choice falls on the differential equation

$$-\frac{d^2u}{dx^2} = f(x), \qquad 0 \leq x \leq 1. \tag{25}$$

This is a linear equation for the unknown function u, with inhomogeneous term f. There is some arbitrariness left in the problem because any combination $C + Dx$ could be added to any solution u and the sum would constitute another solution; the second derivative of $C + Dx$ contributes nothing. Therefore the uncertainty left by these two arbitrary constants C and D will be removed by adding a "boundary condition" at each end of the interval:

$$u(0) = 0, \qquad u(1) = 0. \tag{26}$$

The result is a *two-point boundary-value problem*, describing not a transient but a steady-state phenomenon—the temperature distribution in a rod, for example, with ends fixed at $0°$ and with a distributed heat source $f(x)$.

Remember that our goal is to produce a problem that is discrete, or finite-dimensional—in other words, a problem in linear algebra. For that reason we cannot accept more than a finite amount of information about f, say its values at the equally spaced points $x = h$, $x = 2h$, ..., $x = nh$. And what we compute will be approximate values u_1, ..., u_n for the true solution u at these same points. At the ends $x = 0$ and $x = 1 = (n + 1)h$, we are already given the correct boundary values $u_0 = 0$, $u_{n+1} = 0$.

The first question is, How do we replace the derivative d^2u/dx^2? Since every derivative is a limit of difference quotients, it can be approximated by stopping at a finite stepsize, and not permitting h (or Δx) to approach zero. For the first derivative there are several alternatives:

$$\frac{du}{dx} \approx \frac{u(x + h) - u(x)}{h} \text{ or } \frac{u(x) - u(x - h)}{h} \text{ or } \frac{u(x + h) - u(x - h)}{2h} . \tag{27}$$

The last, because it is symmetric about x, is actually the most accurate. For the second derivative there is just one possible combination that uses only the values at x and $x \pm h$:

$$\frac{d^2u}{dx^2} \approx \frac{u(x + h) - 2u(x) + u(x - h)}{h^2} . \tag{28}$$

It also has the merit of being symmetric about x. To repeat, the right side approaches the true value of d^2u/dx^2 as $h \to 0$, but we have to stop at a finite h.

At a typical meshpoint $x = jh$, the differential equation $-d^2u/dx^2 = f(x)$ is now replaced by this discrete analogue (28); after multiplying through by h^2,

$$-u_{j+1} + 2u_j - u_{j-1} = h^2 f(jh). \tag{29}$$

There are n equations of exactly this form, one for every value $j = 1, \ldots, n$. The first and last equations include the expressions u_0 and u_{n+1}, which are not unknowns—their values are the boundary conditions, and they are shifted to the right side of the equation and contribute to the inhomogeneous terms (or at least they might, if they were not known to equal zero). It is easy to understand (29) as a steady-state equation, in which the flows $(u_j - u_{j+1})$ coming from the right and $(u_j - u_{j-1})$ coming from the left are balanced by the loss of $h^2 f(jh)$ at the center.

The structure of the n equations (29) can be better visualized in matrix 'form $Au = b$. We shall choose $h = \frac{1}{6}$, or $n = 5$:

$$\begin{bmatrix} 2 & -1 & & & \\ -1 & 2 & -1 & & \\ & -1 & 2 & -1 & \\ & & -1 & 2 & -1 \\ & & & -1 & 2 \end{bmatrix} \begin{bmatrix} u_1 \\ u_2 \\ u_3 \\ u_4 \\ u_5 \end{bmatrix} = h^2 \begin{bmatrix} f(h) \\ f(2h) \\ f(3h) \\ f(4h) \\ f(5h) \end{bmatrix} . \qquad (30)$$

From now on, *we will work with Eq.* (30), and it is not essential to look back at the source of the problem. What matters is that we have produced a class of coefficient matrices whose order n can be very large, but which are obviously very far from random. The matrix A possesses several special properties, and the one that interests us most is this: ***the matrix is tridiagonal***. All its nonzero entries lie on the main diagonal and the two adjacent diagonals. In other words, the entries are zero whenever the row and column numbers differ by more than one: $a_{ij} = 0$ if $|i - j| > 1$. The presence of all these zeros will bring a tremendous simplification to Gaussian elimination.

There are two other properties of the matrix A which we hesitate to discuss here, not because they are unimportant but because they are almost too important. We can at least say what they are, and how they affect the elimination algorithm:

(1)   The matrix A is *symmetric*. This means that each entry a_{ij} on one side of the main diagonal equals its "mirror image" a_{ji} on the other side: $a_{ij} = a_{ji}$. In other words, the first column of A agrees exactly with the first row, the second column equals the second row, and so on. One consequence is obvious: A computer has only to store the lower triangular part of A since any entry above the main diagonal can be recovered immediately from the corresponding entry below.

The most important point, however, is that this symmetry is preserved by the elimination process (see the exercises). Therefore, *it is not necessary to compute the changes that occur above the main diagonal during elimination.* Provided we keep track of the results below the diagonal, and of course also of the pivots, the upper triangular part follows automatically. Both the storage and the computation time are cut virtually in half by the symmetry of the matrix.

We call attention to another result of the same kind, and we first introduce the superscript T for the *transpose* of a matrix: If A is any matrix, then exchanging the rows for the columns produces its transpose A^T. Symmetry means that $A^T = A$, and it has the following consequence:

1M If A is symmetric, and if it can be factored in the standard form $A = LDU$, then the columns of the lower triangular L are the same as the rows of the upper triangular U. L is the mirror image, or transpose, of U: $L = U^T$ and $U = L^T$.

Proof Start with the factorization $A = LDU$, and look to see what happens if we exchange the rows for the columns. On the left side we still have A; that is the property of symmetry we are depending on. On the right side, this transposing process will produce the separate transposes of L, D, and U—but *multiplied in reverse order* (just as for inverses). In other words, we arrive at the formula $A^T = U^T D^T L^T$. The first factor U^T, formed by exchanging the rows of an upper triangular U for its columns, is lower triangular. The other factors D^T and L^T are diagonal and upper triangular. Therefore what we now have seems to be a new "LDU factorization" of $A^T = A$. But the factors of any matrix are uniquely determined (p. 23). Therefore $L = U^T$, $D = D^T$, and $U = L^T$, as we wanted to prove.

We will return to transposes in Section 3.1; this proof was a very sudden introduction to a very simple idea.

The second extra property of our finite difference matrix is:

(2) The matrix A is *positive definite*. This property is defined (for symmetric matrices only) in Chapter 6; the definition there has nothing to do with elimination. But later on we shall find an equivalent definition of positive-definiteness, which has everything to do with elimination: *A symmetric matrix A is positive definite exactly when all its pivots d_i are positive*. This means that row exchanges will not be necessary to avoid a zero pivot; elimination can proceed in the natural order, without spoiling the symmetry by any exchanges of rows.

In practice, we know that there is a further question to be asked about the pivots. They are nonzero, but are they sufficiently large? The answer is another consequence of positive-definiteness, and computationally the most important one: Row exchanges are not required in practical computations with positive-definite matrices. The factors L, D, and U cannot be out of scale with A. For our matrix it will be possible to look at the factorization and confirm this prediction.

Remark This does not mean there is no danger of roundoff error; the ill-conditioned matrix

$$A = \begin{bmatrix} 1 & 1 \\ 1 & 1.0001 \end{bmatrix}$$

in the previous section was actually symmetric and positive-definite, and so was the Hilbert matrix in the exercises. Positive-definiteness does mean that

Gaussian elimination without row exchanges is numerically as stable as the matrix itself.

Leaving aside these extra properties of A, we return to the fact that A is tridiagonal. What effect does this have on elimination? To start, suppose we carry out the first stage of the elimination process and produce zeros below the first pivot:

$$
\begin{bmatrix}
2 & -1 & & & \\
-1 & 2 & -1 & & \\
& -1 & 2 & -1 & \\
& & -1 & 2 & -1 \\
& & & -1 & 2
\end{bmatrix}
\rightarrow
\begin{bmatrix}
2 & -1 & & & \\
0 & \frac{3}{2} & -1 & & \\
& -1 & 2 & -1 & \\
& & -1 & 2 & -1 \\
& & & -1 & 2
\end{bmatrix}.
$$

Compared with a general 5 by 5 matrix, there were two major simplifications:

(a)   There was *only one nonzero entry* below the pivot; normally, multiples of the first row would have to be subtracted from every other row.

(b)   This one operation was carried out on a very short row. After the correct multiple $l_{21} = -\frac{1}{2}$ was determined, *only a single multiplication–subtraction* was required to the right of the pivot.

Thus the first step was very much simplified by the zeros in the first row and column. Furthermore, *the tridiagonal pattern is preserved during elimination* (in the absence of row exchanges!).

(c)   The second stage of elimination, as well as every succeeding stage, also admits the simplifications (a) and (b).

We can summarize the final result in several ways. The most revealing is to look at the LDU factorization of A:

$$
A =
\begin{bmatrix}
1 & & & & \\
-\frac{1}{2} & 1 & & & \\
& -\frac{2}{3} & 1 & & \\
& & -\frac{3}{4} & 1 & \\
& & & -\frac{4}{5} & 1
\end{bmatrix}
\begin{bmatrix}
\frac{2}{1} & & & & \\
& \frac{3}{2} & & & \\
& & \frac{4}{3} & & \\
& & & \frac{5}{4} & \\
& & & & \frac{6}{5}
\end{bmatrix}
\begin{bmatrix}
1 & -\frac{1}{2} & & & \\
& 1 & -\frac{2}{3} & & \\
& & 1 & -\frac{3}{4} & \\
& & & 1 & -\frac{4}{5} \\
& & & & 1
\end{bmatrix}.
$$

The observations (a)–(c) can be expressed as follows: *The L and U factors of a tridiagonal matrix are bidiagonal.* These factors have more or less the same structure of zeros as A itself. Note too that L and U are transposes of one another, as was expected from the symmetry, and that the pivots d_i are all positive.† The pivots are obviously converging to a limiting value of $+1$, as n gets large. Such matrices make a computer very happy.

---

† The product of the pivots will later be identified as the **determinant** of A: det $A = 6$.

These simplifications lead to a complete change in the usual operation count. At each elimination stage only two operations are needed—a division by the pivot, and a multiplication–subtraction off the diagonal. There are about n such stages. Therefore *in place of $n^3/3$ operations we need only $2n$*; the computation is quicker by orders of magnitude. And the same is true of back-substitution; instead of $n^2/2$ operations we again need only $2n$. Thus *the number of operations for a tridiagonal system is proportional to n*, not to a higher power of n. Tridiagonal systems $Ax = b$ can be solved almost instantaneously.

Suppose, more generally, that A is a *band matrix*; all its nonzero entries lie within a band running parallel to the main diagonal. Its "bandwidth," or if you prefer its half-bandwidth, is w; a_{ij} is zero whenever $|i - j| \geq w$ (Fig. 1.2). This allows $2w - 1$ nonzero diagonals in all—the main diagonal, and

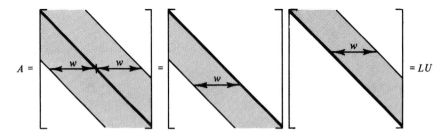

Fig. 1.2.   A band matrix and its factors.

$w - 1$ additional diagonals on each side. The bandwidth is $w = 1$ for a diagonal matrix, $w = 2$ for a tridiagonal matrix, and $w = n$ for a full matrix.

For any band matrix, the observations (a)–(c) still apply. At the first stage of elimination, there are only $w - 1$ zeros to be created below the pivot; the entries further down are already zero. Furthermore, the first row has only w nonzero entries, so that w operations are sufficient to create each zero. Therefore the first stage requires $w(w - 1)$ operations, and after this stage we still have a matrix of the same bandwidth w. Since there are about n stages, elimination on a band matrix must require about $w^2 n$ operations. Again the operation count is proportional to n, and now we see that it is proportional also to the square of w. As w approaches n, the matrix becomes full, and the operation count again is roughly n^3. A more exact count depends on the fact that, once elimination reaches the lower right corner of A, the bandwidth is no longer w; there is not room for that many bands. The precise number of divisions and multiplication–subtractions that produce L, D, and U (without assuming a symmetric A) is $P = \frac{1}{3}w(w - 1)(3n - 2w + 1)$. For a full matrix, which has $w = n$, we recover $P = \frac{1}{3}n(n - 1)(n + 1)$.† To summarize: A band matrix A has triangular factors L and U that lie within the same band, and both elimination and back-substitution are very fast.

† We are happy to confirm that this P is a whole number; since $n - 1$, n, and $n + 1$ are consecutive integers, one of them must be divisible by 3.

This is our last operation count, but we must emphasize the main point. For a matrix like the finite-difference example A, the inverse is a full matrix. Therefore, in solving $Ax = b$, *we are actually much worse off knowing A^{-1} than knowing L and U.* Multiplying A^{-1} by b will take n^2 steps, whereas $4n$ are sufficient to solve $Lc = b$ and $Ux = c$—the forward elimination and back-substitution that produce $x = U^{-1}c = U^{-1}L^{-1}b = A^{-1}b$.

We hope this example has served two purposes: to reinforce the reader's understanding of the elimination sequence (which we now assume to be perfectly understood!) and to provide a genuine example of the kind of large linear system that is actually met in practice. In the next chapter we turn to the "theoretical" structure of a linear system $Ax = b$—the existence and the uniqueness of x.

EXERCISE 1.6.1   Modify the example in the text by changing from $a_{11} = 2$ to $a_{11} = 1$, and find the LDU factorization of this new tridiagonal matrix.

EXERCISE 1.6.2   For the symmetric matrix

$$A = \begin{bmatrix} a & b & c \\ b & d & e \\ c & e & f \end{bmatrix},$$

apply elimination to produce zeros below the pivot a in the first column, and verify that after this stage the lower 2 by 2 matrix is still symmetric. By the same construction, symmetry persists after every stage of elimination.

EXERCISE 1.6.3   Find the 5 by 5 matrix A that approximates

$$-\frac{d^2u}{dx^2} = f(x), \qquad \frac{du}{dx}(0) = \frac{du}{dx}(1) = 0,$$

replacing the boundary conditions by $u_0 = u_1$ and $u_6 = u_5$. Prove that your matrix, applied to the constant vector $[1, 1, 1, 1, 1]$, yields zero; A is a singular matrix. Analogously, show that if $u(x)$ is a solution of the continuous problem, then so is $u(x) + 1$; the two boundary conditions do not remove the uncertainty in the term $C + Dx$, and the solution is not unique.

EXERCISE 1.6.4   With $h = \frac{1}{4}$ and $f(x) = 4\pi^2 \sin 2\pi x$, the difference equation (29) is

$$\begin{bmatrix} 2 & -1 & 0 \\ -1 & 2 & -1 \\ 0 & -1 & 2 \end{bmatrix} \begin{bmatrix} u_1 \\ u_2 \\ u_3 \end{bmatrix} = \frac{\pi^2}{4} \begin{bmatrix} 1 \\ 0 \\ -1 \end{bmatrix}.$$

Solve for the u_i, and find their error in comparison with the true solution $u = \sin 2\pi x$ at $x = \frac{1}{4}$, $x = \frac{1}{2}$, and $x = \frac{3}{4}$.

REVIEW EXERCISES

1.1   For the matrices

$$A = \begin{bmatrix} 1 & 0 \\ 1 & 1 \end{bmatrix}$$

and

$$B = \begin{bmatrix} 1 & 1 \\ 0 & 1 \end{bmatrix},$$

compute $A + B$ and AB.

1.2   For the same matrices, find A^{-1}, B^{-1}, and $(AB)^{-1}$.

1.3   Solve by elimination and back-substitution:

$$2u - 3v \qquad = 8$$
$$4u - 5v + w = 15$$
$$2u \qquad + 4w = 1.$$

1.4   Factor the previous coefficient matrix into $A = LU$.

1.5   Given a system of three equations, what matrix E has the effect of subtracting the second equation from the third?

1.6   What 3 by 3 matrix P has the effect of exchanging the first equation for the third?

1.7   What 3 by 3 matrix multiplies the second equation by -1, and leaves the other two unchanged?

1.8   Decide by elimination whether there is a solution to

$$u + v + w = 0$$
$$u + 2v + 3w = 0$$
$$3u + 5v + 7w = 1.$$

1.9   Solve by elimination and row exchange:

$$u + v - w = 2$$
$$3u + 3v + w = 2$$
$$u \qquad + w = 0.$$

1.10   Factor

$$A = \begin{bmatrix} 1 & 4 & 0 \\ 0 & 1 & 0 \\ 0 & 3 & 1 \end{bmatrix}$$

into LU.

1.11   Find the inverse of

$$\begin{bmatrix} 2 & 3 \\ 3 & 4 \end{bmatrix}.$$

1.12   Is a 3 by 3 matrix necessarily invertible if no row is a multiple of another row? Give an example.

1.13   Invert

$$A = \begin{bmatrix} 1 & 1 & 0 \\ 0 & 1 & 1 \\ 0 & 0 & 1 \end{bmatrix}$$

by the Gauss–Jordan method.

1.14   Explain why the inverse of a triangular matrix is still triangular.

1.15   If E is 2 by 2 and it adds the first equation to the second, what does E^{50} do? Write down E, E^{50}, and $50E$.

1.16   Write down a 2 by 2 system with infinitely many solutions.

THE THEORY OF SIMULTANEOUS LINEAR EQUATIONS

2.1 ■ VECTOR SPACES AND SUBSPACES

The previous chapter described how elimination could proceed, one entry at a time, to simplify a linear system $Ax = b$. Fortunately, it is not only the calculation of x that is made easier by this algorithm; the more theoretical questions about its existence and its uniqueness are answered at the same time. In fact, in order to understand rectangular as well as square matrices, we need to devote one more section to the mechanics of elimination. Then that circle of ideas will be complete. But the concentration on individual entries produces only one kind of understanding of a linear system, and our chief object in this chapter is to achieve also a different and deeper understanding of the problem.

For this we need the concept of a **vector space**. We want to introduce that idea now, not by the usual list of axioms but by an example of three equations in two unknowns:

$$\begin{bmatrix} 1 & 0 \\ 5 & 4 \\ 2 & 4 \end{bmatrix} \begin{bmatrix} u \\ v \end{bmatrix} = \begin{bmatrix} b_1 \\ b_2 \\ b_3 \end{bmatrix}. \tag{1}$$

If there were more unknowns than equations, we might expect to find one or even infinitely many solutions (although that is not always so). In the present case there are more equations than unknowns, $m > n$, and we must expect that *usually there will be no solution*. The system will be solvable only for certain right-hand sides, in fact for a very "thin" subset of all possible three-dimensional vectors b. We want to find that subset.

One way of describing this subset is so simple that it is easy to overlook.

2A The system $Ax = b$ is solvable if and only if the vector b can be expressed as a combination of the columns of A.

This description involves nothing more than a restatement of the system $Ax = b$, writing it in the following way:

$$u \begin{bmatrix} 1 \\ 5 \\ 2 \end{bmatrix} + v \begin{bmatrix} 0 \\ 4 \\ 4 \end{bmatrix} = \begin{bmatrix} b_1 \\ b_2 \\ b_3 \end{bmatrix}. \tag{2}$$

These are the same three equations in two unknowns. But now the problem is seen to be this: Find weights u and v that, multiplying the first and second columns, produce the vector b. The system is solvable exactly when such weights exist, and the weights (u, v) are the solution x.†

Thus, the subset of attainable right-hand sides b is *the set of all combinations of the columns of* A. One possible right side is the first column itself; the weights are $u = 1$ and $v = 0$. Another possibility is the second column: $u = 0$ and $v = 1$. A third is the right side $b = 0$; the weights are $u = 0$, $v = 0$ (and with these trivial weights $b = 0$ will be attainable no matter what the coefficient matrix is).

Now we have to consider all combinations of the two columns, and we describe the result geometrically: $Ax = b$ *can be solved if and only if b lies in the plane that is spanned by the two column vectors* (Fig. 2.1). These vectors are

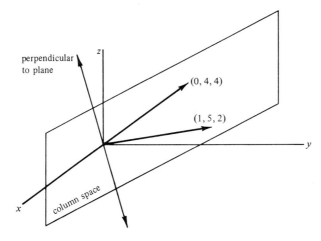

Fig. 2.1. The column space, a plane in three-dimensional space.

† We use the word *weights* as an alternative to *coefficients*; there is no implication that the weights are positive.

line segments connecting the origin $(0, 0, 0)$ to the points $(1, 5, 2)$ and $(0, 4, 4)$; the plane is determined by these two lines. The plane is by no means limited to vectors with positive components; the weights $u = -1$, $v = 1$ produce the possibility $b = (-1, -1, 2)$.

This plane illustrates one of the most fundamental ideas in the theory of linear algebra: It is a vector space. By this we mean a collection of vectors that satisfies the following two requirements:

> (i)   if we add any vectors b and b' that lie in the space, their sum $b + b'$ again lies in the space;
> (ii)   if we multiply any vector b in the space by any scalar constant c, the multiple cb again lies in the space.

In other words, the set of vectors in our plane is *closed under vector addition, and closed under multiplication by scalars.*† The plane is a subset of the set of all possible three-component vectors (b_1, b_2, b_3), and this larger set is also a vector space; it is the full three-dimensional space. Our plane is actually a **subspace,** in other words *a subset that is a vector space in its own right.* We call it the **column space** of the matrix A because it is generated by the columns.

It is easy to recognize the two extreme possibilities for the column space of a matrix. One comes from the n by n matrix of zeros, $A = 0$; its column space contains only one vector, namely $(0, \ldots, 0)$. This is still a (nonempty!) vector space since addition and scalar multiplication are entirely permissible within it. The sum $0 + 0$ is in the space, and so are all multiples $c0$. *This is the smallest possible vector space.* The other extreme comes, for example, from the n by n identity matrix $A = I$. Its column space is the full n-dimensional space; every column vector with n components can be produced from the columns of the identity. We intend these components to be real numbers, *the space of all real n-vectors will be denoted by R^n.* In Figure 2.1, the plane is a subspace of $\mathbf{R}^3$.

The distinction between a *subset* and a *subspace* is best made clear by examples; we give some now and more later. In each case, the question to be decided is whether or not requirements (i) and (ii) are satisfied.

EXAMPLE 1   Let S consist of all multiples of a given vector v, say the vector $(1, 4, -2)$. These multiples fill out *a line in three-dimensional space* $\mathbf{R}^3$, and this line goes through the origin; one multiple of v is $0v$. Such a subset S is actually a subspace; the sum of any two vectors on the line, or any multiple of such a vector, is again on the line.

EXAMPLE 2   Let S contain only the two vectors $(0, 0, 0)$ and $(1, 0, 0)$. Then it fails requirement (ii) because the multiple $5(1, 0, 0)$ is not in S. And it also

---

† A proper definition of a vector space would give the rules which addition and scalar multiplication must satisfy; for example $b + b' = b' + b$. These rules are satisfied in the n-dimensional space $\mathbf{R}^n$, so we are (barely) legitimate if we stick to $\mathbf{R}^n$ and its subspaces.

fails requirement (i): We are allowed to choose both b and b' equal to $(1, 0, 0)$, and ask if $b + b'$ is in S; it is not.

EXAMPLE 3   Let S consist of all vectors on a line running parallel to the x axis; the x component is arbitrary, but we fix $y = 3$ and $z = 4$. This subset again violates both (i) and (ii), even though it is so straight. The sum of two vectors on the line will have $y = 6$ and $z = 8$; it will be on another line. Furthermore, the zero vector is not in S; requirement (ii) allows the scalar multiple to be zero, so that *every subspace must contain the zero vector*.

Now we complete this introduction. Geometrically, there is hardly anything left to be done because all the crucial facts are clear from Fig. 2.1:

(i)   The attainable right sides b, in our system $Ax = b$, form a *two-dimensional column space*; the two columns of A generate the plane, and no fewer than two vectors could possibly do so.

(ii)   The vectors through the origin that are perpendicular to the plane all fall on a single line; these vectors also make up a subspace, and it is one dimensional.†

From the geometry, these facts should seem more or less evident. But so far we have no *algebraic* way of expressing them, or verifying them, or generalizing them to a system of m equations in n unknowns (in other words, to n columns in an m-dimensional space). This is one of our two specific goals in understanding the system $Ax = b$.

The second goal in this chapter is "dual" to the first. We are concerned not only with which right sides b are attainable, but also with the set of all solutions x that attain them. The right side $b = 0$ always allows the particular solution $x = 0$, but there may be infinitely many other solutions. (In case there are more unknowns than equations $(n > m)$, we shall prove that $Ax = 0$ *must have* other solutions than just the trivial one $x = 0$.) **The set of solutions to $Ax = 0$ is itself a vector space—the** *nullspace of* A. It is a subspace of $\mathbf{R}^n$ just as the column space of attainable b's was a subspace of $\mathbf{R}^m$.

It is not hard to solve $Ax = 0$, and thereby find the nullspace, for the example given above:

$$\begin{bmatrix} 1 & 0 \\ 5 & 4 \\ 2 & 4 \end{bmatrix} \begin{bmatrix} u \\ v \end{bmatrix} = \begin{bmatrix} 0 \\ 0 \\ 0 \end{bmatrix}.$$

The first equation gives $u = 0$, and the second one then forces $v = 0$. In this

---

† You must notice that the word "dimensional" is used in two different ways. We speak about an n-dimensional *vector*, meaning a vector with n components, or in other words a member of $\mathbf{R}^n$; and then on p. 62 we define the dimension of a *subspace*. A line in $\mathbf{R}^4$ is a one-dimensional subspace whose points are four-dimensional vectors.

case the nullspace consists of a single point, the zero vector; the only combination of the columns that produces $b = 0$ on the right-hand side is the combination with $u = v = 0$.

The situation is changed if we add a third column that is a combination of the other two:

$$B = \begin{bmatrix} 1 & 0 & 1 \\ 5 & 4 & 9 \\ 2 & 4 & 6 \end{bmatrix}.$$

The column space of B is the same as that of A because the new column also lies in the plane of Fig. 2.1; it is just the sum of the two column vectors we started with. But the nullspace of this new matrix B contains the vector with components $1, 1, -1$, or any multiple of that vector:

$$\begin{bmatrix} 1 & 0 & 1 \\ 5 & 4 & 9 \\ 2 & 4 & 6 \end{bmatrix} \begin{bmatrix} c \\ c \\ -c \end{bmatrix} = \begin{bmatrix} 0 \\ 0 \\ 0 \end{bmatrix}.$$

Therefore the nullspace of B is the line containing all points $x = c$, $y = c$, $z = -c$, where c ranges from $-\infty$ to ∞. The line goes through the origin, as any subspace must. And this one-dimensional nullspace has a perpendicular space (a plane), which is directly related to the rows of the matrix, and is of special importance.

To summarize: We want to be able, for any system $Ax = b$, to find all attainable right-hand sides b, and all solutions to $Ax = 0$. This means that we shall compute the dimensions of the subspaces introduced above, and a convenient set of vectors to generate them. We hope to end up by understanding all four of the subspaces that are intimately related to each other and to A—the column space of A, the nullspace of A, and their two perpendicular spaces.

EXERCISE 2.1.1 Show that the requirements (i) and (ii) for a vector space are genuinely independent by constructing:

(a) a subset of two-dimensional space closed under vector addition and even subtraction but not under scalar multiplication;
(b) a subset of two-dimensional space closed under scalar multiplication but not under vector addition.

EXERCISE 2.1.2 Which of the following subsets of $\mathbf{R}^3$ are actually subspaces?

(a) The plane of vectors with first component $b_1 = 0$.
(b) The plane of vectors b with $b_1 = 1$.
(c) The vectors b with $b_1 b_2 = 0$ (this is the union of two subspaces, the plane $b_1 = 0$ and the plane $b_2 = 0$).
(d) The solitary vector $b = (0, 0, 0)$.
(e) All combinations of two given vectors $u = (1, 1, 0)$ and $v = (2, 0, 1)$.
(f) The vectors (b_1, b_2, b_3) that satisfy $b_3 - b_2 + 3b_1 = 0$.

EXERCISE 2.1.3 Verify that the nullspace of A is a subspace; in other words, the solutions x to any homogeneous system $Ax = 0$ are closed under addition and scalar multiplication. Give a counterexample in the case $b \neq 0$.

THE SOLUTION OF *m* EQUATIONS IN *n* UNKNOWNS ■ 2.2

The elimination process is by now very familiar for square matrices, and one example will be enough to illustrate the new possibilities that arise when the matrix is rectangular. The elimination itself goes forward without major changes, but when it comes to reading off the solution by back-substitution, there are some differences.

Perhaps, even before the example, we should illustrate the possibilities by looking at the scalar equation $ax = b$. This is a "system" of only one equation in one unknown. A moment's thought produces the three alternatives:

(i) If $a \neq 0$, then for any b there exists a solution $x = b/a$, and this solution is unique. This is the **nonsingular** case (of a 1 by 1 invertible matrix a).

(ii) If $a = 0$ and $b = 0$, there are infinitely many solutions; any x satisfies $0x = 0$. This is the **underdetermined** case; a solution exists, but it is not unique.

(iii) If $a = 0$ and $b \neq 0$, there is no solution to $0x = b$. This is the **inconsistent** case.

For square matrices all these alternatives may occur. With a rectangular matrix we cannot have (i), existence and also uniqueness, for *every* b.

Now we choose a less obvious example, ignoring at first the right side b and working exclusively with the 3 by 4 matrix

$$A = \begin{bmatrix} 1 & 3 & 3 & 2 \\ 2 & 6 & 9 & 5 \\ -1 & -3 & 3 & 0 \end{bmatrix}.$$

The pivot $a_{11} = 1$ is nonzero, and the usual elementary operations will produce zeros in the first column below this pivot:

$$A \rightarrow \begin{bmatrix} 1 & 3 & 3 & 2 \\ 0 & 0 & 3 & 1 \\ 0 & 0 & 6 & 2 \end{bmatrix}.$$

The second pivot has become zero, and therefore we look below it for a nonzero entry—intending to carry out a row exchange. In this case the entry below it is also zero. If the original matrix were square, this would signal that the matrix was singular. With a rectangular matrix, we must expect trouble anyway, and there is no reason to terminate the elimination. All we can do is to *go on to the next column*, where the pivot is nonzero. Subtracting twice the second row from the third, we arrive at

$$U = \begin{bmatrix} 1 & 3 & 3 & 2 \\ 0 & 0 & 3 & 1 \\ 0 & 0 & 0 & 0 \end{bmatrix}.$$

Strictly speaking, we then proceed to the fourth column; there we meet another zero in the pivot position, and nothing can be done. The forward stage of elimination is complete.

The final form U is upper triangular, or more accurately it is *upper trapezoidal*; all nonzero entries u_{ij} lie in the area on and above the main diagonal, $i \leq j$. For a rectangular matrix this area may have the shape of a trapezoid. It is easy to see that the nonzero entries actually are confined to a kind of "staircase" or "echelon" form, which is shown in Fig. 2.2. The nonzero pivots

Fig. 2.2.   The nonzero entries of a typical echelon matrix U.

are clearly marked, whereas the other starred entries may or may not be zero. We can summarize in words what the figure illustrates:

(i)   The nonzero rows come first—otherwise there would have been row exchanges—and the pivots are the first nonzero entries in those rows.

(ii)   Below each pivot is a column of zeros, obtained by elimination.

(iii)   Each pivot lies to the right of the pivot in the row above; this produces the echelon pattern.

EXERCISE 2.2.1   The echelon form in Fig. 2.2 arises when some of the candidates for pivots are zero (those in columns 3 and 5–8). If no pivots are zero, draw the corresponding picture of a 5 by 9 echelon form U. Show by example that the sum of two echelon matrices might fail to be in echelon form, even in the 2 by 2 case; the pivots might cancel each other.

Since we started with A and ended with U, the excitable reader is certain to ask: Are these matrices connected by a lower triangular L, $A = LU$, as before? There is no reason why not since the elimination steps have not changed; each step still subtracts a multiple of one row from a row beneath it. Furthermore, the inverse of each step is also accomplished just as before, by adding back the multiple that was subtracted, and these inverses come in an order that permits us to record them directly in L:

$$L = \begin{bmatrix} 1 & 0 & 0 \\ 2 & 1 & 0 \\ -1 & 2 & 1 \end{bmatrix}.$$

The reader should verify that $A = LU$, and note that L is not rectangular but *square*. It is a matrix of the same order $m = 3$ as the number of rows in A and U.

The only operation not required by our example, but needed in general, is an exchange of rows. As in Section 1.5, this would introduce a permutation matrix P. If all row exchanges are carried out on A before the beginning of elimination, the final result matches the one in the first chapter. In fact, since we are now agreed to keep going to the next column when no pivots are available in a given column, there is no need to assume that A is nonsingular. Here is the main theorem:

> **2B** To any m by n matrix A there correspond a permutation matrix P, a lower triangular matrix L with unit diagonal, and an upper trapezoidal matrix U with its nonzero entries in echelon form, such that $PA = LU$.

Our goal is now to read off the solutions (if any) to $Ax = b$.

Suppose we start with the homogeneous case, $b = 0$. Then, since the row operations will have no effect on the zeros on the right side of the equation, $Ax = 0$ is simply reduced to $Ux = 0$:

$$Ux = \begin{bmatrix} 1 & 3 & 3 & 2 \\ 0 & 0 & 3 & 1 \\ 0 & 0 & 0 & 0 \end{bmatrix} \begin{bmatrix} u \\ v \\ w \\ y \end{bmatrix} = \begin{bmatrix} 0 \\ 0 \\ 0 \end{bmatrix}.$$

The unknowns u, v, w, and y go into two groups. One group is made up of the **basic variables**, those that correspond to a **column with a nonzero pivot**; the first and third columns contain the nonzero pivots, so u and w are the basic variables. The other group is made up of the **free variables**, corresponding to **columns without pivots**; these are the second and fourth columns, so that v and y are free variables.

To find the most general solution to $Ux = 0$ (or equivalently, to $Ax = 0$) we may assign arbitrary values to the free variables. Suppose we call these values simply v and y. The basic variables are then completely determined, and can be computed in terms of the free variables by back-substitution. Proceeding upward,

$$3w + y = 0 \qquad \text{yields} \qquad w = -\tfrac{1}{3}y$$
$$u + 3v + 3w + 2y = 0 \qquad \text{yields} \qquad u = -3v - y.$$

There is a "double infinity" of solutions to the system, with two free and independent parameters v and y. The general solution is a combination

$$x = \begin{bmatrix} -3v - y \\ v \\ -\frac{1}{3}y \\ y \end{bmatrix} = v \begin{bmatrix} -3 \\ 1 \\ 0 \\ 0 \end{bmatrix} + y \begin{bmatrix} -1 \\ 0 \\ -\frac{1}{3} \\ 1 \end{bmatrix}. \tag{3}$$

Please look again at the last form of the solution to $Ax = 0$. The vector $(-3, 1, 0, 0)$ gives the solution when the free variables have the values $v = 1$, $y = 0$, and the last vector is the solution when $v = 0$ and $y = 1$. *All solutions are linear combinations of these two.* This is very much like the situation on p. 45, when all attainable right sides b were combinations of the two columns. Here we are dealing, however, not with the columns of A, but with its null-vectors—the solutions to $Ax = 0$. The vectors no longer have the same length as the columns of A; they have not m but n components, matching the rows of A.

Geometrically, the picture is this: Within the four-dimensional space of all possible vectors x, the solutions to $Ax = 0$ form a two-dimensional subspace—the **nullspace** of A. This nullspace could be described as a "plane"; it is generated by the two vectors $(-3, 1, 0, 0)$ and $(-1, 0, -\frac{1}{3}, 1)$. (The reader will not expect a four-dimensional picture.) The combinations of these two vectors form a set that is closed under addition and scalar multiplication; these operations simply lead to more combinations of the same two vectors, and all these combinations comprise the nullspace.

This is the place to recognize one extremely important theorem. Suppose we start with a matrix that has more columns than rows, $n > m$. Then, since there can be at most m nonzero pivots (there are not rows enough to hold any more), *there must be at least $n - m$ free variables.* There will be even more free variables if, as in our example, some rows of U happen to reduce to zero; but no matter what, at least one of the variables must be free. This variable can be assigned an arbitrary value, leading to the following conclusion:

2C Every homogeneous system $Ax = 0$, if it has more unknowns than equations ($n > m$), has a nontrivial solution: There is a solution x other than the trivial solution $x = 0$.

There must actually be infinitely many solutions, since any multiple cx will also satisfy $A(cx) = 0$. And if there are additional free variables, the nullspace becomes more than just a line in n-dimensional space. *The nullspace is a subspace of the same "dimension" as the number of free variables.*

The inhomogeneous case, $b \neq 0$, is quite different. We return to the original example $Ax = b$, and apply to both sides of the equation the operations that led from A to U. The result is an upper trapezoidal system $Ux = c$:

$$\begin{bmatrix} 1 & 3 & 3 & 2 \\ 0 & 0 & 3 & 1 \\ 0 & 0 & 0 & 0 \end{bmatrix} \begin{bmatrix} u \\ v \\ w \\ y \end{bmatrix} = \begin{bmatrix} b_1 \\ b_2 - 2b_1 \\ b_3 - 2b_2 + 5b_1 \end{bmatrix}. \tag{4}$$

The vector c on the right side, which appeared after the elimination steps, is just $L^{-1}b$ as in the previous chapter.

It is not clear that this system of equations has a solution. The third equation is the one in doubt; its left side is zero, and *the equations are inconsistent unless $b_3 - 2b_2 + 5b_1 = 0$*. In other words, *the set of attainable vectors b is not the whole of the three-dimensional space.* Even though there are more unknowns than equations, there may be no solution. We know, from p. 45, another way of considering the same question: $Ax = b$ can be solved if and only if b lies in the column space of A. This subspace is spanned by the four columns of A (not of U!):

$$\begin{bmatrix} 1 \\ 2 \\ -1 \end{bmatrix}, \quad \begin{bmatrix} 3 \\ 6 \\ -3 \end{bmatrix}, \quad \begin{bmatrix} 3 \\ 9 \\ 3 \end{bmatrix}, \quad \begin{bmatrix} 2 \\ 5 \\ 0 \end{bmatrix}.$$

Evidently, even though there are four vectors, their linear combinations only fill out a plane in three-dimensional space; the second column is just three times the first, and the fourth column equals the first plus some fraction of the third. (Note that these dependent columns, the second and fourth, are exactly the ones without pivots.) The column space can now be described in two completely different ways. On the one hand, it is the plane generated by columns 1 and 3; the other columns lie in that plane, and contribute nothing new. Equivalently, it is the plane composed of all points (b_1, b_2, b_3) that satisfy $b_3 - 2b_2 + 5b_1 = 0$; this is the constraint that must be imposed on b if the system (4) is to be solvable. The reader should be morally convinced that these two planes are the same; all four of the vectors displayed above satisfy this constraint. Geometrically, we shall see that the vector $(5, -2, 1)$ is perpendicular to the plane, and therefore to each column of A.

If we assume that b lies in this plane, and thus belongs to the column space, then the solutions of $Ax = b$ are easy to find. The last equation in the system (4) amounts only to $0 = 0$. To the free variables v and y, we may assign arbitrary values as before. Then the basic variables are still determined by back-substitution:

$$3w + y = b_2 - 2b_1, \qquad \text{or} \qquad w = -\tfrac{1}{3}y + \tfrac{1}{3}(b_2 - 2b_1)$$

$$u + 3v + 3w + 2y = b_1, \qquad \text{or} \qquad u = -3v - y + 3b_1 - b_2.$$

Again there is a double infinity of solutions. Looking at the u and w just computed in terms of v and y, the general solution can be expressed as

$$
x = v \begin{bmatrix} -3 \\ 1 \\ 0 \\ 0 \end{bmatrix} + y \begin{bmatrix} -1 \\ 0 \\ -\frac{1}{3} \\ 1 \end{bmatrix} + \begin{bmatrix} 3b_1 - b_2 \\ 0 \\ \frac{1}{3}(b_2 - 2b_1) \\ 0 \end{bmatrix}. \tag{5}
$$

Compared with the solutions (3), which applied to the homogeneous case $b = 0$, the only difference is the inclusion of the vector $(b_1, 0, \frac{1}{3}(b_2 - 2b_1), 0)$. This is a *particular solution* to $Ax = b$, and the general solution x is a sum of this particular solution and the general solution to $Ax = 0$. Geometrically, the solutions again lie on a plane in four-space, but they do not form a subspace, since the plane does not go through the origin. The origin $x = 0$ is not a solution when $b \neq 0$. The plane is certainly *parallel* to the nullspace we had before, but it is displaced from it; it is shifted along the vector giving the particular solution. We should not expect a subspace, since if $Ax = b$ and $Ax' = b$, the sum of these two solutions will be a vector x'' satisfying $Ax'' = 2b$. The solution set is a subspace only for homogeneous equations.

Note that our particular solution is only one of the points lying on the plane of all solutions. Any other point would have done equally well as a particular solution; the one we chose corresponded to the particular values $v = 0$, $y = 0$, of the free variables.

We summarize the conclusions reached by applying Gaussian elimination to a rectangular matrix.

2D   Suppose the m by n matrix A is reduced by elementary operations and row exchanges to a matrix U in echelon form. Let there be r nonzero pivots; the last $m - r$ rows of U are zero. Then there will be r basic variables and $n - r$ free variables, corresponding to the columns of U with and without pivots. The nullspace, formed of solutions to $Ax = 0$, has the $n - r$ free variables as independent parameters. If $r = n$, there are no free variables and the nullspace contains only $x = 0$.

Solutions exist for every right side b if and only if $r = m$; with this number of pivots, the matrix U has no zero rows, and $Ux = c$ can be solved by back-substitution. In case $r < m$, U will have $m - r$ zero rows and there are $m - r$ constraints on b in order for $Ax = b$ to be solvable; they appear explicitly in the last $m - r$ rows of $Ux = c$. If one particular solution exists, then every other solution differs from it by a vector in the nullspace of A.

The number r is called the *rank* of the matrix A.

EXERCISE 2.2.2   Construct the smallest system you can with more unknowns than equations, but no solution.

EXERCISE 2.2.3   Compute an LU factorization for

$$A = \begin{bmatrix} 1 & 2 & 0 & 1 \\ 0 & 1 & 1 & 0 \\ 1 & 2 & 0 & 1 \end{bmatrix}.$$

Determine a set of basic variables and a set of free variables, and find the general solution to $Ax = 0$. Write it in a form similar to (3). What is the rank of A?

EXERCISE 2.2.4   For the matrix

$$A = \begin{bmatrix} 0 & 1 & 4 & 0 \\ 0 & 2 & 8 & 0 \end{bmatrix},$$

determine the echelon form U, the basic variables, the free variables, and the general solution to $Ax = 0$. Then apply elimination to $Ax = b$, with components b_1 and b_2 on the right side; find the conditions for $Ax = b$ to be consistent (that is, to have a solution) and find the general solution in the same form as Eq. (5). What is the rank of A?

EXERCISE 2.2.5   Carry out the same steps, with b_1, b_2, b_3, b_4 on the right side, for the transposed matrix

$$A = \begin{bmatrix} 0 & 0 \\ 1 & 2 \\ 4 & 8 \\ 0 & 0 \end{bmatrix}.$$

EXERCISE 2.2.6   Write the general solution to

$$\begin{bmatrix} 1 & 2 & 2 \\ 2 & 4 & 5 \end{bmatrix} \begin{bmatrix} u \\ v \\ w \end{bmatrix} = \begin{bmatrix} 1 \\ 4 \end{bmatrix}$$

as the sum of a particular solution to $Ax = b$ and the general solution to $Ax = 0$, as in (5).

EXERCISE 2.2.7   Describe the set of attainable right sides b for

$$\begin{bmatrix} 1 & 0 \\ 0 & 1 \\ 2 & 3 \end{bmatrix} \begin{bmatrix} u \\ v \end{bmatrix} = \begin{bmatrix} b_1 \\ b_2 \\ b_3 \end{bmatrix},$$

by finding the constraints on b that turn the third equation into $0 = 0$ (after elimination). What is the rank?

EXERCISE 2.2.8   Find the value of c which makes it possible to solve

$$\begin{aligned} u + v + 2w &= 2 \\ 2u + 3v - w &= 5 \\ 3u + 4v + w &= c. \end{aligned}$$

An optional remark   In many texts the elimination process does not stop at U, but continues until the matrix is in a still simpler "row-reduced echelon form." The difference is that all pivots are normalized to $+1$, by dividing each row by a constant, and zeros are produced not only below but also above every nonzero pivot. For the matrix A used in the text, this form would be

$$\begin{bmatrix} 1 & 3 & 0 & 1 \\ 0 & 0 & 1 & \frac{1}{3} \\ 0 & 0 & 0 & 0 \end{bmatrix}.$$

The row-reduced echelon form of a nonsingular square matrix is exactly the identity matrix I; it is an instance of *Gauss–Jordan elimination* (p. 29), instead of the ordinary Gaussian reduction to $A = LU$. Just as Gauss–Jordan is slower in practical calculations with square matrices, and any band structure of the matrix is lost in A^{-1}, this special echelon form requires too many operations to be the first choice on a computer. It does, however, have some theoretical importance as a "canonical form" for A: Regardless of the choice of elementary operations, including row exchanges and row divisions, the final row-reduced echelon form of A is always the same.

2.3 ■ LINEAR INDEPENDENCE, BASIS, AND DIMENSION

By themselves, the numbers m and n give an incomplete picture of the true size of a linear system. The matrix in our example had three rows and four columns, but in fact the third row was only a combination of the first two. After elimination it became a zero row and had no real effect on the homogeneous problem $Ax = 0$. The four columns also failed to be independent, and the column space degenerated into a two-dimensional plane; the second and fourth columns were simple combinations of the first and third.

The important number which is beginning to emerge is the **rank** r. The rank was introduced on p. 54 in a purely computational way, as the number of nonzero pivots in the elimination process—or equivalently, as the number of nonzero rows in the final matrix U. This definition is so mechanical that it could be given to a computer. But it would be wrong to leave it there because the rank has a simple and intuitive meaning: *It counts the number of genuinely independent rows in the matrix A.* We want to give this quantity, and others like it, a definition that is mathematical rather than computational.

The first step is to define **linear independence**. Given a set of vectors $v_1, \ldots, v_k$, we look at their linear combinations $c_1 v_1 + c_2 v_2 + \cdots + c_k v_k$. The trivial combination, with all weights $c_i = 0$, obviously produces the zero vector: $0 v_1 + \cdots + 0 v_k = 0$. The question is whether any other combination produces zero.

2E If all nontrivial combinations of the vectors are nonzero,

$$c_1v_1 + \cdots + c_kv_k \neq 0 \quad \text{unless} \quad c_1 = c_2 = \cdots = c_k = 0, \quad (6)$$

then the vectors $v_1, \ldots, v_k$ are *linearly independent*. Otherwise they are linearly dependent, and one of them is a linear combination of the others.

EXAMPLE 1 If one of the given vectors, say v_1, should happen to be already the zero vector, then the set is certain to be linearly dependent. We may choose $c_1 = 3$ and all other $c_i = 0$; this is a nontrivial combination that produces zero.

EXAMPLE 2 The three rows of the matrix

$$A = \begin{bmatrix} 1 & 3 & 3 & 2 \\ 2 & 6 & 9 & 5 \\ -1 & -3 & 3 & 0 \end{bmatrix}.$$

are linearly dependent; if these rows are v_1, v_2, and v_3, then $5v_1 - 2v_2 + v_3 = 0$. (This is the same as the combination of b_1, b_2, b_3, which had to vanish on the right-hand side in order for the equations to be consistent; otherwise, the third equation of $Ux = c$ would have been impossible.)

EXAMPLE 3 The rows of the n by n identity matrix

$$I = \begin{bmatrix} 1 & 0 & 0 & \cdot \\ 0 & 1 & 0 & \cdot \\ 0 & 0 & \cdot & 0 \\ \cdot & \cdot & 0 & 1 \end{bmatrix}$$

are linearly independent. We give these particular vectors the special notation $e_1, \ldots, e_n$; they are the unit vectors in the coordinate directions,

$$e_1 = (1, 0, \ldots, 0), \quad \ldots, \quad e_n = (0, 0, \ldots, 1). \quad (7)$$

The usual procedure for proving independence is this: *Assume that some linear combination gives zero, and then prove that all the weights c_i in the combination must be zero.* In the present case, we can simply compute

$$c_1e_1 + \cdots + c_ne_n = (c_1, c_2, \ldots, c_n).$$

Assume that this combination is the zero vector; then obviously every $c_i = 0$. Since no nontrivial combination produces zero, the coordinate vectors $e_1, \ldots, e_n$ are independent.

EXAMPLE 4 Suppose U is an n by n upper triangular matrix, with nonzero entries (the pivots) on the diagonal. Then the rows of U are linearly independent. For proof, we start as before by supposing that some combination of the rows produces zero, $c_1v_1 + \cdots + c_nv_n = 0$. Since U is upper triangular, its first row v_1 is the only one with a nonzero first component u_{11}. By looking at the first component of $c_1v_1 + \cdots + c_nv_n = 0$, we conclude that $c_1u_{11} = 0$, and therefore that $c_1 = 0$. From the second components we find $c_1u_{12} + c_2u_{22} = 0$, and therefore (since we just proved $c_1 = 0$, and assumed that the pivot $u_{22} \neq 0$) we have $c_2 = 0$. Each of the weights, in turn, is proved to be zero. The only combination of the rows that produces zero is the trivial one, and the rows must be linearly independent.

The same reasoning applies to the columns of U; with nonzero pivots, those columns are independent. In fact, the reasoning applies to any echelon matrix, provided we pick out the rows (or columns) that contain the pivots:

2F   The r nonzero rows of an echelon matrix U are linearly independent, and so are the r columns that contain nonzero pivots.

We emphasize that the definition of linear independence is "coordinate-free." Given k points in n-dimensional space, either the vectors from the origin to those points can be combined to give zero, or they cannot—regardless of where we place the coordinate axes. A rotation of axes will change the coordinates of the points, but it has no effect on the question of independence or dependence.

On the other hand, given an arbitrary set of vectors $v_1, \ldots, v_k$, the verification of their dependence or independence obviously does require a calculation. Since we have to consider the combinations $c_1v_1 + \cdots + c_kv_k$, the natural step is to form the matrix A whose columns are the given vectors. Then if we write c for the vector of weights $(c_1, \ldots, c_k)$, and look to see how matrix multiplication works,

$$Ac = \begin{bmatrix} | & | & & | \\ v_1 & v_2 & \cdots & v_k \\ | & | & & | \end{bmatrix} \begin{bmatrix} c_1 \\ c_2 \\ \vdots \\ c_k \end{bmatrix} = c_1v_1 + \cdots + c_kv_k.$$

The vectors are dependent if and only if there is a nontrivial solution to $Ac = 0$. This is settled by going through the elimination process; if the rank of A is k, then there will be no free variables and no nullspace (except $c = 0$), and the vectors are linearly independent. If the rank is less than k, then at least one variable is free to be chosen nonzero, and the columns are linearly dependent.

One case is of special importance. Let the vectors have m components, so that A is an m by k matrix. Suppose now that $k > m$. Then it will be impossible for A to have rank k; the number of nonzero pivots can never exceed the number of rows. The rank must be less than k, and a homogeneous system $Ac = 0$ with more unknowns than equations always has solutions $c \neq 0$.

2G   A set of k vectors in $\mathbf{R}^m$ must be linearly dependent if $k > m$.

The reader will recognize this as a disguised form of 2C.

EXAMPLE 5   Consider the three columns of

$$A = \begin{bmatrix} 1 & 2 & 1 \\ 1 & 3 & 2 \end{bmatrix}.$$

There cannot be three independent vectors in $\mathbf{R}^2$, and to find the combination of the columns producing zero we solve $Ac = 0$:

$$A \to U = \begin{bmatrix} 1 & 2 & 1 \\ 0 & 1 & 1 \end{bmatrix}.$$

If we give the value 1 to the free variable c_3, then back-substitution in $Uc = 0$ gives $c_2 = -1$, $c_1 = 1$. With these three weights, the first column minus the second plus the third equals zero.

EXERCISE 2.3.1   Decide whether or not the following vectors are linearly independent, by looking for the coefficients $c_1, \ldots, c_4$:

$$v_1 = \begin{bmatrix} 1 \\ 1 \\ 0 \\ 0 \end{bmatrix}, \quad v_2 = \begin{bmatrix} 1 \\ 0 \\ 1 \\ 0 \end{bmatrix}, \quad v_3 = \begin{bmatrix} 0 \\ 0 \\ 1 \\ 1 \end{bmatrix}, \quad v_4 = \begin{bmatrix} 0 \\ 1 \\ 0 \\ 1 \end{bmatrix}.$$

EXERCISE 2.3.2   Decide the independence or dependence of the rows of A in Example 5, which had dependent columns.

EXERCISE 2.3.3   Prove that if any diagonal element of

$$T = \begin{bmatrix} a & b & c \\ 0 & d & e \\ 0 & 0 & f \end{bmatrix}$$

is zero, the rows are linearly dependent.

EXERCISE 2.3.4   Is it true that if v_1, v_2, v_3 are linearly independent, then also the vectors $w_1 = v_1 + v_2$, $w_2 = v_1 + v_3$, $w_3 = v_2 + v_3$ are linearly independent? (Hint: Assume some combination $c_1 w_1 + c_2 w_2 + c_3 w_3 = 0$, and find which c_i are possible.)

The next step in discussing vector spaces is to define what it means for a set of vectors to *span the space*. We used this term at the beginning of the chapter, when we spoke of the plane that was spanned by the two columns of the matrix, and called this plane the column space. The general definition is simply this:

2H   If a vector space V consists of all linear combinations of the particular vectors $w_1, \ldots, w_l$, then these vectors **span** the space. In other words, every vector v in V can be expressed as some combination of the w's:

$$v = c_1 w_1 + \cdots + c_l w_l \qquad \text{for some coefficients } c_i. \qquad (8)$$

It is permitted that more than one set of coefficients c_i could give the same vector v; the coefficients need not be unique because the spanning set might be excessively large—it could even include the zero vector.

EXAMPLE 6   The vectors $w_1 = (1, 0, 0)$, $w_2 = (0, 1, 0)$, and $w_3 = (-2, 0, 0)$ span a plane (the x-y plane) within three-dimensional space. So would the first two vectors alone, whereas w_1 and w_3 span only a line.

EXAMPLE 7   The *column space* of an m by n matrix is the space that is spanned by the columns. It is a subspace of the full m-dimensional space $\mathbf{R}^m$ (and of course it may be the whole of $\mathbf{R}^m$). The *row space* of A is defined in the same way; it is the subspace of $\mathbf{R}^n$ that is spanned by the rows of A. (We shall think of the rows as members of $\mathbf{R}^n$, even though the n components happen to be written horizontally.) If $m = n$, then both the row space and column space are subspaces of $\mathbf{R}^n$—and they might even be the same subspace.

EXAMPLE 8   The coordinate vectors in $\mathbf{R}^n$,

$$e_1 = (1, 0, \ldots, 0), \quad \ldots, \quad e_n = (0, \ldots, 0, 1)$$

span $\mathbf{R}^n$. For proof, we have only to show how any vector $x = (x_1, x_2, \ldots, x_n)$ can be written as a combination of the e_i. The right weights are the components x_i themselves:

$$x = x_1 e_1 + x_2 e_2 + \cdots + x_n e_n.$$

In this example, we know that in addition to spanning the space, the set $e_1, \ldots e_n$ is also linearly independent. Roughly speaking, *no vectors in the set are wasted*. The matrix of Example 5 had a redundant column, contributing nothing new to the column space, but here the spanning set is of minimal size; if any vector e_i were removed, the remaining vectors would fail to span $\mathbf{R}^n$. Such a set of vectors is called a basis:

21 A *basis* for a vector space is a set of vectors having two properties at once:

    (1)   It is linearly independent.
    (2)   It spans the space.

This combination of properties is absolutely fundamental to the theory of vector spaces. It means that every vector v in the space can be expanded in one and only one way as a combination of the basis vectors, $v = a_1 v_1 + \cdots + a_k v_k$. (It can be expanded because the vectors span the space; if also $v = b_1 v_1 + \cdots + b_k v_k$, then subtraction gives $0 = \sum (a_i - b_i) v_i$, so linear independence forces every difference $a_i - b_i$ to be zero. Therefore the weights in the expansion, when the v_i are a basis, are uniquely determined by the vector v.)

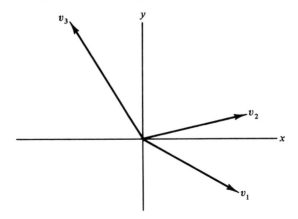

Fig. 2.3.   A spanning set and a basis in $\mathbf{R}^2$.

EXAMPLE 9   Consider the usual x-y plane (Fig. 2.3), which is just $\mathbf{R}^2$. The vector v_1 by itself is linearly independent, but fails to span $\mathbf{R}^2$. The three vectors v_1, v_2, v_3 certainly span $\mathbf{R}^2$, but are not independent. *Any two* of these vectors, say v_1 and v_2, have both properties—they span, and they are independent, so they form a basis. Notice that *a vector space does not have a unique basis.*

EXAMPLE 10   Consider the echelon matrix U corresponding to the matrix A in Example 2:

$$U = \begin{bmatrix} 1 & 3 & 3 & 2 \\ 0 & 0 & 3 & 1 \\ 0 & 0 & 0 & 0 \end{bmatrix}.$$

The four columns span the column space, as always, but they are not independent. There are many possibilities for a basis, but we propose a specific choice: *the columns that contain nonzero pivots* (in this case the first and third, corresponding to the basic variables) *are a basis for the column space.* We noted in 2F that these columns are independent, and it is easy to see that they span the space. In fact, the column space of U is just the x-y plane within $\mathbf{R}^3$.

The first two rows of U—and in general the nonzero rows of any echelon matrix—are a basis for its row space.

EXERCISE 2.3.5   Describe, in words or in a sketch of the x-y plane, the column space and the row space of $A = \begin{bmatrix} 1 & 2 \\ 3 & 6 \end{bmatrix}$. Give a basis for the column space.

EXERCISE 2.3.6   By locating the pivots, find a basis for the column space of

$$U = \begin{bmatrix} 0 & 1 & 4 & 3 \\ 0 & 0 & 2 & -2 \\ 0 & 0 & 0 & 0 \\ 0 & 0 & 0 & 0 \end{bmatrix}.$$

Express each column that is not in the basis as a combination of the basic columns. Find also a basis for the row space (which is a different subspace of $\mathbf{R}^4$).

EXERCISE 2.3.7   Suppose we think of each 2 by 2 matrix as a "vector." Although these are not vectors in the usual sense, we do have rules for adding matrices and multiplying by scalars, and the set of matrices is closed under these operations. Find a basis for this vector space. What subspace is spanned by the set of all echelon matrices U?

EXERCISE 2.3.8   Find two different bases for the subspace of all vectors in $\mathbf{R}^3$ whose first two components are equal.

EXERCISE 2.3.9   Find a counterexample to the following statement: If $v_1, \ldots, v_4$ is a basis for the vector space $\mathbf{R}^4$, and if W is a subspace, then some subset of the v's will form a basis for W.

In spite of the fact that there is no unique choice of basis since infinitely many different possibilities would do equally well, there *is* something common to all of these choices. It is a property that is intrinsic to the space itself:

2J   Any two bases for a vector space V contain the same number of vectors. This number, which is shared by all bases and expresses the number of "degrees of freedom" of the space, is called the **dimension** of V.

Of course we have to prove this fact, that all possible bases contain the same number of vectors. First, we ask the reader to look back at some of the examples, and notice their dimension. The x-y plane in Fig. 2.3 had two vectors in every basis; its dimension is 2. More generally, the dimension of $\mathbf{R}^n$ is n—and the coordinate vectors e_i form a convenient basis. The row space of U in Example 10 was a two-dimensional subspace of $\mathbf{R}^4$, and its column space was a two-dimensional subspace of $\mathbf{R}^3$. The zero matrix $A = 0$ would be rather exceptional; both its column space and row space consist only of the zero vector, in $\mathbf{R}^m$ and $\mathbf{R}^n$, respectively. By convention, the empty set is a basis for such spaces, and their dimension is zero.

Theorem 2J, on which the idea of dimension depends, is equivalent to

2K   Suppose that $v_1, \ldots, v_m$ and $w_1, \ldots, w_n$ are both bases for the same vector space V. Then $m = n$.

Proof   Suppose one set is smaller than the other, say $m < n$; we want to arrive at a contradiction. Since the v's form a basis, they must span the space, and every w_j can be written as a combination of the v's:

$$w_j = a_{1j}v_1 + \cdots + a_{mj}v_m = \sum_{i=1}^{m} a_{ij}v_i .$$

Of course we have no way to know the weights a_{ij}, but we can still look at combinations of the w_j :

$$c_1 w_1 + \cdots + c_n w_n = \sum_{j=1}^{n} c_j \left[\sum_{i=1}^{m} a_{ij} v_i \right] = \sum_{i=1}^{m} v_i \left[\sum_{j=1}^{n} a_{ij} c_j \right]. \qquad (9)$$

This rearrangement of the "double sum" is just like adding up all the entries of a matrix—we can sum along each row and add together those subtotals, or alternatively we can sum first along each column.

Consider now the equations $\sum a_{ij} c_j = 0$, for $i = 1, \ldots m$. This is a system of m homogeneous equations in n unknowns, with $m < n$. Therefore it has a solution $c = (c_1, \ldots, c_n)$ other than the trivial solution $c = 0$. Substituting $\sum a_{ij} c_j = 0$ into (9), this same set of weights leads us to

$$c_1 w_1 + \cdots + c_n w_n = 0.$$

Evidently, the vectors w_j are not linearly independent. Since this contradicts the hypothesis that they form a basis, we must give up the possibility that $m < n$.

This proof was the same as the one used on p. 58 to show that every set of $m + 1$ vectors in $\mathbf{R}^m$ must be dependent. In fact we can see that the general result is this: *In a subspace of dimension k, no set of more than k vectors can be linearly independent, and no set of fewer than k vectors can span the space.*

There are other "dual" theorems, of which we mention only one; it permits us to start with a set of vectors that is either too small or too big, and to end up with a basis:

2L Any linearly independent set in V can be extended to a basis, by adding more vectors if necessary.

Any spanning set in V can be reduced to a basis, by discarding vectors if necessary.

EXERCISE 2.3.10 Suppose that the vectors x, y, and z in $\mathbf{R}^3$ satisfy $x + y + z = 0$. What are all the possible dimensions of the subspace they span? Illustrate each possibility by a specific example of x, y, and z.

EXERCISE 2.3.11 For the matrix $A = \begin{bmatrix} 1 & 2 & 1 \\ 0 & 0 & 4 \end{bmatrix}$, extend the set of rows to a basis for $\mathbf{R}^3$, and (separately) reduce the set of columns to a basis for $\mathbf{R}^2$.

EXERCISE 2.3.12 Suppose V is known to have dimension k. Prove, using the dual theorems 2L if you wish, that

(i)   any k independent vectors in V form a basis;
(ii) any k vectors that span V form a basis.

In other words, if the number of vectors is known to be right, either of the two properties of a basis implies the other.

EXERCISE 2.3.13   Find the dimension of the space of 3 by 3 symmetric matrices, as well as a basis.

EXERCISE 2.3.14   Prove that if V and W are three-dimensional subspaces of $\mathbf{R}^5$, then V and W must have a nonzero vector in common. Hint: Start with bases for the two subspaces, making six vectors in all.

One final note about the language of linear algebra. We never use the terms "basis of a matrix" or "rank of a row space" or "dimension of a basis"; these phrases have no meaning. It is the *dimension of the row space* that equals the *rank of the matrix*.

2.4 ■ THE FOUR FUNDAMENTAL SUBSPACES

The previous section dealt with definitions rather than constructions; we know what a basis is, but not how to find one. Now, starting from an explicit description of a subspace, we would like to compute an explicit basis.

Subspaces are generally described in one of two ways. First, we may be given a set of vectors that span the space; this is the case for the row space and the column space, when the rows and columns are specified. Second, we may be given a list of constraints on the subspace; we are told, not which vectors are in the space, but which conditions they must satisfy. The nullspace, for example, consists of all vectors which satisfy $Ax = 0$, and each equation in this system represents such a constraint. In the first kind of description, there may be redundant rows or columns; in the second kind there may be redundant constraints. In neither case it is possible to write down a basis by inspection, and some systematic procedure is necessary.

The reader can guess what that procedure will be: We shall show how to find, from the L and U (and P) which are produced by elimination, a basis for each of the subspaces associated with A. Then, even if it makes this section longer than the others, we have to look at the two extreme cases:

   (i)   When the rank is very small, $r = 1$, the row and column spaces are especially simple.
   (ii)   When the rank is very large, $r = n$ or $r = m$ or $r = m = n$, the matrix has a left-inverse B or a right-inverse C or a two-sided A^{-1}.

To organize the whole discussion, we consider each of the four fundamental subspaces in turn. The four subspaces associated with U are easy to find and our problem will be to connect them to the original matrix A.

1. The row space of A Elimination acts on A to produce an echelon matrix U, and the row space of U is completely straightforward: Its dimension is the rank r, and a basis is given by its r nonzero rows. Fortunately, it is equally easy to deal with A.

2M   The row space of A has the same dimension r as the row space of U, and the same basis, because the two row spaces are the same.

The reason is that each elementary operation leaves the row space unchanged. Each row in the new matrix U is a combination of the original rows in A, so the new row space is contained in the old—while at the same time, because every step can be reversed by another elementary operation, the old row space is also contained in the new.

Note that we did not start with the m rows of A, which span the row space, and discard $m - r$ of them to end up with a basis. According to 2L, we could have done so; but it might be hard to decide which rows to keep and which to discard, so it was easier just to take the nonzero rows of U.

This reasoning also justifies our previous definition of the rank of A. We took r to be the number of nonzero pivots (or nonzero rows) of U, without considering all the different echelon matrices that could be produced by different sequences of elimination steps. We now know that the nonzero rows of every such U will be a basis for the row space of A, and that all bases have the same number of members. Therefore r can be defined in a new way; *the rank is the dimension of the row space.*

2. The nullspace of A   Recall that the original purpose of elimination was to simplify a system of linear equations without changing any of the solutions. The system $Ax = 0$ is reduced to $Ux = 0$, and this process is reversible. *Therefore the nullspace of A is the same as the nullspace of U.* Of the m constraints apparently imposed by the m equations $Ax = 0$, only r are independent. They are specified by any r linearly independent rows of A, or (more clearly) by the r nonzero rows of U. If we choose the latter, it provides a definite way to find a basis for the nullspace:

2N   The nullspace of A (which we denote by $\mathfrak{N}(A)$) is of dimension $n - r$. A basis can be constructed by reducing to the system $Ux = 0$, which has $n - r$ free variables—corresponding to the columns of U that do not contain pivots. Then, in turn, we give to each free variable the value 1, to the other free variables the value 0, and solve $Ux = 0$ by back-substitution for the remaining (basic) variables. The $n - r$ vectors produced in this way are a basis for $\mathfrak{N}(A)$.

In our example (p. 52), the free variables were v and y, and the basis was

$$\begin{array}{cc} \begin{array}{c} v = 1 \\ y = 0 \end{array} \quad x_1 = \begin{bmatrix} -3 \\ 1 \\ 0 \\ 0 \end{bmatrix} ; & \begin{array}{c} v = 0 \\ y = 1 \end{array} \quad x_2 = \begin{bmatrix} -1 \\ 0 \\ -\frac{1}{3} \\ 1 \end{bmatrix} . \end{array}$$

It is easy to see, either for this example or in general, that these vectors x_i must be independent. Any combination $c_1 x_1 + c_2 x_2$ has the value c_1 as its v component, and c_2 as its y component. Therefore the only way to have $c_1 x_1 + c_2 x_2 = 0$ is to have $c_1 = c_2 = 0$. These two vectors also span the nullspace; the general solution is a combination $v x_1 + y x_2$. Thus the $n - r = 4 - 2$ vectors x_i are a basis.

The nullspace is also called the *kernel* of A, and its dimension is the *nullity* of A. Denoting this nullity by $\nu(A)$, the essential fact to remember is its relation to the rank:

$$\nu(A) = \dim \mathfrak{N}(A) = n - r.$$

3. The column space of A First another point of notation; the column space is often called the **range** of A, and denoted by $\mathfrak{R}(A)$. This is consistent with the usual idea of the range of a function f, as the set of all possible values $f(x)$; if $f(x)$ is defined, then x is in the domain and the value $f(x)$ is in the range. In our case, with vectors, the function in question is $f(x) = Ax$. Its domain consists of all x in $\mathbf{R}^n$; its range is all possible vectors Ax. (In other words, all b for which $Ax = b$ can be solved.) We know that this is the same as all possible combinations of the columns; the range is the column space. We plan to keep the useful term *column space*, but also to adopt the shorthand notation $\mathfrak{R}(A)$.†

Our problem is to find a basis for $\mathfrak{R}(A)$, and its dimension. One reasonable idea is this, to let the columns of A be the rows of a new matrix, and work with a row space again. This new matrix is A^T, the *transpose* of A. Since the columns of A are the rows of A^T, the latter must be an n by m matrix; each entry of A has been carried across the main diagonal into the "mirror image" of its original position, and $(A^T)_{ij} = A_{ji}$. At the same time, the rows of A have become the columns of A^T, and another useful notation falls out: *The row space of A is* $\mathfrak{R}(A^T)$, the column space of A^T. This is especially useful because $\mathfrak{R}(A^T)$ is made up of column vectors—the columns of A^T are the rows of A written vertically—so now even the row space adheres to the convention that vectors are column vectors.

Certainly we could reduce A^T to row echelon form, and thereby understand the column space of A. Nevertheless, this is not the idea we want. There are many uses for the transpose, but this is not one of them. We want to avoid introducing a new quantity, the rank of A^T, and carrying out a new reduction to echelon form. If possible, we prefer to compute the dimension of the column space in terms of the original numbers m, n, and r.

It must be emphasized that A does not have the same column space as U. Elimination left the row space and nullspace unchanged, but the columns are

---

† It is a sad accident that *row space* also starts with the same letter; in this book, r stands for rank, and $\mathfrak{R}$ stands for column space.

entirely different; compare the columns of

$$A = \begin{bmatrix} 1 & 3 & 3 & 2 \\ 2 & 6 & 9 & 5 \\ -1 & -3 & 3 & 0 \end{bmatrix} \quad \text{and} \quad U = \begin{bmatrix} 1 & 3 & 3 & 2 \\ 0 & 0 & 3 & 1 \\ 0 & 0 & 0 & 0 \end{bmatrix}.$$

Nevertheless, whenever certain columns of U form a basis for the column space of U, the corresponding columns of A form a basis for the column space of A. The reason is this: $Ax = 0$ if and only if $Ux = 0$. The two systems are equivalent and have the same solutions. Looking at matrix multiplication, $Ax = 0$ expresses a linear dependence among the columns of A, with weights given by the components of x. Therefore every such dependence is matched by a linear dependence $Ux = 0$ among the columns of U, with exactly the same weights. *If a set of columns of A is independent, then the same is true of the corresponding columns of U, and vice versa.*† For both of our matrices A and U, the last column equals the first plus $\frac{1}{3}$ of the third, and the second column is three times the first.

Now, to find a basis for $\mathcal{R}(A)$, we have the simpler task of finding a basis for the column space of U. This has already been done (p. 58), and the conclusion was this: The r columns of U containing nonzero pivots are a basis for the column space of U. We transfer this result over to A, as follows:

20   The dimension of the column space $\mathcal{R}(A)$ equals the rank r, which also equals the the dimension of the row space: ***The number of independent columns equals the number of independent rows.*** A basis for $\mathcal{R}(A)$ is formed by those r columns of A which correspond, over in U, to the columns containing nonzero pivots.

This fact, that the row space and the column space have the same dimension r, is one of the most important theorems in linear algebra. It is often abbreviated as *"row rank = column rank."* It expresses a result that, for a random 10 by 12 matrix, is not at all obvious.

To see once more that the column space of U has dimension r, consider a typical situation of rank $r = 3$. The echelon matrix U has three independent rows:

$$U = \begin{bmatrix} d_1 & * & * & * & * & * \\ 0 & 0 & 0 & d_2 & * & * \\ 0 & 0 & 0 & 0 & 0 & d_3 \\ 0 & 0 & 0 & 0 & 0 & 0 \end{bmatrix}.$$

† I think this is the most subtle argument to appear so far in the book. Fortunately, it is not wasted: The conclusion 20 to which it leads is also the most subtle and most significant so far.

We claim that there are also three independent columns, and no more; the columns have only three nonzero components. Therefore if we can show that the three basic columns—the first, fourth, and sixth—are linearly independent, they must be a basis (for the column space of U, not A!). Suppose that some combination of these basic columns produced zero:

$$c_1 \begin{bmatrix} d_1 \\ 0 \\ 0 \\ 0 \end{bmatrix} + c_2 \begin{bmatrix} * \\ d_2 \\ 0 \\ 0 \end{bmatrix} + c_3 \begin{bmatrix} * \\ * \\ d_3 \\ 0 \end{bmatrix} = \begin{bmatrix} 0 \\ 0 \\ 0 \\ 0 \end{bmatrix}.$$

Working upward in the usual way, c_3 must be zero because the pivot $d_3 \neq 0$, then c_2 must be zero because $d_2 \neq 0$, and finally $c_1 = 0$. This establishes linear independence and completes the proof. Since $Ax = 0$ if and only if $Ux = 0$, we must find that the first, fourth, and sixth columns of A—whatever the original matrix A was, which we do not even know in this example—are a basis for $\Re(A)$.

We come to the fourth fundamental subspace, which has been keeping out of sight. Since the first three subspaces were $\Re(A^T)$, $\Re(A)$, and $\Re(A)$, it is not hard to guess that the fourth one must be $\Re(A^T)$.

4. The nullspace of A^T This is a subspace of $\mathbf{R}^m$, consisting of those vectors y such that $A^T y = 0$. Thus the columns of A^T, combined with the weights $y_1, \ldots, y_m$, produce the zero column. Since the columns of A^T are the rows of A, we can transpose $A^T y = 0$ into an equation for row vectors:

$$y^T A = \begin{bmatrix} y_1 & \cdots & y_m \end{bmatrix} \begin{bmatrix} \\ \\ A \\ \\ \\ \end{bmatrix} = \begin{bmatrix} 0 & \cdots & 0 \end{bmatrix}.$$

Such a row vector y^T is sometimes called a *left nullvector* of A. The rows of A, weighted by $y_1, \ldots, y_m$, produce the zero row.

The dimension of $\Re(A^T)$ is easy to find. For any matrix, the number of basic variables plus the number of free variables must match the total number of columns. In other words,

rank + nullity = dimension of column space + dimension of nullspace

$$= \text{number of columns}. \tag{10}$$

This rule applies equally to A^T, which has m columns and is just as good a matrix as A. But row rank = column rank = r, which leaves

$$r + \dim \Re(A^T) = m. \tag{11}$$

2P The left nullspace $\mathfrak{N}(A^T)$ is of dimension $m - r$.

To find a basis, we start from $PA = LU$ or $L^{-1}PA = U$. The last $m - r$ rows of U are zero, and therefore the last $m - r$ rows of $L^{-1}P$ are a basis for the left nullspace. When they multiply A, they produce zero.

Now we know the dimensions of the four spaces. We can summarize them in a table, and it even seems fair to advertise them as the

---

Fundamental Theorem of Linear Algebra, Part 1

1. $\mathfrak{R}(A^T)$ = row space of A; dimension r
2. $\mathfrak{N}(A)$ = nullspace of A; dimension $n - r$
3. $\mathfrak{R}(A)$ = column space of A; dimension r
4. $\mathfrak{N}(A^T)$ = left nullspace of A; dimension $m - r$.

EXERCISE 2.4.1 True or false: If $m = n$, then the nullspace of A equals the left nullspace.

EXERCISE 2.4.2 Find the dimension and construct a basis for the four subspaces associated with the matrix in Exercise 2.2.4:

$$A = \begin{bmatrix} 0 & 1 & 4 & 0 \\ 0 & 2 & 8 & 0 \end{bmatrix}.$$

EXERCISE 2.4.3 Find the dimension and a basis for the four fundamental subspaces of

$$A = \begin{bmatrix} 1 & 2 & 0 & 1 \\ 0 & 1 & 1 & 0 \\ 1 & 2 & 0 & 1 \end{bmatrix}.$$

EXERCISE 2.4.4 Describe the four subspaces associated with

$$A = \begin{bmatrix} 0 & 1 & 0 & 0 \\ 0 & 0 & 1 & 0 \\ 0 & 0 & 0 & 1 \\ 0 & 0 & 0 & 0 \end{bmatrix}.$$

EXERCISE 2.4.5 Show that if the product of two matrices is the zero matrix, $AB = 0$, then the column space of B is contained in the nullspace of A.

EXERCISE 2.4.6 Explain why $Ax = b$ is solvable if and only if rank $A = $ rank A', where A' is formed from A by adding b as an extra column. Hint: The rank is the dimension of the column space, and the system is solvable if and only if b is in $\mathfrak{R}(A)$.

Matrices of Rank One

One of the basic themes in mathematics is, given something complicated, to show how it can be put together from simple pieces. We have already seen one example of this synthesis, when the lower triangular L was a product of elementary matrices. Now we turn to the rank r as a different standard of simplicity, and introduce the class of matrices which have *rank one*, $r = 1$. The following matrix is typical:

$$A = \begin{bmatrix} 2 & 1 & 1 \\ 4 & 2 & 2 \\ 8 & 4 & 4 \\ -2 & -1 & -1 \end{bmatrix}.$$

Every row is a multiple of the first row, so the row space is one-dimensional. In fact, we can write the whole matrix in the following special way, as *the product of a column vector and a row vector*:

$$A = \begin{bmatrix} 2 & 1 & 1 \\ 4 & 2 & 2 \\ 8 & 4 & 4 \\ -2 & -1 & -1 \end{bmatrix} = \begin{bmatrix} 1 \\ 2 \\ 4 \\ -1 \end{bmatrix} \begin{bmatrix} 2 & 1 & 1 \end{bmatrix}.$$

The product of a 4 by 1 matrix and a 1 by 3 matrix is a 4 by 3 matrix, and this product has rank one. Note that, at the same time, the columns are all multiples of the same column vector; the column space shares the dimension $r = 1$ and reduces to a line.

The same thing will happen for any other matrix of rank one: *It can be factored into the simple form $A = uv^T$*. The rows are all multiples of the same vector v^T, and the columns are all multiples of the same vector u.

We shall show in the last section how to decompose any matrix of rank r into the sum of r matrices of rank one.

EXERCISE 2.4.7   If a, b, and c are given with $a \neq 0$, how must d be chosen so that

$$A = \begin{bmatrix} a & b \\ c & d \end{bmatrix}$$

has rank one? With this choice of d, factor A into uv^T.

EXERCISE 2.4.8   Compute the product AB of the rank one matrices

$$A = \begin{bmatrix} 2 & -2 \\ 4 & -4 \\ 0 & 0 \end{bmatrix} \quad \text{and} \quad B = \begin{bmatrix} 1 & 1 & 2 \\ 3 & 3 & 6 \end{bmatrix}.$$

Writing A and B in the form uv^T and wz^T, verify that their product is a multiple of the matrix uz^T, and that the multiplying factor is the inner product $v^T w$.

EXERCISE 2.4.9   For the previous matrix A, sketch the row space and the nullspace in the x-y plane.

Existence of Inverses

We know already, from 1.5, that if A has both a left-inverse $(BA = I)$ and a right-inverse $(AC = I)$, then the two are equal: $B = B(AC) = (BA)C = C$. Now, from the rank of a matrix, it is easy to decide which matrices actually have these inverses. Roughly speaking, *an inverse exists only when the rank is as large as possible.*

The rank always satisfies $r \leq m$ and $r \leq n$ since an m by n matrix cannot have more than m independent rows or n independent columns. We want to prove that if $r = m$ there is a right-inverse, and if $r = n$ there is a left-inverse. In the first case $Ax = b$ always has a solution, and in the second case the solution (if it exists) is unique. Only a square matrix can have both $r = m$ and $r = n$, and therefore only a square matrix can achieve both existence and uniqueness.

2Q   EXISTENCE: The system $Ax = b$ has **at least** one solution x for every b if and only if the columns span $\mathbf{R}^m$; then $r = m$. In this case there exists an n by m right-inverse C such that $AC = I_m$, the identity matrix of order m. This is possible only if $m \leq n$.

UNIQUENESS: The system $Ax = b$ has **at most** one solution x for every b if and only if the columns are linearly independent; then $r = n$. In this case there exists an n by m left-inverse B such that $BA = I_n$, the identity matrix of order n. This is possible only if $m \geq n$.

In the first case, one possible solution is $x = Cb$, since then $Ax = ACb = b$. But there will be other solutions if there are other right-inverses.

In the second case, if there is a solution to $Ax = b$, it has to be $x = BAx = Bb$. But there may be no solution.†

EXAMPLE   Consider a simple 2 by 3 matrix of rank 2:

$$A = \begin{bmatrix} 4 & 0 & 0 \\ 0 & 5 & 0 \end{bmatrix}.$$

Since $r = m = 2$, the theorem guarantees a right-inverse C:

$$AC = \begin{bmatrix} 4 & 0 & 0 \\ 0 & 5 & 0 \end{bmatrix} \begin{bmatrix} c_{11} & c_{12} \\ c_{21} & c_{22} \\ c_{31} & c_{32} \end{bmatrix} = \begin{bmatrix} 1 & 0 \\ 0 & 1 \end{bmatrix}.$$

† The number of solutions in the "uniqueness case" is 0 or 1, whereas in the "existence case" it is 1 or ∞.

In fact, there are many right-inverses; the last row of C is completely arbitrary. This is a case of existence but no uniqueness:

$$C = \begin{bmatrix} \frac{1}{4} & 0 \\ 0 & \frac{1}{5} \\ c_{31} & c_{32} \end{bmatrix}$$

For an example in the opposite direction, consider the transpose of A. A^T will be a 3 by 2 matrix, still of rank 2, with infinitely many left-inverses:

$$BA^T = \begin{bmatrix} b_{11} & b_{12} & b_{13} \\ b_{21} & b_{22} & b_{23} \end{bmatrix} \begin{bmatrix} 4 & 0 \\ 0 & 5 \\ 0 & 0 \end{bmatrix} = \begin{bmatrix} 1 & 0 \\ 0 & 1 \end{bmatrix}.$$

Now it is the last column of B that is completely arbitrary. In fact, this B is just the transpose of the previous C.

The example suggests a general way of constructing the matrices C and B, and proving the statements in 2Q. We separate the two cases:

(1)   EXISTENCE   We are given that the columns span $\mathbf{R}^m$, so that the rank (dimension of the column space) is m. Then every vector b, including the coordinate vectors $e_1, \ldots, e_m$, is a linear combination of the columns of A. In other words, we can find at least one solution x_i to the system $Ax_i = e_i$, for each $i = 1, \ldots, m$. If C is the n by m matrix whose columns are these solutions $x_1, \ldots, x_m$, it follows column by column that

$$AC = A[x_1 \quad \cdots \quad x_m] = [e_1 \quad \cdots \quad e_m] = I_m.$$

C is the required right-inverse.

(2)   UNIQUENESS   We are given that the n columns of A are linearly independent. Therefore the column rank is n, there are no free variables, the nullspace contains only 0; and if there is any particular solution to $Ax = b$, then that is the only solution. The simplest construction of a left-inverse is to consider A^T. This is an n by m matrix, and by the fundamental theorem it has the same rank $r = n$—which now matches the number of rows. Therefore, the existence part of the theorem applies to A^T; there is a right-inverse, say Q, such that $A^T Q = I_n$. Transposing, $Q^T A = I_n$ and Q^T is the required left-inverse B.

There is a simpler way of constructing these inverses; we write down

$$B = (A^T A)^{-1} A^T \qquad \text{and} \qquad C = A^T (AA^T)^{-1}.$$

Certainly $BA = I$ and $AC = I$, but what is not so obvious is that $(A^T A)^{-1}$ and $(AA^T)^{-1}$ actually exist. Here the conditions on the column space must play their part. We show later in 3G that $A^T A$ is invertible when the rank is n and AA^T is invertible when the rank is m.

It is natural to think of A as providing a transformation from $\mathbf{R}^n$ into $\mathbf{R}^m$: Given any vector x in $\mathbf{R}^n$, it is transformed into the vector Ax in $\mathbf{R}^m$. It is a *linear* transformation because the rules of matrix multiplication ensure that $A(cx + dy) = cAx + dAy$; this is its most important property, and we have devoted Appendix A to the relation between linear transformations and matrices. In the case of "existence," when $r = m$, the transformation is called *onto*; every vector b in $\mathbf{R}^m$ comes from at least one x in $\mathbf{R}^n$, $Ax = b$. The range (column space) is all of $\mathbf{R}^m$. In the case of "uniqueness," when $r = n$, the transformation is called *one-to-one*; each b comes from at most one x in $\mathbf{R}^n$. Nonlinear example: The function $y = x^2$ from $\mathbf{R}^1$ to $\mathbf{R}^1$ is not onto, because the number $y = -4$ does not come from any x; it is not one-to-one, because the number $y = 4$ comes both from $x = +2$ and $x = -2$. By contrast, the function $y = x^3$ is both onto and one-to-one. In this case there is a perfect pairing, a one-to-one correspondence, between real numbers x and their cubes x^3—or, going in the other direction, there is a one-to-one correspondence between real numbers y and their real cube roots $y^{1/3}$. This second transformation is the (two-sided!) inverse of the first. Invertible transformations are identical with one-to-one correspondences—they are simultaneously one-to-one and onto.

A rectangular matrix can have one of these properties without the other. So can a nonlinear function $y(x)$; the exercises ask for examples. But a square matrix is different. If $m = n$, then the matrix A has a left-inverse if and only if it has a right-inverse; *existence implies uniqueness* and *uniqueness implies existence*. Since all right-inverses equal all left-inverses, there can be only one of each: $B = C = A^{-1}$. The condition for this invertibility is that the rank must be as large as possible: $r = m = n$. We can say this in still another way: For a square matrix A of order n to be nonsingular, each of the following conditions is a necessary and sufficient test:

(1)   The columns span $\mathbf{R}^n$, so $Ax = b$ has at least one solution for every b.
(2)   The columns are independent, so $Ax = 0$ has only the solution $x = 0$.

This list can be made much longer, especially if we look ahead to later chapters; every condition in the list is equivalent to every other, and ensures that A is nonsingular.

(3)   The rows of A span $\mathbf{R}^n$.
(4)   The rows are linearly independent.
(5)   Gaussian elimination, with row exchanges, can be accomplished without any zero pivots: $PA = LDU$, with all $d_i \neq 0$.
(6)   There exists a matrix A^{-1} such that $AA^{-1} = A^{-1}A = I$.†
(7)   The determinant of A is not zero.
(8)   Zero is not an eigenvalue of A.
(9) A^TA is positive definite.
(10) AA^T is positive definite.

Here is a typical application. Consider all polynomials $P(t)$ of degree $n - 1$. The only such polynomial that vanishes at n given points $t_1, \ldots, t_n$ is $P(t) \equiv 0$;

---

† This means that *invertible* is a synonym for *nonsingular*. These two words apply only to square matrices.

no other polynomial of degree $n - 1$ can have n roots. This is a statement of uniqueness, and it implies a statement of existence: Given any values $b_1, \ldots, b_n$, there exists a polynomial of degree $n - 1$ interpolating these values: $P(t_i) = b_i$, $i = 1, \ldots, n$. The point is that we are dealing with a square matrix; the number of coefficients in $P(t)$ (which is n) matches the number of equations. In fact, if $P(t) = x_1 + x_2 t + \cdots + x_n t^{n-1}$, the equations $P(t_i) = b_i$ are the same as

$$\begin{bmatrix} 1 & t_1 & t_1^2 & \cdots & t_1^{n-1} \\ 1 & t_2 & t_2^2 & \cdots & t_2^{n-1} \\ \vdots & \vdots & \vdots & \vdots & \vdots \\ 1 & t_n & t_n^2 & \cdots & t_n^{n-1} \end{bmatrix} \begin{bmatrix} x_1 \\ x_2 \\ \vdots \\ x_n \end{bmatrix} = \begin{bmatrix} b_1 \\ b_2 \\ \vdots \\ b_n \end{bmatrix}.$$

The coefficient matrix A is n by n, and is known as *Vandermonde's matrix*. To repeat the argument: Since $Ax = 0$ has only the solution $x = 0$ (in other words $P(t_i) = 0$ is only possible if $P \equiv 0$), it follows that A is nonsingular. Thus $Ax = b$ always has a solution—a polynomial can be passed through any n values b_i at distinct points t_i. Later we shall actually find the determinant of A; it is not zero.

There is one more point. Since A is nonsingular, so is A^T; they have the same rank. Therefore we can solve not only $Ax = b$ for any b, but also $A^T y = c$ for any c. The latter system arises in a quite different application. Suppose we look for an approximate numerical integration formula of the form

$$\int_0^1 f(t) \, dt \approx \sum_1^n y_i f(t_i).$$

The integral of f is to be estimated from the its values $f(t_i)$ at the n particular points t_i. With only n pieces of information to go on, no choice of coefficients y_i can make the formula correct for all functions f; but it can (and in practice it would) be made correct for all polynomials $f = P$ of degree $n - 1$. Requiring correctness for each possibility $f = 1, f = t, \ldots, f = t^{n-1}$, we have

$$\int_0^1 1 \, dt = y_1 + \cdots + y_n$$

$$\int_0^1 t \, dt = y_1 t_1 + \cdots + y_n t_n$$

$$\vdots$$

$$\int_0^1 t^{n-1} \, dt = y_1 t_1^{n-1} + \cdots + y_n t_n^{n-1}.$$

This is exactly the system $c = A^T y$ whose solvability has been proved, and therefore the required coefficients y_i can be found. In a roundabout way, the integration formula exists because polynomials of degree $n - 1$ have only $n - 1$ roots.

EXERCISE 2.4.10 Construct a nonlinear function $y(x)$ that is one-to-one but not onto, and a nonlinear function $z(x)$ that is onto but not one-to-one.

EXERCISE 2.4.11 Explain why existence holds for A if and only if uniqueness holds for the matrix A^T and vice versa.

EXERCISE 2.4.12 Construct all possible left or right inverses for the 1 by 3 matrix $A = \begin{bmatrix} 1 & 1 & 0 \end{bmatrix}$, and for A^T.

ORTHOGONALITY OF VECTORS AND SUBSPACES ■ 2.5

The first step in this section is to find the length of a vector. In two dimensions, this length $\| x \|$ is the hypotenuse of a right triangle (Fig. 2.4a), and was

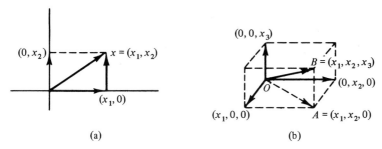

Fig. 2.4.   The length of two- and three-dimensional vectors.

given a long time ago by Pythagoras†: $\| x \|^2 = x_1^2 + x_2^2$.

In three-dimensional space, the vector $x = (x_1, x_2, x_3)$ is the diagonal of a box (Fig. 2.4b), and its length comes from *two* applications of the Pythagoras formula. First, the two-dimensional case takes care of the diagonal $OA = (x_1, x_2, 0)$ which runs across the base, and gives $\overline{OA^2} = x_1^2 + x_2^2$. This forms a right angle with the vertical side $(0, 0, x_3)$, so we may appeal to Pythagoras again (in the plane of OA and AB). The hypotenuse of the triangle OAB is the length $\| x \|$ we want, and it is given by

$$\| x \|^2 = \overline{OA^2} + \overline{AB^2} = x_1^2 + x_2^2 + x_3^2.$$

The generalization to a vector in n dimensions, $x = (x_1, \ldots, x_n)$, is immediate. **The length $\| x \|$ of a vector in R^n is the positive square root of**

$$\| x \|^2 = x_1^2 + x_2^2 + \cdots + x_n^2. \tag{12}$$

Geometrically, this amounts to applying the Pythagoras formula $n - 1$ times, adding one more dimension at each step. In the case of one dimension, $n = 1$, the length is just the absolute value of the only component x_1.

Suppose we are now given two vectors x and y (Fig. 2.5). How can we decide

† Or perhaps by the Egyptians, and then actually proved by the Pythagoreans.

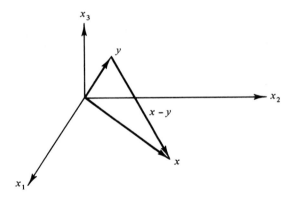

Fig. 2.5.   The plane triangle with sides x, y, and $x - y$.

whether or not they are perpendicular? In other words, what is the test for orthogonality? This is a question that can be answered in the two-dimensional plane by trigonometry; we need the generalization to $\mathbf{R}^n$, but even there we can stay in the plane spanned by x and y. Within this plane, x is orthogonal to y provided they form a right triangle, and we use the Pythagoras formula as a test:

$$|| x ||^2 + || y ||^2 = || x - y ||^2. \tag{13}$$

Applying the formula (12), this condition becomes

$$(x_1{}^2 + \cdots + x_n{}^2) + (y_1{}^2 + \cdots + y_n{}^2) = (x_1 - y_1)^2 + \cdots + (x_n - y_n)^2.$$

The right hand side is exactly

$$(x_1{}^2 + \cdots + x_n{}^2) - 2(x_1 y_1 + \cdots + x_n y_n) + (y_1{}^2 + \cdots + y_n{}^2).$$

Thus equality holds in (13), *and the vectors* x *and* y *are orthogonal, when the* "*cross-product terms*" *give zero:*

$$x_1 y_1 + \cdots + x_n y_n = 0. \tag{14}$$

Notice that this quantity is exactly the same as $x^T y$, the product of a 1 by n matrix (the row vector x^T) with an n by 1 matrix (the column vector y):

$$x^T y = [x_1 \quad \cdots \quad x_n] \begin{bmatrix} y_1 \\ \vdots \\ y_n \end{bmatrix} = x_1 y_1 + \cdots + x_n y_n. \tag{15}$$

Using the notation for summation, it is $\sum x_i y_i$. This combination appears in every discussion of the geometry of n-dimensional space. It is sometimes called the scalar product or dot product of the two vectors, and denoted by (x, y) or $x \cdot y$, but we prefer to call it the inner product and to keep the notation $x^T y$:

2R   The quantity $x^T y$ is the inner product of the (column) vectors x and y in $\mathbf{R}^n$. It is zero if and only if x and y are orthogonal.

Notice that the ideas of length and inner product are connected by $x^Tx = x_1^2 + \cdots + x_n^2 = \| x \|^2$. The only vector with length zero—in other words, the only vector orthogonal to itself—is the zero vector. This vector $x = 0$ is orthogonal to every vector y in $\mathbf{R}^n$.

EXERCISE 2.5.1 Find the lengths and the inner product of $x = (1, 4, 0, 2)^T$ and $y = (2, -2, 1, 3)^T$.

We have shown that *x and y are orthogonal if and only if their inner product is zero.* In the next chapter we discuss the inner product at greater length.†
There we are interested also in nonorthogonal vectors; the inner product gives a natural definition of the cosine in n-space, and determines the angle between any two vectors. In this section, however, the goal is still to understand the four fundamental subspaces—and the property we are after is orthogonality.

First, there is a simple connection between independence and orthogonality: *If the nonzero vectors $v_1, \ldots, v_k$ are mutually orthogonal* (every vector is orthogonal to every other), *then they are linearly independent.*

Proof Suppose $c_1v_1 + \cdots + c_kv_k = 0$. To show that a typical coefficient such as c_1 must be zero, take the inner product of both sides with v_1 :

$$v_1^T(c_1v_1 + \cdots + c_kv_k) = v_1^T0 = 0. \tag{16}$$

The orthogonality of the v's leaves only one term in (16), $c_1v_1^Tv_1 = 0$. Because the vectors were assumed nonzero, $v_1^Tv_1 \neq 0$ and therefore $c_1 = 0$. The same is true of every c_i, and the only combination of the v's producing zero is the trivial one with all $c_i = 0$. Therefore the vectors are linearly independent.

The most important example of mutually orthogonal vectors is the set of coordinate vectors $e_1, \ldots, e_n$ in $\mathbf{R}^n$. These are the columns of the identity matrix, they form the simplest basis for $\mathbf{R}^n$, and they are *unit vectors*—each of them has length $\| e_i \| = 1$. Geometrically, they point in the direction of the coordinate axes. If this system of vectors is rotated, the result is a new *orthonormal set*, that is, a new system of mutually orthogonal unit vectors. In the plane, this rotation produces the orthonormal pair

$$v_1 = (\cos \theta, \sin \theta), \qquad v_2 = (-\sin \theta, \cos \theta).$$

EXERCISE 2.5.2 Give an example in $\mathbf{R}^2$ of linearly independent vectors that are not mutually orthogonal, proving that the converse of the above theorem is false. Also, give an example of mutually orthogonal vectors that are not independent, because a condition of the theorem is violated.

EXERCISE 2.5.3 According to analytic geometry, two lines in the plane are perpendicular when the product of their slopes is -1. Apply this to the vectors $x = (x_1, x_2)$ and

---

† Or perhaps we should say from another angle.

$y = (y_1 , y_2)$, whose slopes are x_2/x_1 and y_2/y_1 , to derive again the orthogonality condition $x^{\mathrm{T}}y = 0$.

EXERCISE 2.5.4   How do we know that the ith row of B is orthogonal to the jth column of B^{-1}, if $i \neq j$?

EXERCISE 2.5.5   Which pairs are orthogonal among the vectors

$$v_1 = \begin{bmatrix} 1 \\ 2 \\ -2 \\ 1 \end{bmatrix}, \qquad v_2 = \begin{bmatrix} 4 \\ 0 \\ 4 \\ 0 \end{bmatrix}, \qquad v_3 = \begin{bmatrix} 1 \\ -1 \\ -1 \\ -1 \end{bmatrix}?$$

EXERCISE 2.5.6   In $\mathbf{R}^3$ find all vectors that are orthogonal to both $(1, 1, 1)$ and $(1, -1, 0)$. Produce from these vectors a mutually orthogonal system of unit vectors (an orthonormal system) in $\mathbf{R}^3$.

Orthogonal Subspaces

We come next to the orthogonality of two subspaces. In ordinary three-dimensional space, subspaces are represented by lines or planes through the origin—and, in the extreme cases, by the origin alone or the whole space. The subspace $\{0\}$ is orthogonal to all subspaces. A line can be orthogonal either to another line or to a plane, but a plane cannot be orthogonal to a plane,† and the full space $\mathbf{R}^3$ is orthogonal only to $\{0\}$. In n dimensions, the basic definition is this:

25   Two subspaces V and W of the same space $\mathbf{R}^n$ are called orthogonal if every vector v in V is orthogonal to every vector w in W: $v^{\mathrm{T}}w = 0$ for all v and w.

EXAMPLE   Suppose V is the plane spanned by $v_1 = (1, 0, 0, 0)$ and $v_2 = (1, 1, 0, 0)$, and W is the line spanned by $w_1 = (0, 0, 4, 5)$. Then since w_1 is orthogonal to both of the v's, the line W will be orthogonal to the whole plane V.

EXERCISE 2.5.7   In the previous example, find a w_2 so that the plane W spanned by w_1 and w_2 is still orthogonal to V. Also find a v_3 so that the three-dimensional subspace spanned by v_1, v_2, v_3 is orthogonal to the original line W.

EXERCISE 2.5.8   If V and W are orthogonal subspaces, show that the only vector they have in common is the zero vector: $V \cap W = \{0\}$.

---

† I have to admit that the front wall and side wall of a room look very much like perpendicular planes in $\mathbf{R}^3$. But by our definition, that is not so! There are lines v and w in the two walls that do not meet at a right angle.

Now we explain our reason for introducing orthogonality. Of the four fundamental subspaces associated with a matrix, recall that two are subspaces of $\mathbf{R}^n$—the nullspace $\mathfrak{N}(A)$ and the row space $\mathfrak{R}(A^T)$—and the other two lie in $\mathbf{R}^m$. The most important fact about these spaces, except for their dimensions, is that they are orthogonal:

2T   For any m by n matrix A, the nullspace $\mathfrak{N}(A)$ and the row space $\mathfrak{R}(A^T)$ are orthogonal subspaces of $\mathbf{R}^n$. Similarly, the left nullspace $\mathfrak{N}(A^T)$ and the column space $\mathfrak{R}(A)$ are orthogonal subspaces of $\mathbf{R}^m$.

First Proof   Suppose w is any vector in the nullspace $\mathfrak{N}(A)$. Then $Aw = 0$, and this system of m equations can be written out more fully as

$$Aw = \begin{bmatrix} \text{row } 1 & \cdots \\ \text{row } 2 & \cdots \\ \vdots \\ \text{row } m & \cdots \end{bmatrix} \begin{bmatrix} w_1 \\ w_2 \\ \vdots \\ w_n \end{bmatrix} = \begin{bmatrix} 0 \\ 0 \\ \vdots \\ 0 \end{bmatrix}. \tag{17}$$

The first equation means that a certain inner product is zero; the vector w is orthogonal to the first row of A, or more properly (to keep the column vector convention) to the first column of A^T. The second equation states the orthogonality of w to the second column of A^T. Continuing in this way, w is orthogonal to every column of A^T. Therefore w is orthogonal to the whole space spanned by these columns, in other words to every v in the column space of A^T. This is true for every w in the nullspace, and therefore $\mathfrak{N}(A) \perp \mathfrak{R}(A^T)$.

The second statement in the theorem, that $\mathfrak{N}(A^T) \perp \mathfrak{R}(A)$, is just the first statement applied to A^T. (The first statement was proved for any matrix whatsoever, including the transpose of any given A.) Alternatively, we can start with any y in the left nullspace, and read off from

$$y^T A = \begin{bmatrix} y_1 & \cdots & y_m \end{bmatrix} \begin{bmatrix} \text{c} & & \text{c} \\ \text{o} & & \text{o} \\ \text{l} & & \text{l} \\ \text{u} & \cdots & \text{u} \\ \text{m} & & \text{m} \\ \text{n} & & \text{n} \\ 1 & & n \end{bmatrix} = \begin{bmatrix} 0 & \cdots & 0 \end{bmatrix}$$

the fact that y is orthogonal to every column. Therefore it is orthogonal to every combination of the columns, that is, every y in $\mathfrak{N}(A^T)$ is orthogonal to every w in $\mathfrak{R}(A)$.

Second Proof   We want to establish the same result by a more coordinate-free argument. The contrast between the two proofs should be useful to the reader, as a specific example of an "abstract" versus a "concrete" method of reasoning. I wish I were sure which is the clearer and more permanently understood.

Suppose w is in $\mathfrak{N}(A)$ and v is in $\mathfrak{R}(A^T)$. Then $Aw = 0$, and v is of the form $v = A^T x$ for some vector x. (In the concrete notation, v is a combination of the columns of A^T, and $x_1, \ldots, x_m$ are the weights in this combination.) Therefore

$$w^T v = w^T(A^T x) = (w^T A^T)x = (Aw)^T x = 0^T x = 0. \tag{18}$$

EXAMPLE   Suppose A is the matrix in Exercise 2.4.2:

$$A = \begin{bmatrix} 0 & 1 & 4 & 0 \\ 0 & 2 & 8 & 0 \end{bmatrix} \rightarrow U = \begin{bmatrix} 0 & 1 & 4 & 0 \\ 0 & 0 & 0 & 0 \end{bmatrix}.$$

The second column is basic, and the other three variables are free. Therefore if we set each free variable equal to one, in turn, and solve $Ux = 0$, we find three vectors that are a basis for the nullspace of A:

$$\begin{bmatrix} 1 \\ 0 \\ 0 \\ 0 \end{bmatrix}, \quad \begin{bmatrix} 0 \\ -4 \\ 1 \\ 0 \end{bmatrix}, \quad \begin{bmatrix} 0 \\ 0 \\ 0 \\ 1 \end{bmatrix}.$$

These are all perpendicular to the rows of A, as 2T guarantees.

The column space of A is one dimensional (row rank = column rank) and it is spanned by the one basic column $\begin{bmatrix} 1 \\ 2 \end{bmatrix}$. On the other hand, the left null-space is found by combining the rows of A to produce the zero row in U: $(-2)(\text{row } 1) + (1)(\text{row } 2) = 0$. Therefore $y^T = (-2, 1)$ is in the left null-space, and as predicted it is orthogonal to the column space:

$$\begin{bmatrix} -2 & 1 \end{bmatrix} \begin{bmatrix} 1 \\ 2 \end{bmatrix} = 0.$$

Now 1 have to ask for your patience about one more point. It is certainly the truth that the nullspace is perpendicular to the row space—but it is not the whole truth. $\mathfrak{N}(A)$ does not contain just some of the vectors orthogonal to the row space, *it contains every such vector*. The nullspace was formed from *all* solutions to $Ax = 0$.

2U   Definition.   Given a subspace V of $\mathbf{R}^n$, the space of all vectors orthogonal to V is called the ***orthogonal complement*** of V, and denoted by $V^\perp$.†

Using this terminology, the nullspace $\mathfrak{N}(A)$ is the orthogonal complement of $\mathfrak{R}(A^T)$: $\mathfrak{N}(A) = (\mathfrak{R}(A^T))^\perp$. At the same time, the opposite relation also holds: The row space $\mathfrak{R}(A^T)$ contains all vectors that are orthogonal to the

† Suggested pronunciation: "V perp."

nullspace. This is not so obvious from the construction, since in solving $Ax = 0$ we started with the row space and found all x that were orthogonal to it; now we are going in the opposite direction. Suppose, however, that some vector z in $\mathbf{R}^n$ is orthogonal to the nullspace but is outside the row space. Then adding z as an extra row of A would enlarge the row space without changing the nullspace. But we know that there is a fixed formula: $r + (n - r) = n$, or

$$\dim(\text{row space}) + \dim(\text{nullspace}) = \text{number of columns}.$$

Since the last two numbers are unchanged when the new row z is added, it is impossible for the first one to change either. We conclude that every vector orthogonal to the nullspace is already in the row space: $\mathfrak{R}(A^T) = (\mathfrak{N}(A))^\perp$.

The same reasoning applied to A^T produces the dual result: *The left nullspace* $\mathfrak{N}(A^T)$ *and the column space* $\mathfrak{R}(A)$ *are orthogonal complements of one another in* $\mathbf{R}^m$. This completes the second half of the fundamental theorem of linear algebra. The first half gave the dimensions of the four subspaces, including the fact that row rank = column rank, and now we know that they are not only perpendicular, they are orthogonal complements.

2V *Fundamental Theorem of Linear Algebra*, Part 2

$$\mathfrak{N}(A) = (\mathfrak{R}(A^T))^\perp, \qquad \mathfrak{R}(A^T) = \mathfrak{N}(A)^\perp,$$

$$\mathfrak{N}(A^T) = (\mathfrak{R}(A))^\perp, \qquad \mathfrak{R}(A) = (\mathfrak{N}(A^T))^\perp.$$

The last equality means: $Ax = b$ has a solution if and only if b is orthogonal to $\mathfrak{N}(A^T)$; b is in the column space if and only if it is orthogonal to every solution y of the transposed homogeneous equation $A^Ty = 0$.

We must emphasize that two subspaces V and W can be orthogonal without being orthogonal complements of one another. In three-space, the line V spanned by $(1, 0, 0)$ is orthogonal to the line W spanned by $(0, 0, 1)$, but V does not equal $W^\perp$. The orthogonal complement of W is a two-dimensional subspace, containing all vectors of the form $(x_1, x_2, 0)$. The line V can be only a part of $W^\perp$ because its dimension is too small. If the dimensions are right, however, then two orthogonal subspaces are necessarily orthogonal complements. That was the case for the row space and nullspace. Furthermore, if $W = V^\perp$, then this ensures that the dimensions are right, and automatically $V = W^\perp$. The space is simply being decomposed into two perpendicular parts V and W, as in Fig. 2.6.

The theorem that goes with the picture is this:

If V and W are subspaces of $\mathbf{R}^n$, then any one of the following conditions forces them to be orthogonal complements of one another:

(1) $W = V^\perp$ (*W consists of all vectors orthogonal to V*).
(2) $V = W^\perp$ (*V consists of all vectors orthogonal to W*).
(3) *V and W are orthogonal, and* $\dim V + \dim W = n$.

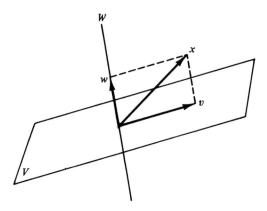

Fig. 2.6.   Orthogonal decomposition of the space $\mathbf{R}^3$.

Assuming any of these three equivalent conditions, every vector x can be split in exactly one way into a sum $x = v + w$, with v in V and w in W. These components, the projections of x onto V and W, are orthogonal: $v^{\mathrm{T}}w = 0$.

Permit us to summarize the previous section and this one. Together, they give a very complete picture of the true effect of a matrix A. The previous section determined the *dimensions* of the four fundamental subspaces; in particular, the row space $\mathfrak{R}(A^{\mathrm{T}})$ and the column space $\mathfrak{R}(A)$ share the same dimension r. This section determined the *orientation* of these four spaces; two of them are orthogonal complements in $\mathbf{R}^n$, and the other two are orthogonal complements in $\mathbf{R}^m$. The true effect of any matrix A is illustrated (in a very schematic way) by Fig. 2.7. An arbitrary x is split into $x_r + x_n$, and A trans-

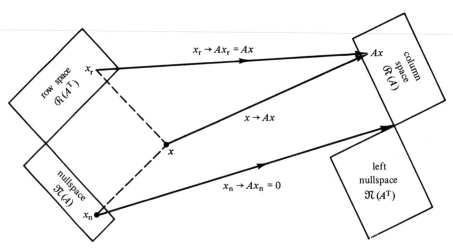

Fig. 2.7.   The action of a matrix A.

forms the row space component x_r into a vector $Ax_r = Ax$ in the column space, while it transforms the nullspace component x_n into zero.†

2W The mapping from row space to column space is actually nonsingular, or invertible; every vector b in the column space comes from one and only one vector x in the row space.

Proof If b is in the column space, it is some combination Ax of the columns. Split x into $x_r + x_n$, with x_r in the row space and x_n in the nullspace. Then $Ax_r = Ax_r + Ax_n = Ax = b$, so a suitable x_r in the row space has been found. If there were another vector x_r', also in the row space and also with $Ax_r' = b$, then $A(x_r - x_r') = b - b = 0$. This puts $x_r - x_r'$ in both the nullspace and the row space, which makes it orthogonal to itself. Therefore it is zero, and $x_r = x_r'$.

Every matrix A is invertible when it is properly understood as a transformation of a certain r-dimensional subspace onto another one, $\mathfrak{R}(A^T)$ *onto* $\mathfrak{R}(A)$. Acting on the orthogonal complement $\mathfrak{N}(A)$, A is the zero matrix. In the same way, A^T is an invertible transformation in the reverse direction, from $\mathfrak{R}(A)$ onto $\mathfrak{R}(A^T)$. That does not mean that A^T is the inverse of A; A^T moves the spaces around correctly, but not the individual vectors. In fact A^T transforms Ax into A^TAx, whereas the inverse of A—or, in the singular case discussed in Section 3.4, the *pseudoinverse* of A—transforms Ax back into x.

EXERCISE 2.5.9 Find the orthogonal complement of the plane spanned by the vectors $(1, 1, 2)$ and $(1, 2, 3)$, by taking these to be the rows of A and solving $Ax = 0$. Remember that the complement is a whole line.

EXERCISE 2.5.10 Construct a homogeneous equation in three unknowns whose solutions are the linear combinations of the vectors $(1, 1, 2)$ and $(1, 2, 3)$. This is the reverse of the previous exercise, but of course the two problems are really the same.

EXERCISE 2.5.11 The fundamental theorem of linear algebra is often stated in the form of *Fredholm's alternative*: For any A and b, one and only one of the following systems has a solution:

$$(1) \quad Ax = b \qquad (2) \quad A^Ty = 0, \quad y^Tb \neq 0.$$

In other words, either b is in the column space $\mathfrak{R}(A)$ or there is a y in $\mathfrak{N}(A^T)$ such that $y^Tb \neq 0$. By splitting b into a column space component and a left nullspace component, suggest a suitable y.

EXERCISE 2.5.12 Find a basis for the nullspace of

$$A = \begin{bmatrix} 1 & 0 & 2 \\ 1 & 1 & 4 \end{bmatrix},$$

† We did not really know how to draw two orthogonal subspaces of dimension r and $n - r$. If you already understand these dimensions, and the orthogonality, do not allow Fig. 2.7 to confuse you!

and verify that it is orthogonal to the row space. Given $x = (3, 3, 3)$, split it into a row space component x_r and a nullspace component x_n.

EXERCISE 2.5.13 Show that the orthogonal complement of $V + W$ is the intersection of the orthogonal complements of V and W: $(V + W)^\perp = V^\perp \cap W^\perp$. (The sum of two subspaces is defined on page 87.)

EXERCISE 2.5.14 Illustrate the action of A^T by a picture corresponding to Fig. 2.7, sending $\mathcal{R}(A)$ back to the row space and the left nullspace to zero.

Incidence Matrices and Kirchhoff's Laws

If direct current is flowing around the network in Fig. 2.8, then the only

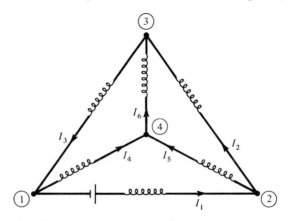

Fig. 2.8. A network with four nodes and six branches.

way to maintain equilibrium is to satisfy both of Kirchhoff's laws:

(1) *At every node, the sum of incoming currents equals the sum of outgoing currents.* For example, $I_3 = I_1 + I_4$ at node 1.

(2) *Around every closed loop, the sum of the voltage drops is zero.* If the drop is E_k in the direction indicated for I_k, then around the triangle of nodes 1–4–3 the second law requires $E_4 + E_6 + E_3 = 0$.

We want to express these laws in terms of matrix algebra. Notice that they involve only "graph theory"—they depend on the way the nodes are linked by the branches, and on the directions of the arrows, but not on the sizes of the resistors (or batteries) in the network.† The connections between the nodes are completely described by the incidence matrix of the graph. This matrix has a row for every node and a column for every branch; in each column the nonzero entries $+1$ and -1 indicate the nodes that begin and end the branch.

---

† In other words, Kirchhoff's laws are "equilibrium conditions," which have to be supplemented by Ohm's law $E = IR$.

The incidence matrix for Fig. 2.8 is

$$
M = \begin{bmatrix}
1 & 0 & -1 & 1 & 0 & 0 \\
-1 & 1 & 0 & 0 & 1 & 0 \\
0 & -1 & 1 & 0 & 0 & -1 \\
0 & 0 & 0 & -1 & -1 & 1
\end{bmatrix}
\begin{matrix}
\text{node 1} \\
\text{node 2} \\
\text{node 3} \\
\text{node 4}
\end{matrix}
$$

Suppose we look along the first row of M, corresponding to node 1. There is a -1 for the incoming branch 3 and there are $+1$'s for the outgoing branches 1 and 4. Remembering that Kirchhoff's first law gave $I_3 = I_1 + I_4$, it translates immediately into matrix form: *If I is the column vector formed from the six currents, then $MI = 0$.* This is a system of four equations in six unknowns.

To translate the second law, fix the potential of node 1 at $p_1 = 0$, and define p_i at every other node by requiring the voltage drop from node i to node j (if there is a branch in that direction) to equal the difference in potentials, $E = p_i - p_j$. The second law guarantees that the total drop around a circuit is zero, so we come back to the same potential we started with. In terms of M, the drop $p_i - p_j$ across any branch comes from multiplying the potential vector p by the column corresponding to that branch; from its construction, that column has $+1$ at node i and -1 at node j. In other words, *the voltage drops E are the components of $M^T p$.*

Now the laws can be stated very concisely: *I is in the nullspace of M, and E is in its row space.* Since these two spaces are orthogonal for any matrix M, and we have proved "Tellegen's theorem" in circuit theory: $E^T I = 0$.

EXAMPLE Suppose a 20-V (volt) battery is connected into the first branch, producing a current I_1. From the symmetry, half of that current will flow back through node 3 and half through node 4, with no effect on the sixth branch: $I_6 = 0$. If the common value of the resistors is $R = 5$, then we can verify that the basic current will be $I_1 = 2$, and the return currents will be $I_2 = I_3 = -I_4 = I_5 = 1$. (Note: The arrows were drawn at random, just to fix a direction along each branch, so $I_4 = -1$ means that the flow is against the arrow.) These currents provide a drop of $E = IR = 10$ V across the bottom resistor, and 5 V across each of the resistors 2, 3, 4, 5—which balances the 20-V battery. Therefore Kirchhoff's laws are satisfied and the proposed currents are correct.

EXERCISE 2.5.15 Find the total voltage drops across each branch, as well as the potentials p_2, p_3, p_4—given $p_1 = 0$. Verify that $E^T I = 0$.

EXERCISE 2.5.16 Draw the network whose incidence matrix is

$$
M = \begin{bmatrix}
1 & 0 & 0 & -1 & 1 & 0 \\
-1 & 1 & 0 & 0 & 0 & 1 \\
0 & -1 & 1 & 0 & -1 & 0 \\
0 & 0 & -1 & 1 & 0 & -1
\end{bmatrix}.
$$

With a 6-V battery in the branch 1–3, and unit resistors in all branches, what are the currents I and the voltage drops E?

2.6 ■ PAIRS OF SUBSPACES AND PRODUCTS OF MATRICES

This chapter has pursued one goal exclusively, the understanding of a linear system $Ax = b$. Each new idea and each definition—including vector spaces and linear independence, basis and dimension, rank and nullity, inner product and orthogonality—was introduced as it was needed for this one purpose. Now we want to look back at these same ideas, to find some of the other relationships which were missed.†

Many of them are simple. They arise from considering not just a single subspace or a single matrix A, but the interconnections between two subspaces or two different matrices. The first connection is this:

2X If V and W are both subspaces of a given vector space, then so is their intersection $V \cap W$.

The proof is immediate. Suppose x and x' belong to $V \cap W$, in other words they are vectors both in V and in W. Then, because these are vector spaces in their own right, $x + x'$ and cx are also in both V and W. Therefore the results of addition and scalar multiplication stay within the intersection. Geometrically, the intersection of two planes through the origin (or "hyperplanes" in $\mathbf{R}^n$) is again a subspace. The same will be true of the intersection of several subspaces, or even of infinitely many.

EXAMPLE 1 The intersection of two orthogonal subspaces V and W is the subspace $\{0\}$.

EXAMPLE 2 Regarding the set of all n by n matrices as a vector space, and the sets of upper and lower triangular matrices as subspaces V and W, their intersection is the set of n by n diagonal matrices. This set is certainly a subspace; adding two diagonal matrices, or multiplying by a scalar, leaves us with a diagonal matrix.

EXAMPLE 3 Suppose V is the nullspace of a k by n matrix A, and W is the nullspace of an l by n matrix B. Then $V \cap W$ is the nullspace of the matrix

$$C = \begin{bmatrix} A \\ B \end{bmatrix},$$

† The most important definitions are given right away in 2X and 2Y. Then the rest of the section develops ideas that are more relevant to the theory than to the applications.

formed from the k rows of A and the l rows of B. Proof: x is in the nullspace of C, $Cx = 0$, if and only if both $Ax = 0$ and $Bx = 0$.

EXAMPLE 4   Suppose $v_1, \ldots, v_k$ is a set of vectors in $\mathbf{R}^n$. Consider all subspaces that contain this set; $\mathbf{R}^n$ is one such subspace (every vector space is a subspace of itself), and there may be others. The intersection of all these subspaces is the smallest subspace containing the given vectors; it is the subspace spanned by $v_1, \ldots, v_k$.

Usually, after discussing and illustrating the intersection of two sets, it is natural to look at their union. With vector spaces, however, it is not natural. *The union $V \cup W$ of two subspaces will not in general be a subspace.* Consider for example the x axis and the y axis in the plane. Each axis by itself is a subspace, but their union is not; the sum of $(1, 0)$ and $(0, 1)$ is not on either axis. It is not hard to see that this will always happen, unless one of the subspaces is contained in the other; only then is their union (which coincides with the larger one) again a subspace.

Nevertheless, we do want to combine two subspaces, and therefore in place of their union we turn to their sum.

2Y   If V and W are both subspaces of a given space, then so is their sum $V + W$. It is made up of all possible combinations $x = v + w$, where v is an arbitrary vector in V and w is an arbitrary vector in W.

This is nothing but the space spanned by $V \cup W$. It is the smallest vector space that contains both V and W. The sum of the x axis and the y axis is the whole x-y plane; the sum of a two-dimensional plane and a line not in that plane is a three-dimensional subspace.

EXAMPLE 5   Suppose V and W are orthogonal complements of one another in $\mathbf{R}^n$. Then their sum is $V + W = \mathbf{R}^n$; every x is the sum of its projection v in V and its projection w in W.

EXAMPLE 6   If V is the space of upper triangular matrices, and W is the space of lower triangular matrices, then $V + W$ is the space of all matrices. Every matrix can be written as the sum of an upper and a lower triangular matrix— in many ways, because the diagonals are not uniquely determined.

EXAMPLE 7   If V is the column space of a matrix A, and W is the column space of B, then $V + W$ is the column space of the combined matrix $Q = [A \quad B]$. The dimension of $V + W$ may be less than the combined dimensions of V and W (because the two spaces may overlap), but it is easy to find:

$$\dim(V + W) = \text{rank of } Q. \tag{19}$$

Surprisingly, *the computation of $V \cap W$ is much more subtle.* Suppose we are given the two bases $v_1, \ldots, v_k$ and $w_1, \ldots, w_l$; this time we want a basis for the intersection of the two subspaces. Certainly it is not enough just to check whether any of the v's equal any of the w's; the two spaces could even be identical, $V = W$, and still the bases might be completely different.

The most efficient method is this. Form the same matrix Q whose columns are $v_1, \ldots, v_k, w_1, \ldots, w_l$, and compute its nullspace $\mathfrak{N}(Q)$. We shall show that a basis for this nullspace leads to a basis for $V \cap W$, and that the two spaces have the same dimension:

$$\dim(V \cap W) = \dim \mathfrak{N}(Q). \tag{20}$$

This leads to a formula which is important in its own right. Adding Eqs. (19) and (20),

$$\dim(V + W) + \dim(V \cap W) = \text{rank of } Q + \text{nullity of } Q.$$

From our computations with the four fundamental subspaces, we know that the rank plus the nullity equals the number of columns. In this case Q has $k + l$ columns, and since $k = \dim V$ and $l = \dim W$, we are led to the following conclusion:

$$\dim(V + W) + \dim(V \cap W) = \dim V + \dim W. \tag{21}$$

Not a bad formula.

EXAMPLE 8 The spaces V and W of upper and lower triangular matrices both have dimension $n(n + 1)/2$. The space $V + W$ of all matrices has dimension n^2, and the space $V \cap W$ of diagonal matrices has dimension n. As predicted by (21), $n^2 + n = n(n + 1)/2 + n(n + 1)/2$.

We now look at the proof of (20). For once in this book, the interest is less in the actual computation than in the technique of proof. It is the only time we will use the trick of understanding one space by matching it with another, and it leads to the useful formula (21). Note first that the nullspace of Q is a subspace of $\mathbf{R}^{k+l}$, whereas $V \cap W$ is a subspace of $\mathbf{R}^m$: we have to prove that these two spaces have the same dimension. The trick is to show that these two subspaces are perfectly matched by the following correspondence:

Given any vector x in the nullspace of Q, write the equation $Qx = 0$ in terms of the columns as follows:

$$x_1 v_1 + \cdots + x_k v_k + x_{k+1} w_1 + \cdots + x_{k+l} w_l = 0,$$

or

$$x_1 v_1 + \cdots + x_k v_k = -x_{k+1} w_1 - \cdots - x_{k+l} w_l.$$

The left side of this equation is obviously in V, being a combination of the v_k, and the right side is in W. Since the two are equal, they represent a vector y

in $V \cap W$. This provides the correspondence between the vector x in $\mathfrak{N}(Q)$ and the vector y in $V \cap W$. It is easy to check that the correspondence preserves addition and scalar multiplication: If x corresponds to y and x' to y', then $x + y$ corresponds to $x' + y'$ and cx corresponds to cx'. Furthermore, every y in $V \cap W$ comes from one and only one x in $\mathfrak{N}(Q)$ (Exercise 2.6.5).

This is a perfect illustration of an *isomorphism* between two vector spaces. The spaces are different, but *for all algebraic purposes they are exactly the same.* They match completely: Linearly independent sets correspond to linearly independent sets, and a basis in one corresponds to a basis in the other. So their dimensions are equal, which completes the proof of (20) and (21). This is the kind of result an algebraist is after, to identify two different mathematical objects as being fundamentally the same. It is a fact that any two spaces with the same scalars and the same (finite) dimension are always isomorphic, but this is too general to be very exciting. The interest comes in matching two superficially dissimilar spaces, like $\mathfrak{N}(Q)$ and $V \cap W$.

EXAMPLE 9 Suppose V is spanned by the first two columns, and W is spanned by the last two columns, of

$$Q = \begin{bmatrix} 4 & 2 & 0 & 0 \\ 0 & 3 & -6 & 0 \\ 0 & 0 & 0 & 1 \end{bmatrix}.$$

This matrix is already in echelon form, $Q = U$, and the free variable (the column without a pivot) is the third. Therefore the first, second, and fourth columns span the column space $V + W$—which, having dimension three, must be all of $\mathbf{R}^3$. To find $V \cap W$, we compute the nullspace of Q. Let the free variable be $x_3 = 1$, and solve for the basic variables by back-substitution: $x_4 = 0$, $x_2 = 2$, $x_1 = -1$. This solution to $Qx = 0$ is $x = (-1, 2, 1, 0)^T$, and our one-to-one correspondence between $\mathfrak{N}(Q)$ and $V \cap W$ matches it with

$$y = (-1)\ (\text{column } 1) + (2)\ (\text{column } 2)$$

$$= (-1) \begin{bmatrix} 4 \\ 0 \\ 0 \end{bmatrix} + (2) \begin{bmatrix} 2 \\ 3 \\ 0 \end{bmatrix} = \begin{bmatrix} 0 \\ 6 \\ 0 \end{bmatrix}.$$

This is a basis for the (one-dimensional) intersection $V \cap W$.

EXERCISE 2.6.1 Within the space of all 4 by 4 matrices, let V be the subspace of tridiagonal matrices and W the subspace of upper triangular matrices. Describe the subspace $V + W$, whose members are the upper Hessenberg matrices, and the subspace $V \cap W$. Verify formula (21).

EXERCISE 2.6.2 Suppose that $V \cap W = \{0\}$; then (21) yields $\dim(V + W) = \dim V + \dim W$. Show that each x in $V + W$ can be written in only one way as $x = v + w$, with

v in V and w in W; in other words, assume that also $x = v' + w'$, and prove that $v = v'$ and $w = w'$. In this situation, with $V \cap W = \{0\}$, $V + W$ is called the *direct sum* of V and W, and is sometimes written $V \oplus W$. Orthogonal subspaces always produce direct sums, and so do any two subspaces that share only the zero vector.

EXERCISE 2.6.3 Suppose V is spanned by the column vectors with components $(1, 1, 0, 1)$ and $(1, 2, 0, 0)$. Find a subspace W such that, in the notation of the previous exercise, $V \oplus W = \mathbf{R}^4$.

EXERCISE 2.6.4 Find a basis for the sum $V + W$ of the space V spanned by $v_1 = (1, 1, 0, 0)$, $v_2 = (1, 0, 1, 0)$ and the space W spanned by $w_1 = (0, 1, 0, 1)$, $w_2 = (0, 0, 1, 1)$. Find also the dimension of $V \cap W$ and a basis for it.

EXERCISE 2.6.5 Verify the statement that "every y in $V \cap W$ comes from one and only one x in $\mathfrak{N}(Q)$"—by describing, for a given y, how to find the only possible x.

The Fundamental Spaces for Products AB

We turn from pairs of subspaces to products of matrices. As always, it is not the individual entries of AB that are particularly interesting; they probably have no similarity to the entries of A and B. Instead, it is at the level of vectors—the rows or columns of a matrix, rather than its entries—that some properties of A and B may be inherited by their product AB. And it is not even so much the individual rows or columns, as the subspaces they span; these subspaces reflect the properties of the whole matrix at once.

Our basic question is this: What are the relationships between the four fundamental subspaces associated with A, the four associated with B, and the four associated with the product AB? All these matrices may be rectangular, and there are four principal relationships:

(i)   *The nullspace of* AB *contains the nullspace of* B:
$$\mathfrak{N}(AB) \supseteq \mathfrak{N}(B).$$

(ii)   *The column space of* AB *is contained in the column space of* A:
$$\mathfrak{R}(AB) \subseteq \mathfrak{R}(A).$$

(iii)   *The left nullspace of* AB *contains the left nullspace of* A:
$$\mathfrak{N}((AB)^{\mathrm{T}}) \supseteq \mathfrak{N}(A^{\mathrm{T}}).$$

(iv)   *The row space of* AB *is contained in the row space of* B:
$$\mathfrak{R}((AB)^{\mathrm{T}}) \subseteq \mathfrak{R}(B^{\mathrm{T}}).$$

The proof is extremely simple.

(i)   If $Bx = 0$, then $ABx = 0$. Therefore every x in $\mathfrak{N}(B)$ is also in $\mathfrak{N}(AB)$, and $\mathfrak{N}(B) \subseteq \mathfrak{N}(AB)$.

(ii)   Suppose b is in the column space of AB, or in other words, $ABx = b$ for some vector x. Then also b is in the column space of A; $Ay = b$ for some vector y, namely the vector $y = Bx$. Thus $\mathfrak{R}(AB) \subseteq \mathfrak{R}(A)$.

Another proof comes directly from matrix multiplication: Each column of AB is a linear combination of the columns of A.

(iii)   Since $(AB)^T = B^T A^T$, the third statement coincides with the first—after the matrix A in (i) is replaced by B^T, and B is replaced by A^T.

(iv)   The fourth statement coincides with the second, after the same replacements. It also follows directly from matrix multiplication: the ith row of AB is a combination of the rows of B, with weights coming from the ith row of A. Therefore the row space of AB is contained in the row space of B.

EXERCISE 2.6.6   Show by example that the nullspace of AB need not contain the nullspace of A, and the column space of AB is not necessarily contained in the column space of B.

Corollary   *The rank r and nullity ν satisfy*

$$r(AB) \leq r(A) \qquad and \qquad r(AB) \leq r(B) \tag{22}$$

$$\nu(AB) \geq \nu(B). \tag{23}$$

Since the column space of AB is contained in the column space of A, and the rank is the dimension of the column space, it is immediate that $r(AB) \leq r(A)$. Similarly, it follows from (iv) that $r(AB) \leq r(B)$, and the inequality (23) for the nullity follows from (i). Note that we do not try to prove $\nu(AB) \geq \nu(A)$, which cannot be guaranteed for all rectangular matrices.

EXERCISE 2.6.7   Show, with matrices that are full of zeros, that $\nu(AB)$ may be less than $\nu(A)$.

There is one specific application that we have in mind for these relationships. It starts with an arbitrary m by n matrix A and its factorization into $PA = LU$ (or $A = P^{-1}LU$). Remember that the last $m - r$ rows of U are all zero; suppose we throw them away to produce an r by n matrix $\bar{U}$. (In case the rank r equals m, nothing is thrown away and $\bar{U} = U$.) Now look at matrix multiplication:

$$A = (P^{-1}L)U = (P^{-1}L) \begin{bmatrix} \bar{U} \\ 0 \end{bmatrix}. \tag{24}$$

The last $m - r$ columns of $P^{-1}L$ are only multiplying the zero rows at the bottom of U, so we might as well throw away these columns too. We will call the resulting matrix $\bar{L}$ (even though it should really be $P^{-1}\bar{L}$); it is formed from the first r columns of $P^{-1}L$. So far, our observation amounts to nothing more than this: *Throwing away zeros in equation* (24) *leads to a new factorization $A = \bar{L}\bar{U}$.*

To this matrix product we apply the relationship (ii) for column spaces of products: The column space of $A = \bar{L}\bar{U}$ is contained in the column space of $\bar{L}$. We know that the column space of A has dimension r. Since $\bar{L}$ has only r columns, its column space cannot be any larger, and *therefore the two column spaces are the same.* $\bar{L}$ shares the same column space as A, and $\bar{U}$ shares the same row space.

EXERCISE 2.6.8   Factor the matrix

$$A = \begin{bmatrix} 0 & 1 & 4 & 0 \\ 0 & 2 & 8 & 0 \end{bmatrix}$$

into $A = \bar{L}\bar{U}$, and verify that the columns of $\bar{L}$ are a basis for the column space of A.

EXERCISE 2.6.9   Repeat the previous exercise for a matrix A that requires a permutation P:

$$A = \begin{bmatrix} 0 & 0 \\ 1 & 2 \\ 4 & 8 \\ 0 & 0 \end{bmatrix}.$$

EXERCISE 2.6.10   Multiplying each column of $\bar{L}$ by the corresponding row of $\bar{U}$ decomposes any $A = \bar{L}\bar{U}$ into the *sum of r matrices of rank one*. Construct $\bar{L}$ and $\bar{U}$, and the splitting into r matrices of rank one, for the following matrix of rank two:

$$A = \begin{bmatrix} 1 & 3 & 3 & 2 \\ 2 & 6 & 9 & 5 \\ -1 & -3 & 3 & 0 \end{bmatrix}.$$

Finally, recall that a *submatrix* C is formed by striking out some (or none) of the rows of A, and some (or none) of its columns. It is not hard to guess a limitation on the rank of C.

2Z   Suppose A is an m by n matrix of rank r. Then:

   (i)   Every submatrix C is of rank $\leq r$.
   (ii)   At least one r by r submatrix is of rank exactly r.

Proof   We shall reduce A to C in two stages. The first keeps the number of columns intact, and removes only the rows that are not wanted in C. The row space of this intermediate matrix B is obviously contained in the row space of A, so that $\text{rank}(B) \leq \text{rank}(A) = r$. At the second stage B is reduced to C by excluding the unwanted columns. Therefore the column space of C is contained in the column space of B, and $\text{rank}(C) \leq \text{rank}(B) \leq r$. This establishes (i).

To prove (ii), suppose that B is formed from r independent rows of A. Then the row space of B is of dimension r; $\text{rank}(B) = r$, and the column space of B must also have dimension r. Suppose next that C is formed from r independent columns of B. Then the column space of C has dimension r, and $\text{rank}(C) = r$. This completes the proof of (ii): Every matrix of rank r contains a nonsingular r by r submatrix.

EXAMPLE 10   Consider once more that 3 by 4 matrix of rank 2,

$$A = \begin{bmatrix} 1 & 3 & 3 & 2 \\ 2 & 6 & 9 & 5 \\ -1 & -3 & 3 & 0 \end{bmatrix} \quad \text{with submatrix} \quad C = \begin{bmatrix} 1 & 3 \\ 2 & 9 \end{bmatrix}.$$

Every 3 by 3 submatrix of A is singular, but C is not.

This theorem does not deserve to be overemphasized. Superficially, it resembles a theorem that *is* important—the one next to Fig. 2.7, and numbered 2W, at the end of the previous section. There we proved that every A is an invertible transformation from its r-dimensional row space to its r-dimensional column space. Those spaces, and that transformation, give total information about A; the whole matrix can be reassembled once the transformation is known. Here it is only a question of finding an invertible submatrix C, and there is nothing special about the one that is chosen; there may be, and in the example there are, many other invertible submatrices of order r. The only thing is, we do get a new and equivalent definition of rank: It is *the order of the largest nonsingular submatrix.*

EXERCISE 2.6.11   For the matrix $A = \begin{bmatrix} 1 & 2 & 3 \\ 3 & 6 & 9 \end{bmatrix}$, find the largest invertible submatrix and the rank.

We now give a series of rather abstract exercises, to provide some practice with proofs.

EXERCISE 2.6.12   Suppose A is m by n and B is n by m, with $n < m$. Prove that their product AB is singular.

EXERCISE 2.6.13   Prove that $\text{rank}(A + B) \leq \text{rank}(A) + \text{rank}(B)$.

EXERCISE 2.6.14   If $\mathfrak{N}(AB) = \mathfrak{N}(B)$, assume that some vector y is in both $\mathfrak{R}(B)$ and $\mathfrak{N}(A)$ and prove that $y = 0$.

EXERCISE 2.6.15   If A is square and invertible, show that AB has the same nullspace, the same row space, and the same rank as B itself. Hint: Apply the relationship (i) from 2Z also to the product of A^{-1} and AB.

EXERCISE 2.6.16   Consider the four-dimensional vector space V of all 2 by 2 matrices B. Suppose A is one particular matrix, and suppose V is transformed into itself by the following rule: Every 2 by 2 matrix B is transformed into AB. If A is invertible, then this transformation has no nullspace; the only matrix B that is transformed into zero, $AB = 0$, is the matrix $B = 0$.

   (i)   If $A = \begin{bmatrix} 1 & 0 \\ 2 & 0 \end{bmatrix}$, describe the nullspace of the transformation—the space of all B such that $AB = 0$—and compute its dimension (which is the nullity of the transformation, not of A!).

   (ii)   Describe also the range, or "column space," of this transformation—the space of all matrices AB, where $A = \begin{bmatrix} 1 & 0 \\ 2 & 0 \end{bmatrix}$ and B varies? Compute the dimension of this range (which is the rank of the transformation, not of A). Verify that rank + nullity = dimension of V.

REVIEW EXERCISES

2.1   Find a basis for the subspace of $\mathbf{R}^4$ in which $x_1 = x_2 = x_3$.

2.2   What is the echelon form U of

$$A = \begin{bmatrix} 1 & 2 & 0 & 2 & 1 \\ -1 & -2 & 1 & 1 & 0 \\ 1 & 2 & -3 & -7 & -2 \end{bmatrix}?$$

What are the dimensions of its four fundamental subspaces?

2.3   By giving a basis, describe a two-dimensional subspace of $\mathbf{R}^3$ that contains none of the coordinate vectors $(1, 0, 0)$, $(0, 1, 0)$, $(0, 0, 1)$.

2.4   Find the rank and the nullspace of

$$A = \begin{bmatrix} 0 & 0 & 1 \\ 0 & 0 & 1 \\ 1 & 1 & 1 \end{bmatrix}.$$

2.5   Construct a matrix whose nullspace is spanned by $[1 \ 0 \ 1]^{\mathrm{T}}$.

2.6   True or false, with counterexample if false:

   (i)   If the vectors $x_1, \ldots, x_m$ span a subspace S, then dim $S = m$.
  (ii)   The intersection of two subspaces of a vector space cannot be empty.
 (iii)   If $Ax = Ay$, then $x = y$.
 (iv)   The row space of A has a unique basis that can be computed by reducing A to echelon form.
  (v)   If a square matrix A has independent columns, so does A^2.

2.7   Find bases for the four fundamental subspaces associated with

$$A = \begin{bmatrix} 1 & 2 \\ 3 & 6 \end{bmatrix}, \qquad B = \begin{bmatrix} 0 & 0 \\ 0 & 0 \end{bmatrix}, \qquad C = \begin{bmatrix} 1 & 1 & 0 & 0 \\ 0 & 1 & 0 & 1 \end{bmatrix}.$$

2.8   What is the most general solution to $u + v + w = 1$, $u - w = 2$?

2.9   Is there a matrix whose row space contains $[1 \ 1 \ 1]^{\mathrm{T}}$ and whose nullspace contains $[1 \ 0 \ 0]^{\mathrm{T}}$?

2.10   Find all vectors which are perpendicular to $(1, 4, 4, 1)$ and $(2, 9, 8, 2)$.

2.11   Do the vectors $(1, 2, 1)$, $(0, 1, 3)$, and $(3, 7, 6)$ lie on the same plane in $\mathbf{R}^3$? If so, find a vector orthogonal to that plane.

2.12   The system $Ax = b$ has a solution if and only if b is orthogonal to which of the four fundamental subspaces?

2.13   Find all solutions to the systems

$$\begin{bmatrix} 1 & 1 & 1 \\ 2 & 1 & 1 \\ 3 & 1 & 1 \end{bmatrix} \begin{bmatrix} x_1 \\ x_2 \\ x_3 \end{bmatrix} = \begin{bmatrix} 2 \\ 3 \\ 4 \end{bmatrix} \qquad \text{and} \qquad \begin{bmatrix} 1 & 1 & 1 \\ 2 & 1 & 1 \\ 3 & 1 & 1 \end{bmatrix} \begin{bmatrix} x_1 \\ x_2 \\ x_3 \end{bmatrix} = \begin{bmatrix} 2 \\ 3 \\ 5 \end{bmatrix}.$$

2.14 Do the vectors $(1, 1, 3)$, $(2, 3, 6)$, and $(1, 4, 3)$ form a basis for $\mathbf{R}^3$?

2.15 Find matrices A for which the number of solutions to $Ax = b$ is

    (i)   0 or 1, depending on b;

    (ii)   1 or ∞, depending on b;

    (iii)   0 or ∞, depending on b;

    (iv)   1, regardless of b.

2.16 In the previous exercise, how must r be related to m and n in each case?

[3]

ORTHOGONAL PROJECTIONS AND LEAST SQUARES

3.1 ■ INNER PRODUCTS AND TRANSPOSES

We know already that the inner product of two vectors x and y is the number $x^T y$. So far we have been interested only in whether or not that inner product is zero—in other words, whether or not the two vectors are orthogonal. Now we want to allow also the possibility of inner products that are not zero, and angles that are not right angles—and to understand the relationship of the inner product to the angle. Also, still in this first section, we would like to clarify the connection between inner products and transposes. In the last chapter the transpose was constructed by flipping over a matrix as if it were some kind of pancake. We have to do better than that.

If we try also to summarize the rest of the chapter, there is no way to avoid the fact that *the orthogonal case is by far the most important*. Suppose we are given a point b in n-dimensional space, and we want to find its distance to a given line, say the line in the direction of the vector a. We are looking along that line for the point p closest to b. Then, as you know, the line connecting b to p (the dotted line in Fig. 3.1) is perpendicular to the original vector a. This fact will allow us to find the closest point p, and to compute its distance from b. Even though the given vectors a and b are not orthogonal, the solution to the problem automatically brings in orthogonality.

The situation is the same when, instead of a line in the direction of a, we are given a plane—or more generally any subspace S of $\mathbf{R}^n$. Again the problem is to find the point p on that subspace that is closest to b, and again *this point p*

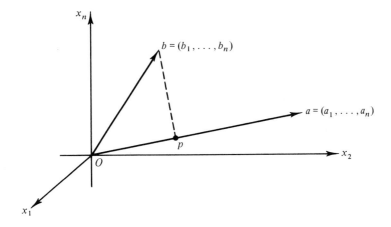

Fig. 3.1. Projections in n-dimensional space.

is the projection of b onto the subspace. When we draw a perpendicular from b to S, p is the point where the perpendicular meets the subspace. Geometrically speaking, this is a simple solution to a very natural problem about distances between points b and subspaces S. But there are some questions that need to be asked:

(1)   Does this problem actually arise in practical applications?

(2)   Analytically, if the subspace S is described by specifying a basis (or even just some set of vectors spanning it), is there a definite formula for the point p?

(3)   Computationally, is there a numerically stable way to use this formula and to calculate p?

The answer to the first two questions is certainly yes. Our problem, described so far only in geometrical terms, is exactly the problem of *least squares solution to an overdetermined system.* The vector b represents the data, given by a series of experiments or questionnaires, and they contain too many errors to be found in the given subspace. In other words, when we try to write b as a vector in the subspace S, it cannot be done—the equations we meet are inconsistent, and have no solution. Therefore the least squares method selects the point p as the best choice possible. There can be no doubt of the importance of this application.†

The second question, to find a formula for p, is very easy when the subspace is a line. We shall project one vector onto another in several different ways, in this section and the next, and relate this projection to the earlier question of inner products and angles. Fortunately, the formula for p remains fairly simple when we project onto a higher dimensional subspace, provided we are given a

† In economics and statistics it is called *regression analysis.*

basis. This is by far the most important case; it corresponds to a least squares problem with several parameters to be chosen, and it is solved in Section 3.2. Then there remain two more possibilities which need special attention:

(a)   If S is described by a set of vectors that span it, but which are linearly dependent, this degeneracy leads to a breakdown in the formula for p. The point p is still uniquely determined, by being closest to b, but it can be expressed in many different ways as a combination of the linearly dependent spanning set. Therefore a further rule is required to pick out one special combination, and this rule leads us—in Section 3.4—to the pseudoinverse of a matrix.

(b)   Suppose, in a much more favorable case, that the vectors describing S are not only independent but mutually orthogonal. Then the formula for p becomes particularly simple—it virtually reduces to the easy case of projection onto a line—and so do the numerical calculations. This suggests the idea of preparing the basis beforehand in such a way that its members are orthogonal to one another. Any basis can be converted to an orthogonal basis, and Section 3.3 describes the simple "Gram–Schmidt process" which accomplishes this orthogonalization.

It remains to answer question 3 above, whether or not least squares approximation can be made numerically stable. This question is not so easy, for the following reason: The subspace S is specified by a set of vectors that span it, and if those vectors are extremely close to being linearly dependent, then numerically it is virtually impossible to decide whether or not they *are* dependent. In other words, the dimension of S (the number of independent vectors) is very unstable, and that makes the pseudoinverse, and the point p itself, both unstable. In a less delicate case, the "normal equations" in Section 3.2 are the simplest—and either Gram–Schmidt orthogonalization or the singular value decomposition in Section 3.4 is the safest—of several different ways to calculate p.

Inner Products and the Schwarz Inequality

Now we return to pick up again the discussion of inner products and angles. You will soon see that it is not the angle, but *the cosine of the angle*, that is related more directly to inner products. Therefore we first look back to trigonometry, that is to the two-dimensional case, in order to find that relationship. (See Fig. 3.2.) Suppose α is the angle that the vector a makes with the x axis. Remembering that $\| a \|$ is the length of the vector a, which is the hypotenuse in the triangle OaQ, the sine and cosine of α are defined by

$$\sin \alpha = \frac{a_2}{\| a \|}, \qquad \cos \alpha = \frac{a_1}{\| a \|}.$$

The same is true for b and its corresponding angle β; the sine is $b_2/\| b \|$, and

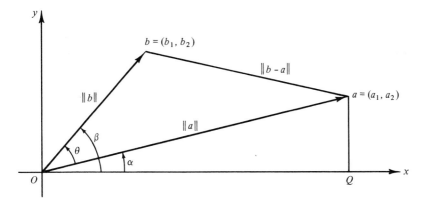

Fig. 3.2.   The cosine of the angle $\theta = \beta - \alpha$.

the cosine is $b_1/\| \, b \, \|$. Now, since θ is just $\beta - \alpha$, its cosine comes from a trigonometric identity which no one could forget:

$$\cos \theta = \cos \beta \cos \alpha + \sin \beta \sin \alpha = \frac{a_1 b_1 + a_2 b_2}{\| \, a \, \| \, \| \, b \, \|}. \tag{1}$$

The numerator in this formula is exactly the inner product of b and a, and gives the relationship we are looking for:

3A   The cosine of the angle between any two vectors is

$$\cos \theta = \frac{a^{\mathrm{T}} b}{\| \, a \, \| \, \| \, b \, \|}. \tag{2}$$

Notice that the formula is dimensionally correct; if we double the length of b, then both numerator and denominator are doubled, and the cosine is unchanged. Reversing the sign of b, on the other hand, reverses the sign of $\cos \theta$—and changes the angle by 180°.

Remark   There is another law of trigonometry, the law of cosines, that leads directly to the same result. It is not quite so unforgettable as the formula in (1), but it relates the lengths of the sides of any triangle:

$$\| \, b - a \, \|^2 = \| \, b \, \|^2 + \| \, a \, \|^2 - 2 \| \, b \, \| \, \| \, a \, \| \cos \theta. \tag{3}$$

When θ is a right angle, we are back to Pythagoras. But regardless of θ, the expression $\| \, b - a \, \|^2$ can be expanded as $(b - a)^{\mathrm{T}}(b - a)$, and (3) becomes

$$b^{\mathrm{T}} b - 2 a^{\mathrm{T}} b + a^{\mathrm{T}} a = b^{\mathrm{T}} b + a^{\mathrm{T}} a - 2 \| \, b \, \| \, \| \, a \, \| \cos \theta.$$

Canceling the terms that appear on both sides of this equation, you recognize formula (2) for the cosine. In fact this proves the cosine formula in n dimensions since we only have to worry about the plane triangle Oab.

Now we want to find the projection point p. This point must be some multiple $p = \bar{x}a$ of the given vector a—every point on the line is a multiple of a—and the problem is to compute the coefficient $\bar{x}$. All that we need for this computation is the geometrical fact that the line from b to the closest point $p = \bar{x}a$ is perpendicular to the vector a:

$$(b - \bar{x}a) \perp a, \quad \text{or} \quad a^T(b - \bar{x}a) = 0, \quad \text{or} \quad \bar{x} = \frac{a^Tb}{a^Ta}.$$

3B   The projection p of the point b onto the line spanned by the vector a is given by

$$p = \frac{a^Tb}{a^Ta}a. \tag{4}$$

The distance (squared) from the point to the line is therefore

$$\left\| b - \frac{a^Tb}{a^Ta}a \right\|^2 = b^Tb - 2\frac{(a^Tb)^2}{a^Ta} + \left(\frac{a^Tb}{a^Ta}\right)^2 a^Ta$$

$$= \frac{(b^Tb)(a^Ta) - (a^Tb)^2}{(a^Ta)}. \tag{5}$$

This allows us to redraw Fig. 3.1 and to include the correct formula for the point p (Fig. 3.3).

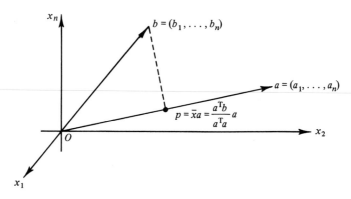

Fig. 3.3.   The projection of b onto a.

This formula has a remarkable corollary, which is probably the most important inequality in mathematics. (It includes the famous one about arithmetic and geometric means as a special case, and it is equivalent—see Exercise 3.1.1—to the so-called triangle inequality.) The result itself seems to be almost an accidental consequence of formula (5) for the distance between a point and a line, that is, the distance between b and $\bar{x}a$. This distance, and certainly the square of this distance, cannot avoid being greater than or equal to zero. Therefor the numerator in (5) must be nonnegative: $(b^Tb)(a^Ta) - (a^Tb)^2 \geq 0$. If we add $(a^Tb)^2$ to both sides, and then take square roots, the conclusion can be rewritten in the following way:

3C   Any two vectors satisfy the Schwarz inequality

$$|a^Tb| \leq \|a\| \|b\|. \tag{6}$$

Remark   According to formula (2), the ratio between the two sides of the Schwarz inequality is exactly $|\cos\theta|$. Since all cosines lie in the interval $-1 \leq \cos\theta \leq 1$, this gives another proof of (6)—and in some ways a more easily understood proof, because cosines are so familiar. But we have to insist on the simplicity of our proof, which simply amounted to doing the mechanical calculation in Eq. (5). The expression on the left of (5) is nonnegative; and it will stay nonnegative even when we later introduce some new possibilities for the lengths and inner products of vectors. Therefore the expression on the right side of (5) is also nonnegative, and without any appeal to trigonometry, the Schwarz inequality is proved.†

One final observation: Equality in (6) holds if and only if b is a multiple of a. In this case b is identical with the point p, and the distance (5) between the point and the line is zero.

The Transpose of a Matrix

Now we turn to transposes. Up to now, A^T has been defined simply by reflecting A across its main diagonal; the rows of A become the columns of A^T, and vice versa. In other words, the entry in row i and column j of A^T is the (j, i) entry of A:

$$(A^T)_{ij} = (A)_{ji}. \tag{7}$$

There is a deeper significance to the transpose, which comes from its close connection to inner products. In fact this connection can be used to give a new and much more "abstract" definition of the transpose:

3D   The transpose A^T can be defined by the following property: The inner product of Ax with y equals the inner product of x with A^Ty. Formally, this simply means that

$$(Ax)^Ty = x^TA^Ty = x^T(A^Ty). \tag{8}$$

† The name of Cauchy is also attached to this inequality $|a^Tb| \leq \|a\| \|b\|$, and the Russians even refer to it as the Cauchy–Schwarz–Buniakowsky inequality! Mathematical historians seem to agree that Buniakowsky's claim is genuine.

This definition has two purposes:

(i) It tells us how, when we measure the inner product in a different way, to make the proper change in the transpose. This becomes significant in the case of complex numbers; the new inner product is in Section 5.5.

(ii) It allows us to calculate the transpose of a product AB, without getting drowned by all the subscripts on the individual entries in A and B. The rule is exactly parallel to the familiar one for inverses, $(AB)^{-1} = B^{-1}A^{-1}$.

3E The transpose of AB is the product of the transposes in reverse order:

$$(AB)^{\mathrm{T}} = B^{\mathrm{T}}A^{\mathrm{T}}. \tag{9}$$

Proof The inner product of $(AB)x$ with y must equal, according to 3D, the inner product of x with $(AB)^{\mathrm{T}}y$. On the other hand, it is also the inner product of $A(Bx)$ with y, which by 3D is the inner product of Bx with $A^{\mathrm{T}}y$, which again by 3D is the inner product of x with $B^{\mathrm{T}}(A^{\mathrm{T}}y)$. Comparing these identities, we find $(AB)^{\mathrm{T}} = B^{\mathrm{T}}A^{\mathrm{T}}$. Or perhaps you prefer a concrete example:

$$AB = \begin{bmatrix} 1 & 4 \\ 0 & 3 \end{bmatrix}\begin{bmatrix} 0 & 1 & 1 \\ 2 & 0 & 1 \end{bmatrix} = \begin{bmatrix} 8 & 1 & 5 \\ 6 & 0 & 3 \end{bmatrix},$$

and

$$B^{\mathrm{T}}A^{\mathrm{T}} = \begin{bmatrix} 0 & 2 \\ 1 & 0 \\ 1 & 1 \end{bmatrix}\begin{bmatrix} 1 & 0 \\ 4 & 3 \end{bmatrix} = \begin{bmatrix} 8 & 6 \\ 1 & 0 \\ 5 & 3 \end{bmatrix}.$$

There is one more matrix that is needed. For projections onto a single line, the number $a^{\mathrm{T}}a$ appeared in the denominator. For projections onto a subspace, this number turns into a matrix $A^{\mathrm{T}}A$—and its rank can be determined in advance.

3F For any m by n matrix A of rank r, the product $A^{\mathrm{T}}A$ is a symmetric matrix and its rank is also r.

Remember that a *symmetric matrix is one that equals its transpose*. Therefore we compute the transpose of $A^{\mathrm{T}}A$ by rule (9):

$$(A^{\mathrm{T}}A)^{\mathrm{T}} = A^{\mathrm{T}}(A^{\mathrm{T}})^{\mathrm{T}}. \tag{10}$$

But when we transpose the matrix A twice, we are back to A again. Therefore the right side of (10) is nothing but $A^{\mathrm{T}}A$, and the equation says that it equals its transpose. In other words, $A^{\mathrm{T}}A$ is symmetric.

To find the rank, we shall show that A and $A^{\mathrm{T}}A$ have exactly the same nullspace. Then since the rank plus the dimension of the nullspace always equals the number of columns—$r + (n - r) = n$, and both A and $A^{\mathrm{T}}A$ have n columns—it follows immediately that the rank is the same for both matrices. First if x is in the nullspace of A, then $Ax = 0$ and $A^{\mathrm{T}}Ax = A^{\mathrm{T}}0 = 0$, so that x is in the nullspace of $A^{\mathrm{T}}A$. To go in the other direction, start by supposing that $A^{\mathrm{T}}Ax = 0$, and take the inner product with x:

$$x^{\mathrm{T}}A^{\mathrm{T}}Ax = 0, \quad \text{or} \quad \| Ax \|^2 = 0, \quad \text{or} \quad Ax = 0.$$

Thus x is in the nullspace of A; the two nullspaces are identical.

There is a particular case which is the most common, and most important. It occurs when the columns of A are linearly independent, so that the rank is $r = n$. Then, according to 3F, the n by n matrix $A^{\mathrm{T}}A$ is also of rank n; and therefore it must be invertible.

3G If A has linearly independent columns, so that $r = n$, then $A^{\mathrm{T}}A$ is a square, symmetric, and invertible matrix.

It is easy to give an example:

$$\text{if} \quad A = \begin{bmatrix} 1 & 4 \\ 2 & 0 \\ 3 & 1 \end{bmatrix}, \quad \text{then} \quad A^{\mathrm{T}}A = \begin{bmatrix} 14 & 7 \\ 7 & 17 \end{bmatrix}.$$

Both matrices have independent columns, rank equal to two, and no nullspace to speak of.

EXERCISE 3.1.1 (a) Given any two positive numbers x and y, choose the vector b equal to $(\sqrt{x}, \sqrt{y})$, and choose $a = (\sqrt{y}, \sqrt{x})$. Apply the Schwarz inequality to compare the arithmetic mean of x and y with their geometric mean.

(b) Suppose we start with a vector from the origin to the point x, and then add a vector of length $\| y \|$ connecting x to $x + y$. The third side of the triangle goes directly from the origin to $x + y$, and *the triangle inequality asserts that this distance cannot be greater than the sum of the first two*:

$$\| x + y \| \leq \| x \| + \| y \|.$$

After squaring both sides, reduce this to the Schwarz inequality.

EXERCISE 3.1.2 Verify Pythagoras' law for the triangle Obp in Figure 3.3, using (5) for the length of the side bp.

EXERCISE 3.1.3 Find the point p on the ray connecting the origin to $a = (1, 1, 1)$ closest to the point $b = (2, 4, 4)$. Find also the point closest to a on the line through b.

EXERCISE 3.1.4 Explain why the Schwarz inequality becomes an equality in case the points a and b lie on the same line through the origin, and only in that case. What if they lie on opposite sides of the origin?

EXERCISE 3.1.5 In n dimensions, what angle does the vector $(1, 1, \ldots, 1)$ make with the coordinate axes?

EXERCISE 3.1.6 The Schwarz inequality has a one-line proof if a and b are normalized ahead of time to be unit vectors:

$$| a^{\mathsf{T}}b | = | \sum a_j b_j | \leq \sum | a_j | \, | b_j | \leq \sum \frac{| a_j |^2 + | b_j |^2}{2} = \frac{1}{2} + \frac{1}{2} = \| a \| \, \| b \|.$$

The exercise is to justify the middle step.

EXERCISE 3.1.7 By transposing $AA^{-1} = I$, we discover that *the transpose of the inverse is the inverse of the transpose*: $(A^{-1})^{\mathsf{T}} = (A^{\mathsf{T}})^{-1}$. Show that if A is symmetric, so is A^{-1}.

EXERCISE 3.1.8 Construct 2 by 2 symmetric matrices A and B whose product AB is not symmetric. Note that if A commutes with B, then the product does remain symmetric: $(AB)^{\mathsf{T}} = B^{\mathsf{T}}A^{\mathsf{T}} = BA = AB$.

EXERCISE 3.1.9 If A has rank r, then so do the three matrices A^{T} (row rank = column rank), $A^{\mathsf{T}}A$ (by 3F), and AA^{T} (3F applied to A^{T}). Give an example to show that AA^{T} might nevertheless fail to be invertible even when $A^{\mathsf{T}}A$ is.

EXERCISE 3.1.10 The methane molecule CH_4 is arranged as if the carbon atom were at the center of a regular tetrahedron with the four hydrogen atoms at the vertices. If vertices are placed at $(0, 0, 0)$, $(1, 1, 0)$, $(1, 0, 1)$, and $(0, 1, 1)$—note that all six edges have length $\sqrt{2}$, so the tetrahedron is regular—what is the cosine of the angle between the rays going from the center $(\frac{1}{2}, \frac{1}{2}, \frac{1}{2})$ to the vertices? (The bond angle itself is about $109.5°$, an old friend of chemists.)

3.2 ■ PROJECTIONS ONTO SUBSPACES AND LEAST SQUARES
APPROXIMATIONS

Up to this point, a system $Ax = b$ either has a solution or not. If b is not in the column space $\mathfrak{R}(A)$, the system is inconsistent and Gaussian elimination fails. This is almost certain to be the case for a system of m equations, $m > 1$, in only one unknown. For example, the simultaneous equations

$$2x = b_1$$
$$3x = b_2 \tag{11}$$
$$4x = b_3$$

will be solvable only if the right-hand sides b_i are in the ratio $2:3:4$. The solution x is certainly unique if it exists, but it will exist only if b is on the same line as the vector

$$a = \begin{bmatrix} 2 \\ 3 \\ 4 \end{bmatrix}.$$

In spite of their unsolvability, inconsistent equations arise in practice and have to be solved. One possibility is to determine x from a part of the system, and ignore the rest; this is hard to justify if all m equations come from the same source. Rather than expecting no error in some equations and large errors in the others, it is more reasonable to choose x so as to minimize the *average error in the m equations*. There are many ways to define such an average, but the most convenient is to use the sum of squares

$$E^2 = (2x - b_1)^2 + (3x - b_2)^2 + (4x - b_3)^2. \qquad (12)$$

If there is an exact solution to $ax = b$, the minimum error is $E = 0$. In the more likely case that b is not proportional to a, the function E^2 will be a parabola with its minimum at the point where

$$\frac{dE^2}{dx} = 2[(2x - b_1)2 + (3x - b_2)3 + (4x - b_3)4] = 0. \qquad (13)$$

Solving for x, the least squares solution of the system $ax = b$ is

$$\bar{x} = \frac{2b_1 + 3b_2 + 4b_3}{2^2 + 3^2 + 4^2} = \frac{2b_1 + 3b_2 + 4b_3}{29}. \qquad (14)$$

It is not hard to find the general formula, given any $a \neq 0$ and any right side b. First of all, the error E is nothing but the length of the vector $ax - b$,

$$E = \| ax - b \| = [(a_1x - b_1)^2 + \cdots + (a_mx - b_m)^2]^{1/2}. \qquad (15)$$

Squaring, the parabola is

$$E^2 = (ax - b)^{\mathrm{T}}(ax - b) = a^{\mathrm{T}}ax^2 - 2a^{\mathrm{T}}bx + b^{\mathrm{T}}b. \qquad (16)$$

This has a minimum at the point where

$$\frac{dE^2}{dx} = 2a^{\mathrm{T}}ax - 2a^{\mathrm{T}}b = 0.$$

3H The least squares solution to a problem $ax = b$ in one unknown is

$$\bar{x} = \frac{a^{\mathrm{T}}b}{a^{\mathrm{T}}a}. \qquad (17)$$

Geometrically, this solution is identical with the projection: $p = \bar{x}a$ is the point on the line through a closest to b.

You see that we keep coming back to the geometrical interpretation of a least squares problem—to minimize a distance. In fact, by differentiating the parabola E^2 in (16) and setting the derivative to zero, we have used calculus to confirm the geometry of the previous section; the line connecting b to p

must be perpendicular to the line in the direction of a, and this gives the correct $\bar{x}$:

$$a^\mathrm{T}(b - \bar{x}a) = a^\mathrm{T}b - \frac{a^\mathrm{T}b}{a^\mathrm{T}a}\,a^\mathrm{T}a = 0. \tag{18}$$

As a side remark, we notice the degenerate case $a = 0$. All multiples of a are zero, and the line is only a point. Therefore $p = 0$ is the only candidate for the projection of b onto the vector a. But the formula (17) for $\bar{x}$ becomes a meaningless $0/0$, and correctly reflects the fact that the multiple $\bar{x}$ is left completely undetermined. In fact, all values of x give the same error $E = \|\, 0x - b \,\|$, so E^2 is a horizontal line instead of a parabola, and there is no unique minimizing point $\bar{x}$. One purpose of the pseudoinverse in Section 3.4 is to assign some definite value to $\bar{x}$; in this case it would assign $\bar{x} = 0$, which at least seems a more "symmetric" choice than any other number.

EXERCISE 3.2.1   Suppose we observe a patient's weight on four different occasions, with the results $b_1 = 150$, $b_2 = 153$, $b_3 = 150$, $b_4 = 151$. What is the best value, in the least squares sense, which we can assign to this weight?

EXERCISE 3.2.2   Find the best least squares solution $\bar{x}$ to $3x = 10$, $4x = 5$.

Least Squares Problems with Several Variables

Now we are ready for the next step, to project b onto a subspace rather than just a line. This geometrical problem arises in the following way. Suppose we start again from a system $Ax = b$, but this time let A be an m by n matrix—instead of permitting only one unknown, with a single column vector a, the matrix has n columns. We shall still imagine that the number m of observations is larger than the number n of unknowns, so it must be expected that the system $Ax = b$ will be inconsistent. *Probably there will not exist a choice of x that perfectly fits the data b; or, in other words, probably the vector b will not be a combination of the columns of A.*

Again the problem is to choose $\bar{x}$ so as to minimize the error, and again this minimization will be done in the least squares sense. The error is $E = \|\, Ax - b \,\|$, and *this is exactly the distance from b to the point Ax in the column space of A.* (Remember that Ax is the linear combination of the columns using the coefficients $x_1, \ldots, x_n$.) Therefore searching for the least squares solution $\bar{x}$, which will minimize the error E, is the same as locating the point $p = A\bar{x}$ that is closer to b than any other point in the column space.

We may use either geometry or calculus to determine $\bar{x}$, and we prefer the appeal of geometry; p must be the "projection of b onto the column space," and *the error vector $A\bar{x} - b$ must be perpendicular to that space* (Fig. 3.4). This perpendicularity to a space is expressed as follows. Each vector in the column space of A is a linear combination of the columns, with some coefficients $y_1, \ldots, y_n$. In other words, it is a vector of the form Ay. For all choices

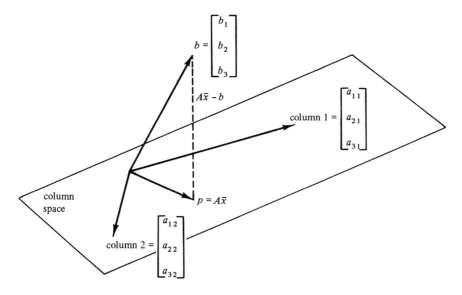

Fig. 3.4. Projection onto the column space of a 3 by 2 matrix.

of y, these vectors in the plane must be perpendicular to the error vector $A\bar{x} - b$:

$$(Ay)^{\mathrm{T}}(A\bar{x} - b) = 0, \quad \text{or} \quad y^{\mathrm{T}}[A^{\mathrm{T}}A\bar{x} - A^{\mathrm{T}}b] = 0. \tag{19}$$

This is true for every y, and there is only one way in which it can happen: The vector in brackets has to be the zero vector, $A^{\mathrm{T}}A\bar{x} - A^{\mathrm{T}}b = 0$. The geometry has led us directly to the fundamental equations of least squares theory.

3I The least squares solution to an inconsistent system $Ax = b$ of m equations in n unknowns satisfies

$$A^{\mathrm{T}}A\bar{x} = A^{\mathrm{T}}b. \tag{20}$$

These are known as the *"normal equations."* If the columns of A are linearly independent, then by 3G the matrix $A^{\mathrm{T}}A$ is invertible, and the unique least squares solution is

$$\bar{x} = (A^{\mathrm{T}}A)^{-1}A^{\mathrm{T}}b. \tag{21}$$

The projection of b onto the column space is therefore

$$p = A\bar{x} = A(A^{\mathrm{T}}A)^{-1}A^{\mathrm{T}}b. \tag{22}$$

The matrix $B = (A^{\mathrm{T}}A)^{-1}A^{\mathrm{T}}$ which appears in (21), $\bar{x} = Bb$, is one of the left-inverses of $A : BA = (A^{\mathrm{T}}A)^{-1}A^{\mathrm{T}}A = I$. Such a left-inverse was guaranteed by the uniqueness half of Theorem 2Q because the columns of A are linearly independent.

We choose a numerical example to which we can apply either our intuition or formula (22). Suppose that the two columns of A and the vector b are given by

$$A = \begin{bmatrix} 1 & 2 \\ 1 & 5 \\ 0 & 0 \end{bmatrix}, \qquad b = \begin{bmatrix} 4 \\ 3 \\ 9 \end{bmatrix}.$$

The column space of A is easy to visualize since both columns end with a zero; it is just the x-y plane within the full three-dimensional space. The projection of b onto this plane will not change the x and y components, which are 4 and 3; but the z component will disappear, and $p = (4,\ 3,\ 0)^{\mathrm{T}}$. This picture is confirmed by formula (22):

$$A^{\mathrm{T}}A = \begin{bmatrix} 1 & 1 & 0 \\ 2 & 5 & 0 \end{bmatrix} \begin{bmatrix} 1 & 2 \\ 1 & 5 \\ 0 & 0 \end{bmatrix} = \begin{bmatrix} 2 & 7 \\ 7 & 29 \end{bmatrix}, \qquad (A^{\mathrm{T}}A)^{-1} = \frac{1}{9}\begin{bmatrix} 29 & -7 \\ -7 & 2 \end{bmatrix}$$

$$p = \begin{bmatrix} 1 & 2 \\ 1 & 5 \\ 0 & 0 \end{bmatrix} \frac{1}{9} \begin{bmatrix} 29 & -7 \\ -7 & 2 \end{bmatrix} \begin{bmatrix} 1 & 1 & 0 \\ 2 & 5 & 0 \end{bmatrix} \begin{bmatrix} 4 \\ 3 \\ 9 \end{bmatrix} = \begin{bmatrix} 4 \\ 3 \\ 0 \end{bmatrix}.$$

We can also look directly at the inconsistent system

$$u + 2v = 4$$

$$u + 5v = 3$$

$$0u + 0v = 9.$$

In this case, the best we can do is to solve the first two equations (giving the components $\bar{u}$ and $\bar{v}$ of $\bar{x}$) and ignore the third equation; the error in that equation is bound to be 9.

Note that when b is actually in the column space, which means that it can be written as some combination $b = Ax$ of the columns, the projection simplifies to

$$p = A(A^{\mathrm{T}}A)^{-1}A^{\mathrm{T}}Ax = Ax = b. \tag{23}$$

The closest point p is just b itself—which is obvious.

At the other extreme, the vector b might be orthogonal to all the columns of A. In this case b not only fails to lie in the subspace, it is actually perpendicular to it. Then the geometrical picture suggests that the closest point on the subspace is at the origin; b has a zero component along the subspace, so its projection is $p = 0$. This is confirmed when we compute $\bar{x}$: The columns of A are the rows of A^{T}; and if b is orthogonal to them all, then $A^{\mathrm{T}}b = 0$. Therefore

$$\bar{x} = (A^{\mathrm{T}}A)^{-1}A^{\mathrm{T}}b = 0, \qquad \text{and} \qquad p = A\bar{x} = 0.$$

It is also worth checking that in the special case of projection onto a line, we get back to the earlier formula (17). A becomes just a column vector a, A^TA is the number a^Ta, and $\bar{x}$ in (21) is a^Tb/a^Ta as we expect.

EXERCISE 3.2.3   From the normal equations, find the best least squares solution to the inconsistent system

$$\begin{bmatrix} 1 & 0 \\ 0 & 1 \\ 1 & 1 \end{bmatrix} \begin{bmatrix} u \\ v \end{bmatrix} = \begin{bmatrix} 1 \\ 1 \\ 0 \end{bmatrix}.$$

Do the same for the system $x = 1$, $x = 3$, $x = 5$, or

$$\begin{bmatrix} 1 \\ 1 \\ 1 \end{bmatrix} [x] = \begin{bmatrix} 1 \\ 3 \\ 5 \end{bmatrix}.$$

EXERCISE 3.2.4   (a)   Let

$$A = \begin{bmatrix} 1 & 0 \\ 0 & 1 \\ 1 & 1 \end{bmatrix}, \qquad x = \begin{bmatrix} u \\ v \end{bmatrix}, \qquad b = \begin{bmatrix} 1 \\ 3 \\ 4 \end{bmatrix}.$$

Write out $E^2 = \| Ax - b \|^2$ in full, and set to zero its derivatives with respect to u and v. Compare the resulting equations with $A^TA\bar{x} = A^Tb$, confirming that calculus as well as geometry can be used to derive the normal equations. These equations come directly from minimizing E^2.

(b)   Find the solution $\bar{x}$ and the projection $p = A\bar{x}$ of b onto the column space.

(c)   Verify that $b - p$ is perpendicular to the columns of A.

Projection Matrices

Our computations have shown that the closest point to b is $p = A(A^TA)^{-1}A^Tb$. *This formula expresses in matrix terms the geometrical construction of a perpendicular line from b to the column space of A.* The matrix that describes this construction is called a ***projection matrix***, and it will be denoted by P:

$$P = A(A^TA)^{-1}A^T. \tag{24}$$

This matrix projects any vector b onto the column space of A.† In other words, $p = Pb$ is the component of b in the column space, and the error $b - Pb$ is the component in the orthogonal complement. (Or, as it seems natural to say, $I - P$ is also a projection matrix; it projects any vector b onto the orthogonal complement, and the projection is $(I - P)b = b - Pb$.) In short, we have a

---

† There may be a risk of confusion with permutation matrices, also denoted by P. But the risk should be small since we will try never to let both appear on the same page.

matrix formula for splitting a vector into two perpendicular components; Pb is in the column space $\Re(A)$, and the other component $(I - P)b$ is in the left nullspace $\Re(A^T)$—which is the orthogonal complement of the column space.

These projection matrices can be understood either from a geometric or from an algebraic standpoint. They are a family of matrices with very special properties, and in fact they will later be used as the fundamental building blocks for all symmetric matrices. Therefore we pause for a moment, before returning to complete the applications to least squares, in order to identify these properties.

3J   The projection matrix $P = A(A^TA)^{-1}A^T$ has two basic properties:

  (i)   It is idempotent: $P^2 = P$.
  (ii) It is symmetric: $P = P^T$.

Conversely, any matrix with these two properties represents a projection onto the column space of P.

Proof   It is easy to see geometrically why $P^2 = P$, since if we start with any b, the vector Pb lies in the subspace we are projecting onto. Therefore when we project again, producing $P(Pb)$ or P^2b, nothing is changed; the vector is already in the subspace, and $Pb = P^2b$ for every b. Algebraically, the same conclusion follows from

$$P^2 = A(A^TA)^{-1}A^TA(A^TA)^{-1}A^T = A(A^TA)^{-1}A^T = P.$$

To prove that P is also symmetric, we multiply the transposes in reverse order and use the identity $(B^{-1})^T = (B^T)^{-1}$ of Exercise 3.1.7, with $B = A^TA$:

$$P^T = (A^T)^T((A^TA)^{-1})^TA^T = A((A^TA)^T)^{-1}A^T = A(A^TA)^{-1}A^T = P.$$

Or, to give a more geometric proof, imagine any vectors b and c. If b is projected in the usual way to produce Pb, and c is projected onto the orthogonal complement to produce $(I - P)c$, then these two projections are perpendicular:

$$(Pb)^T(I - P)c = b^TP^T(I - P)c = 0.$$

Since this is true for every b and c, we conclude that

$$P^T(I - P) = 0, \qquad \text{or} \qquad P^T = P^TP, \qquad \text{or} \qquad P = (P^TP)^T = P^TP.$$

Thus $P^T = P$, and P is symmetric.

For the converse, we have to deduce from properties (i) and (ii) that P is a projection matrix onto its column space. This space consists of all combinations Pc of the columns of P. For any vector b, the vector Pb is certainly in this space; matrix multiplication tells us that Pb is a combination of the columns of P. Furthermore, we can show that *the error vector $b - Pb$ is orthogonal to the space*: For any Pc in the space, properties (i) and (ii) imply

$$(b - Pb)^TPc = b^T(I - P)^TPc = b^T(P - P^2)c = 0. \qquad (25)$$

Since $b - Pb$ is orthogonal to the space, it is the perpendicular we wanted, and P is the projection.

EXERCISE 3.2.5 Suppose we are given a basis u_1, u_2 for a subspace S of three-dimensional space and a vector b that is outside S:

$$u_1 = \begin{bmatrix} 1 \\ 1 \\ 0 \end{bmatrix}, \qquad u_2 = \begin{bmatrix} 1 \\ 0 \\ 1 \end{bmatrix}, \qquad b = \begin{bmatrix} 0 \\ 2 \\ 1 \end{bmatrix}.$$

Find the projection matrix P for the subspace S by constructing the matrix A that has u_1, u_2 as its columns. Compute the projection of b onto S, and its projection onto the orthogonal complement $S^{\perp}$.

EXERCISE 3.2.6 (a) Show that if P is a projection matrix, so that it has properties (i) and (ii), then $I - P$ also has these properties.
(b) Show that if P_1 and P_2 are projections onto subspaces S_1 and S_2, and if $P_1 P_2 = P_2 P_1 = 0$, then $P = P_1 + P_2$ is also a projection. Construct a 2 by 2 example.

EXERCISE 3.2.7 If P is the projection matrix onto a line in the x-y plane, draw a figure to describe the effect of the "reflection matrix" $H = I - 2P$. Explain both geometrically and algebraically why $H^2 = I$.

EXERCISE 3.2.8 Show that if u has unit length, then the rank one matrix $P = uu^{\mathrm{T}}$ is a projection matrix: It has properties (i) and (ii). By choosing $u = a/\| a \|$, P becomes the projection onto the line through a, and Pb is the point $p = \bar{x}a$: Rank-one projections correspond exactly to least squares problems in one unknown.

EXERCISE 3.2.9 What 2 by 2 matrix projects the x-y plane onto the y axis?

Least Squares Fitting of Data

Suppose we do a series of experiments, and expect the output y to be pretty much a linear function of the input t, $y = C + Dt$. For example:

(1) At a number of time intervals, we measure the distance to a satellite which is on its way to Mars. In this case t is the time, y is the distance, and unless the motor was left on or gravity is strong, the satellite should move with nearly constant velocity v: $y = y_0 + vt$.

(2) We may vary the load that is applied to a structure, and measure the strain it produces. In this experiment t is the load, and y is the reading from the strain gauge. Unless the load is so great that the material becomes plastic, a linear relationship $y = C + Dt$ is normal in the theory of elasticity.

(3) In economics and business, there are complicated interconnections between the costs of production and the volume produced and the prices and the profits. Nevertheless, within a certain range these connections may not be far

from linear. Suppose, for example, that t_1 copies of a magazine are printed at an actual cost y_1; the next week, there are t_2 copies at a cost y_2; and so on. Then the printer may predict his costs in a future week by assuming that $y = C + Dt$, and estimating C and D from the figures he already has. The coefficient D—the cost of each additional copy, or marginal production cost—will often be more critical to his decisions than the overhead cost C.

The question is, How does one compute the coefficients C and D from the results of these experiments? If the relationship is truly linear, and there is no experimental error, then there is no problem; two measurements of y at different values of t will determine the line $y = C + Dt$ and all further measurements will lie on this line. But if there *is* error, and the additional points fail to land on the line, then we must be prepared to "average" all the experiments and find an optimal line—which is not to be confused with the line on which we have been projecting the vector b in the previous pages! In fact, since there are two unknowns C and D to be determined, we shall be involved with projections onto a two-dimensional subspace. The least squares problem comes directly from the experimental results

$$C + Dt_1 = y_1$$

$$C + Dt_2 = y_2 \tag{26}$$
$$\vdots$$
$$C + Dt_m = y_m .$$

This is an overdetermined system; and if errors are present, it will have no solution. We emphasize that the unknown vector x has two components C and D:

$$\begin{bmatrix} 1 & t_1 \\ 1 & t_2 \\ \vdots & \vdots \\ 1 & t_m \end{bmatrix} \begin{bmatrix} C \\ D \end{bmatrix} = \begin{bmatrix} y_1 \\ y_2 \\ \vdots \\ y_m \end{bmatrix} , \qquad \text{or} \qquad Ax = b. \tag{27}$$

The best solution in the least squares sense is the one that minimizes the sum of the squares of the errors; we choose $\bar{C}$ and $\bar{D}$ to minimize

$$E^2 = \| b - Ax \|^2 = (y_1 - C - Dt_1)^2 + \cdots + (y_m - C - Dt_m)^2.$$

In matrix terminology, we choose $\bar{x}$ so that the point $p = A\bar{x}$ is as close as possible to b. Of all straight lines $y = C + Dt$, we are choosing the one that best fits the data (Fig. 3.5). On the graph, the errors are the vertical distances $y - C - Dt$ to the straight line; these are the distances that are squared, summed, and minimized.

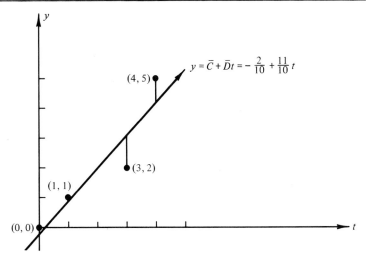

Fig. 3.5. Straight-line approximation.

EXAMPLE   Suppose we are given the four measurements marked on the figure:

$$y = 0 \quad \text{at} \quad t = 0, \quad y = 1 \quad \text{at} \quad t = 1,$$
$$y = 2 \quad \text{at} \quad t = 3, \quad y = 5 \quad \text{at} \quad t = 4.$$

Note that the values of t are not required to be equally spaced; the experimenter may choose any convenient values (even negative values, if the experiment permits them) without any effect on the mathematical formulation. The overdetermined system $Ax = b$, for this set of measurements, is

$$\begin{bmatrix} 1 & 0 \\ 1 & 1 \\ 1 & 3 \\ 1 & 4 \end{bmatrix} \begin{bmatrix} C \\ D \end{bmatrix} = \begin{bmatrix} 0 \\ 1 \\ 2 \\ 5 \end{bmatrix}.$$

We shall need to form $A^{\mathrm{T}}A$ and its inverse:

$$A^{\mathrm{T}}A = \begin{bmatrix} 4 & 8 \\ 8 & 26 \end{bmatrix}, \quad (A^{\mathrm{T}}A)^{-1} = \frac{1}{20}\begin{bmatrix} 13 & -4 \\ -4 & 2 \end{bmatrix}.$$

Then the least squares solution $\bar{x} = (A^{\mathrm{T}}A)^{-1}A^{\mathrm{T}}b$ is calculated from

$$\bar{x} = \begin{bmatrix} \bar{C} \\ \bar{D} \end{bmatrix} = \frac{1}{20}\begin{bmatrix} 13 & -4 \\ -4 & 2 \end{bmatrix}\begin{bmatrix} 1 & 1 & 1 & 1 \\ 0 & 1 & 3 & 4 \end{bmatrix}\begin{bmatrix} 0 \\ 1 \\ 2 \\ 5 \end{bmatrix} = \begin{bmatrix} -\frac{2}{10} \\ \frac{11}{10} \end{bmatrix}.$$

The best straight line is $y = -\frac{2}{10} + \frac{11}{10}t$.

Remark   It makes no special difference to the mathematics of the least squares method that we are fitting the data by straight lines. In many experiments there is no reason to expect a linear relationship, and it would be crazy to look for one. Suppose for example that we are handed some radioactive material. The output y will be the reading on a Geiger counter at various times t. We may know that we are holding a mixture of two radioactive chemicals, and we may know their half-lives (or rates of decay), but we do not know how much of each is in our hands. If these two unknown amounts are C and D, then the Geiger counter readings would behave like the sum of two exponentials† (and not like a straight line):

$$y = Ce^{-\lambda t} + De^{-\mu t}. \tag{28}$$

In practice, because radioactivity jumps out in discrete amounts and at irregularly spaced times, the law (28) is not reflected exactly by the counter. Instead, we make a series of readings $y_1, \ldots, y_m$ at different times $t_1, \ldots, t_m$, and the relationship (28) will only be approximately satisfied:

$$Ce^{-\lambda t_1} + De^{-\mu t_1} \approx y_1,$$
$$\vdots \tag{29}$$
$$Ce^{-\lambda t_m} + De^{-\mu t_m} \approx y_m.$$

If there are more than two readings, $m > 2$, then in all likelihood we cannot solve these as equations for C and D. But the least squares principle will give optimal values $\bar{C}$ and $\bar{D}$.

The situation would be completely different if we knew the exact amounts C and D, and were trying to discover the decay rates λ and μ. This is a problem in *nonlinear least squares*, and it is very much harder. We would still form E^2, the sum of the squares of the errors, and minimize it. But E^2 will not be a quadratic polynomial in λ and μ, and setting its derivatives to zero will not give linear equations for the optimal $\bar{\lambda}$ and $\bar{\mu}$. In the exercises, we stay with linear least squares.

EXERCISE 3.2.10   Show that the best least squares fit to a set of measurements $y_1, \ldots, y_m$ by a horizontal line—in other words, by a constant function $y = C$—is their average

$$C = \frac{y_1 + \cdots + y_m}{m}.$$

(Compare with Exercise 3.2.1.) In statistical terms, the choice $\bar{y}$ that minimizes $E^2 = (y_1 - y)^2 + \cdots + (y_m - y)^2$ is the *mean* of the sample, and the resulting E^2 is the *variance* σ^2.

---

† In statistics or economics, this corresponds to two goods that are being shipped out or destroyed with probabilities given by Poisson's law. In the theory of populations, λ and μ will be negative if the birthrate exceeds the deathrate.

EXERCISE 3.2.11   Find the best straight line fit to the following measurements, and sketch your solution:

$$y = 2 \quad \text{at} \quad t = -1, \quad y = 0 \quad \text{at} \quad t = 0,$$

$$y = -3 \quad \text{at} \quad t = 1, \quad y = -5 \quad \text{at} \quad t = 2.$$

EXERCISE 3.2.12   Suppose that instead of a straight line, we fit the data in the previous exercise by a parabola: $y = C + Dt + Et^2$. In the inconsistent system $Ax = b$ that comes from the four measurements, what are the coefficient matrix A, the unknown vector x, and the data vector b? You need not compute $\bar{x}$.

ORTHOGONAL BASES, ORTHOGONAL MATRICES, AND ■ 3.3 GRAM–SCHMIDT ORTHOGONALIZATION

We have tried to explain the importance of orthogonality in solving least squares problems, and will continue to do so. But orthogonality has an importance, and an appeal to the intuition, that goes much deeper than least squares. Every time I think of the x-y plane, or of three-dimensional space, the imagination adds a set of coordinate axes to the picture. They provide a point of reference, which we call the origin. But more than that, *the coordinate axes that the imagination constructs are always orthogonal*. In choosing a basis for the x-y plane—which is the same thing as choosing a set of coordinate axes—we tend to choose an orthogonal basis.

If the idea of a basis is one of the key steps in connecting the geometry of a vector space to the algebra, then the specialization to an orthogonal basis is not very far behind. We need a basis in order to convert each geometric construction into an algebraic calculation, and we need an orthogonal basis in order to make these calculations simple. There is even a further specialization, which makes the basis just about optimal: We start with a set of mutually orthogonal vectors, and normalize them all to become unit vectors. This just means that each v in the set is divided by its own length, and replaced by $v/\| v \|$. This step changes an *orthogonal basis into an orthonormal basis*.

3K   A basis $v_1, \ldots, v_k$ is called ***orthonormal*** if

$$v_i^{\mathsf{T}} v_j = \begin{cases} 0 & \text{whenever} \quad i \neq j, \quad \text{giving the orthogonality} \\ \\ 1 & \text{whenever} \quad i = j, \quad \text{giving the normalization.} \end{cases} \tag{30}$$

The most important example is the *standard basis*. For the x-y plane or for $\mathbf{R}^n$, we not only imagine perpendicular coordinate axes but we also mark out on each axis a vector of unit length:

$$e_1 = \begin{bmatrix} 1 \\ 0 \\ 0 \\ \vdots \\ 0 \end{bmatrix}, \qquad e_2 = \begin{bmatrix} 0 \\ 1 \\ 0 \\ \vdots \\ 0 \end{bmatrix}, \qquad \cdots, \qquad e_n = \begin{bmatrix} 0 \\ 0 \\ 0 \\ \vdots \\ 1 \end{bmatrix}.$$

This is by no means the only orthonormal basis; we can rotate the whole set of axes without changing the right angles at which they meet. These rotation matrices, or orthogonal matrices, will be introduced below. On the other hand, if we are thinking not about $\mathbf{R}^n$ but about one of its subspaces, the standard vectors e_i might well lie outside that subspace. In this case, it is not so clear that even one orthonormal basis can be found. But we shall show that there does always exist such a basis, and that it can be constructed in a simple way out of any basis whatsoever: This construction, which converts a skew set of axes into a perpendicular set, is known as *Gram–Schmidt orthogonalization*.

To summarize, the three topics basic to this section are:

(1) The least squares solution of $Ax = b$, when the columns of A are orthonormal.

(2) The definition and properties of orthogonal matrices.

(3) The Gram–Schmidt process and its interpretation as a new matrix factorization.

Projections and Least Squares: The Orthonormal Case

Suppose A is an m by n matrix, and suppose its columns are orthonormal. Then these columns are certain to be independent, so we already know the projection matrix onto the column space, and the least squares solution $\bar{x}$: $P = A(A^{\mathrm{T}}A)^{-1}A^{\mathrm{T}}$ and $\bar{x} = (A^{\mathrm{T}}A)^{-1}A^{\mathrm{T}}b$. Not only are these formulas valid in the presence of orthonormal columns, but they become extremely simple: *The matrix $A^{\mathrm{T}}A$ is the identity*.

3L If the columns of A are orthonormal, then

$$A^{\mathrm{T}}A = \begin{bmatrix} \text{---} a_1^{\mathrm{T}} \text{---} \\ \text{---} a_2^{\mathrm{T}} \text{---} \\ \hline \\ \text{---} a_n^{\mathrm{T}} \text{---} \end{bmatrix} \begin{bmatrix} \Big| & \Big| & & \Big| \\ a_1 & a_2 & & a_n \\ \Big| & \Big| & & \Big| \end{bmatrix} = \begin{bmatrix} 1 & 0 & \cdot & 0 \\ 0 & 1 & \cdot & 0 \\ \cdot & \cdot & \cdot & \cdot \\ 0 & 0 & \cdot & 1 \end{bmatrix} = I. \quad (31)$$

This immediately improves the algebra, since P and $\bar{x}$ are changed to

$$P = AA^{\mathrm{T}}, \quad \text{and} \quad \bar{x} = A^{\mathrm{T}}b. \quad (32)$$

And of course the geometry ought to be improved at the same time; a simple algebraic formula should have a simple geometric interpretation. *When the coordinate axes are perpendicular, projection onto the space is simplified into projection onto each axis* (Fig. 3.6). The projection matrix becomes just $P = a_1 a_1^T + \cdots + a_n a_n^T$:

$$p = AA^Tb = \begin{bmatrix} | & & | \\ a_1 & & a_n \\ | & & | \end{bmatrix} \begin{bmatrix} a_1^Tb \\ \cdot \\ \cdot \\ \cdot \\ a_n^Tb \end{bmatrix} = a_1 a_1^Tb + \cdots + a_n a_n^Tb. \qquad (33)$$

The usual coupling term $(A^TA)^{-1}$ has disappeared, and p is the sum of the n separate projections.

We now have the five equations which are basic to this chapter, and it might be useful to collect them all in one place:

1. $Ax = b$, the given equation, probably inconsistent;
2. $A^TA\bar{x} = A^Tb$, the normal equations for $\bar{x}$;
3. $p = A\bar{x}$, the projection of b onto the column space of A;
4. $P = A(A^TA)^{-1}A^T$, the projection matrix giving $p = Pb$;
5. $\bar{x} = A^Tb$ and $P = AA^T = a_1 a_1^T + \cdots + a_n a_n^T$, the special case in which A has orthonormal columns.

EXAMPLE 1 The following case is simple but typical: Suppose we project a point $b = (x, y, z)$ onto the x-y plane. Obviously its projection is $p = (x, y, 0)$,

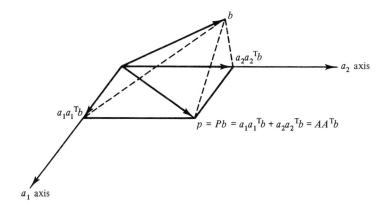

Fig. 3.6. Projection onto a plane = sum of projections onto orthonormal a_1 and a_2.

and this is the sum of the separate projections onto the x and y axes:

$$a_1 = \begin{bmatrix} 1 \\ 0 \\ 0 \end{bmatrix} \quad \text{and} \quad a_1a_1{}^{\mathrm{T}}b = \begin{bmatrix} x \\ 0 \\ 0 \end{bmatrix}; \quad a_2 = \begin{bmatrix} 0 \\ 1 \\ 0 \end{bmatrix} \quad \text{and} \quad a_2a_2{}^{\mathrm{T}}b = \begin{bmatrix} 0 \\ y \\ 0 \end{bmatrix}.$$

The overall projection matrix is

$$P = a_1a_1{}^{\mathrm{T}} + a_2a_2{}^{\mathrm{T}} = \begin{bmatrix} 1 & 0 & 0 \\ 0 & 1 & 0 \\ 0 & 0 & 0 \end{bmatrix}, \quad \text{and} \quad P\begin{bmatrix} x \\ y \\ z \end{bmatrix} = \begin{bmatrix} x \\ y \\ 0 \end{bmatrix}.$$

EXAMPLE 2   There is one case in which fitting a straight line leads to orthogonal columns. If measurements b_1, b_2, and b_3 are taken at times symmetrically placed around $t = 0$, say at $t_1 = -1$, $t_2 = 0$, and $t_3 = 1$, then the attempt to fit $y = C + Dt$ leads to three equations in two unknowns:

$$\begin{array}{ccc} C + Dt_1 = b_1 \\ C + Dt_2 = b_2 \\ C + Dt_3 = b_3 \end{array} \quad \text{or} \quad \begin{bmatrix} 1 & -1 \\ 1 & 0 \\ 1 & 1 \end{bmatrix}\begin{bmatrix} C \\ D \end{bmatrix} = \begin{bmatrix} b_1 \\ b_2 \\ b_3 \end{bmatrix}.$$

The two columns are orthogonal. They are not orthonormal, but that requires only a simple normalization; we factor out their lengths $\sqrt{3}$ and $\sqrt{2}$, and change the unknowns to $c = C\sqrt{3}$ and $d = D\sqrt{2}$:

$$Ax = \begin{bmatrix} 1/\sqrt{3} & -1/\sqrt{2} \\ 1/\sqrt{3} & 0 \\ 1/\sqrt{3} & 1/\sqrt{2} \end{bmatrix}\begin{bmatrix} c \\ d \end{bmatrix} = \begin{bmatrix} b_1 \\ b_2 \\ b_3 \end{bmatrix}.$$

Now $A^{\mathrm{T}}A$ is the identity, and the least squares solution can be written down at sight:

$$\bar{x} = A^{\mathrm{T}}b = \begin{bmatrix} (b_1 + b_2 + b_3)/\sqrt{3} \\ (-b_1 + b_3)/\sqrt{2} \end{bmatrix} = \begin{bmatrix} \bar{c} \\ \bar{d} \end{bmatrix}. \tag{34}$$

In other words, the coefficients C and D in the optimal straight line are

$$C = \frac{c}{\sqrt{3}} = \frac{b_1 + b_2 + b_3}{3}, \qquad D = \frac{d}{\sqrt{2}} = \frac{-b_1 + b_3}{2}.$$

Each of the components C and D now has a significance of its own. C is the *average*, or *mean*, of the data; it gives the best fit by a horizontal line, while Dt is the best fit by a straight line through the origin. *Because the columns are orthogonal, the sum of these two separate pieces is the best fit by any straight line whatsoever.*

Remark   Studying the special properties of an orthonormal basis often has a valuable side effect: An ordinary basis is understood more clearly than ever, by recognizing the properties that it does not have. The imagination is wrong if it thinks of every basis as orthonormal. The same is true for the special property just discovered about projections, that in the orthonormal case they are the sum of n one-dimensional projections. Intuition expects that a vector can be reproduced as the sum of its components along the coordinate axes, but *this depends on the orthogonality of the axes*. If the x axis is replaced by the line $y = x$ and $(0, 1)$ is projected onto this line and onto the y axis, the sum of these two projections is far from the original $(0, 1)$.

EXERCISE 3.3.1   (a)   Write down the four equations for fitting $y = C + Dt$ to the data

$$y = -4 \quad \text{at} \quad t = -2, \qquad y = -3 \quad \text{at} \quad t = -1,$$

$$y = -1 \quad \text{at} \quad t = 1, \qquad y = 0 \quad \text{at} \qquad t = 2.$$

Show that the columns are orthogonal and normalize them. What are the unknowns c and d in the new problem $Ax = b$?
   (b)   Find the optimal straight line, draw a graph, and write down the error E^2.
   (c)   Interpret the fact that the error is zero in terms of the original system of four equations in two unknowns: Where is the right side b with relation to the column space, and what is its projection p?

EXERCISE 3.3.2   Project $b = (0, 3, 0)$ onto each of the orthonormal vectors $a_1 = (\frac{2}{3}, \frac{2}{3}, -\frac{1}{3})$ and $a_2 = (-\frac{1}{3}, \frac{2}{3}, \frac{2}{3})$, and then find its projection p onto the plane of a_1 and a_2 .

EXERCISE 3.3.3   Find also the projection of $b = (0, 3, 0)$ onto $a_3 = (\frac{2}{3}, -\frac{1}{3}, \frac{2}{3})$, add up the three one-dimensional projections, and interpret the result. Why is $P = a_1 a_1{}^\mathrm{T} + a_2 a_2{}^\mathrm{T} + a_3 a_3{}^\mathrm{T}$ equal to the identity?

Orthogonal Matrices

An orthogonal matrix is simply a *square matrix with orthonormal columns*.†
We shall use the letter Q to denote an orthogonal matrix, and $q_1 , \ldots , q_n$ to denote its columns. One essential fact which has already been discovered, namely $Q^\mathrm{T}Q = I$, is only a different way of saying that the columns are orthonormal:

$$Q^\mathrm{T}Q = \begin{bmatrix} \text{---} q_1{}^\mathrm{T}\text{---} \\ \hline \\ \hline \\ \text{---} q_n{}^\mathrm{T}\text{---} \end{bmatrix} \begin{bmatrix} \bigg| & & \bigg| \\ q_1 & \cdots & q_n \\ \bigg| & & \bigg| \end{bmatrix} = \begin{bmatrix} 1 & & \\ & \ddots & \\ & & 1 \end{bmatrix} = I. \qquad (35)$$

Equation (35) is a repetition of (31) above, where we wrote $A^\mathrm{T}A = I$ and the matrix was not necessarily square.

---

† Perhaps *orthonormal matrix* would have been a better name, but it is too late to change.

What is special about the case when Q is square? The difference is this: A square matrix with independent columns has full rank $r = n$, and is invertible; if Q^T is a left-inverse then it is *the* inverse. In other words Q^T is also a right-inverse, and $QQ^T = I$.

3M   An orthogonal matrix has all the following properties:

$$Q^TQ = I, \quad \text{and} \quad QQ^T = I, \quad \text{and} \quad Q^T = Q^{-1}. \quad (36)$$

Not only does Q have orthonormal columns, as required by the definition of an orthogonal matrix, but also its rows are orthonormal. In other words, if Q is an orthogonal matrix, then so is Q^T.

This last statement, that the rows are orthonormal at the same time, follows immediately from $QQ^T = I$. The multiplication QQ^T takes the inner product of every row of Q with every other row, and since the answer is the identity those rows are orthonormal. We have to note how remarkable that is. The rows point in completely different directions from the columns, but somehow they are automatically perpendicular whenever the columns are.

EXAMPLE 1

$$Q = \begin{bmatrix} \cos\theta & -\sin\theta \\ \sin\theta & \cos\theta \end{bmatrix}, \qquad Q^T = Q^{-1} = \begin{bmatrix} \cos\theta & \sin\theta \\ -\sin\theta & \cos\theta \end{bmatrix}.$$

Q rotates every vector through the angle θ, and Q^T rotates it back through $-\theta$.

EXAMPLE 2   Any permutation P is an orthogonal matrix since the columns are certainly unit vectors and certainly orthogonal—the 1 appears in a different place in each column:

$$\text{if} \quad P = \begin{bmatrix} 0 & 1 \\ 1 & 0 \end{bmatrix}, \qquad \text{then} \quad P^{-1} = P^T = \begin{bmatrix} 0 & 1 \\ 1 & 0 \end{bmatrix},$$

and

$$\text{if} \quad P = \begin{bmatrix} 0 & 1 & 0 \\ 0 & 0 & 1 \\ 1 & 0 & 0 \end{bmatrix}, \qquad \text{then} \quad P^{-1} = P^T = \begin{bmatrix} 0 & 0 & 1 \\ 1 & 0 & 0 \\ 0 & 1 & 0 \end{bmatrix}.$$

In the first case, where only two rows of the identity are exchanged, there is the extra property that $P^{-1} = P$. But this is not the case in general and is not the case in the second example, where three rows were moved. Notice also that $P = \begin{bmatrix} 0 & 1 \\ 1 & 0 \end{bmatrix}$ is not one of the rotations Q of Example 1; there is no value of θ that will produce P. Instead P reflects every point (x, y) into its mirror image (y, x) across the 45° line $y = x$. So we were wrong to suggest earlier that every orthogonal Q represents a pure rotation.

There does remain one property that is shared by the rotations Q and the permutations P, and in fact by every orthogonal matrix. It is the most important and most characteristic property of all:

3N Multiplication by an orthogonal Q preserves lengths,

$$|| Qx || = || x || \qquad \text{for every vector } x, \tag{37}$$

and it also preserves inner products:

$$(Qx)^{\mathrm{T}}(Qy) = x^{\mathrm{T}}y \qquad \text{for all vectors } x \text{ and } y. \tag{38}$$

The proofs are immediate since $(Qx)^{\mathrm{T}}(Qy) = x^{\mathrm{T}}Q^{\mathrm{T}}Qy = x^{\mathrm{T}}Iy = x^{\mathrm{T}}y$. If $y = x$, this equation becomes $|| Qx ||^2 = || x ||^2$, and therefore lengths are preserved whenever inner products are.

EXAMPLE For the "plane rotations" described above,

$$\begin{bmatrix} \cos\theta & -\sin\theta \\ \sin\theta & \cos\theta \end{bmatrix}\begin{bmatrix} x \\ y \end{bmatrix} = \begin{bmatrix} x\cos\theta - y\sin\theta \\ x\sin\theta + y\cos\theta \end{bmatrix}.$$

The length is preserved because

$$(x\cos\theta - y\sin\theta)^2 + (x\sin\theta + y\cos\theta)^2 = x^2 + y^2.$$

EXERCISE 3.3.4 If Q_1 and Q_2 are orthogonal matrices, and therefore satisfy the requirements (36), show that Q_1Q_2 is also orthogonal.

EXERCISE 3.3.5 If u is a unit vector, show that $Q = I - 2uu^{\mathrm{T}}$ is an orthogonal matrix. (It is known as a Householder transformation.) Compute Q explicitly when $u = (1, 1, 1)/\sqrt{3}$.

EXERCISE 3.3.6 Find a third column so that the matrix

$$Q = \begin{bmatrix} 1/\sqrt{3} & 1/\sqrt{2} & \\ 1/\sqrt{3} & 0 & \\ 1/\sqrt{3} & -1/\sqrt{2} & \end{bmatrix}$$

is orthogonal. It must be a unit vector that is orthogonal to the other columns; how much freedom does this leave? Verify that the rows automatically become orthonormal at the same time.

EXERCISE 3.3.7 Show, by forming $v^{\mathrm{T}}v$ directly, that Pythagoras' law holds for any combination $v = x_1q_1 + \cdots + x_nq_n$ of orthonormal vectors:

$$|| v ||^2 = x_1^2 + \cdots + x_n^2.$$

In matrix terms $v = Qx$, so this gives a new and more explicit proof of the fact that lengths are preserved: $|| Qx ||^2 = || x ||^2$.

EXERCISE 3.3.8   Is the following matrix orthogonal:

$$
Q = \begin{bmatrix} \frac{1}{2} & \frac{1}{2} & \frac{1}{2} & \frac{1}{2} \\[4pt] \frac{1}{2} & \frac{1}{2} & -\frac{1}{2} & -\frac{1}{2} \\[4pt] \frac{1}{2} & -\frac{1}{2} & -\frac{1}{2} & \frac{1}{2} \\[4pt] \frac{1}{2} & -\frac{1}{2} & \frac{1}{2} & -\frac{1}{2} \end{bmatrix} ?
$$

Gram–Schmidt Orthogonalization

We have emphasized the special advantages that come with orthonormal columns: To invert Q we only have to transpose it, and to solve the least squares problem $Ax = b$ we only have to form $\bar{x} = A^{T}b$. But in every case we have been forced to start by saying, *"If the columns are orthonormal . . .."* Now we propose to find a way to *make* them orthonormal.

This is a process that can be carried out in advance of any particular application, and it is easy to visualize when only two vectors are involved. We are given two independent vectors a and b, and we want to produce two perpendicular vectors v_1 and v_2. The first one can certainly go in the direction of a, $v_1 = a$. Then the problem is to find a second vector that is perpendicular. But that is exactly the construction with which this chapter began: The vector $b - p$ was perpendicular to a because the component of b in the direction of a (namely p) had been subtracted off. The second axis will lie in the direction of

$$
v_2 = b - p = b - \frac{a^{T}b}{a^{T}a}\, a = b - \frac{v_1^{T}b}{v_1^{T}v_1}\, v_1 . \tag{39}
$$

You can verify at sight that $v_1^{T}v_2 = v_1^{T}b - v_1^{T}b = 0$.

Now suppose, to get the Gram–Schmidt process straight, that there is a third independent vector c. Then the idea remains exactly the same. *We subtract off the components of c in the two directions v_1 and v_2 which have already been settled:*

$$
v_3 = c - \frac{v_1^{T}c}{v_1^{T}v_1}\, v_1 - \frac{v_2^{T}c}{v_2^{T}v_2}\, v_2 . \tag{40}
$$

Again it is automatic that v_3 is perpendicular to v_1 and v_2 : For example,

$$
v_1^{T}v_3 = v_1^{T}c - \frac{v_1^{T}c}{v_1^{T}v_1}\, v_1^{T}v_1 - \frac{v_2^{T}c}{v_2^{T}v_2}\, v_1^{T}\!\overset{0}{\cancel{v_2}} = 0.
$$

We are really subtracting from c its component in the plane of a and b, but it is much easier to use the perpendicular directions v_1 and v_2 already established in that same plane. The projection of c onto the plane is the sum of its projections on the axes v_1 and v_2 , and that is what makes (40) so simple. Notice that v_3 cannot be the zero vector, or c would lie entirely in the plane of a and b, contradicting the linear independence of a, b, and c. Of course the vectors v_i are only orthogonal, and not yet orthonormal. To turn them into unit vectors, we divide each one by its length:

$$q_1 = \frac{v_1}{\| v_1 \|}, \qquad q_2 = \frac{v_2}{\| v_2 \|}, \qquad q_3 = \frac{v_3}{\| v_3 \|}.$$

This orthogonalization process, which produces each new v_i by subtracting off the right multiples of $v_1, \ldots, v_{i-1}$, can be summarized as follows:

30 Any set of independent vectors $a_1, \ldots, a_n$ can be converted into a set of orthogonal vectors by the Gram–Schmidt process: First $v_1 = a_1$, and then each v_i is orthogonal to the preceding $v_1, \ldots, v_{i-1}$:

$$v_i = a_i - \frac{v_1^T a_i}{v_1^T v_1} v_1 - \cdots - \frac{v_{i-1}^T a_i}{v_{i-1}^T v_{i-1}} v_{i-1}. \qquad (41)$$

For every choice of i, the subspace spanned by the original $a_1, \ldots, a_i$ is also spanned by $v_1, \ldots, v_i$. The final vectors $q_i = v_i / \| v_i \|$ are orthonormal.

EXAMPLE Suppose the given vectors are

$$a_1 = \begin{bmatrix} 1 \\ 1 \\ 0 \end{bmatrix}, \qquad a_2 = \begin{bmatrix} 1 \\ 0 \\ 1 \end{bmatrix}, \qquad a_3 = \begin{bmatrix} 0 \\ 1 \\ 1 \end{bmatrix}.$$

Then $v_1 = a_1$, and v_2 is computed as in (39):

$$\frac{a_2^T v_1}{v_1^T v_1} = \frac{1}{2}, \qquad v_2 = a_2 - \frac{1}{2} v_1 = \begin{bmatrix} \frac{1}{2} \\ -\frac{1}{2} \\ 1 \end{bmatrix}.$$

The third perpendicular axis comes from Eq. (40):

$$\frac{a_3^T v_1}{v_1^T v_1} = \frac{1}{2}, \qquad \frac{a_3^T v_2}{v_2^T v_2} = \frac{\frac{1}{2}}{\frac{6}{4}} = \frac{1}{3}, \qquad v_3 = a_3 - \frac{1}{2} v_1 - \frac{1}{3} v_2 = \begin{bmatrix} -\frac{2}{3} \\ \frac{2}{3} \\ \frac{2}{3} \end{bmatrix}.$$

The final orthonormal vectors are

$$q_1 = \frac{v_1}{\| v_1 \|} = \sqrt{\frac{1}{2}} \begin{bmatrix} 1 \\ 1 \\ 0 \end{bmatrix}, \qquad q_2 = \frac{v_2}{\| v_2 \|} = \sqrt{\frac{2}{3}} \begin{bmatrix} \frac{1}{2} \\ -\frac{1}{2} \\ 1 \end{bmatrix},$$

$$q_3 = \frac{v_3}{\| v_3 \|} = \sqrt{\frac{3}{4}} \begin{bmatrix} -\frac{2}{3} \\ \frac{2}{3} \\ \frac{2}{3} \end{bmatrix}.$$

Since this Gram–Schmidt algorithm is simple and direct, there ought to be an equally simple and direct way to write down its final result. We want to explain how that can be done. The situation is exactly comparable to Gaussian elimination, where the elementary operations are chosen in a natural way as we go, but it would be very awkward to list the whole sequence in advance. In that case, the right way to keep a record was found in the factorization $A = LU$. In this case, the Gram–Schmidt process is recorded in a different factorization of the matrix A.

Just as in the case of elimination, the trick is to ask how the original columns a_i can be recovered from the final vectors q_i. If we look at the example, and undo the equations for the v_i, we find

$$a_1 = v_1, \qquad \text{or} \qquad a_1 = \sqrt{2}\, q_1$$

$$a_2 = \tfrac{1}{2} v_1 + v_2, \qquad \text{or} \qquad a_2 = \sqrt{\tfrac{1}{2}}\, q_1 + \sqrt{\tfrac{3}{2}}\, q_2$$

$$a_3 = \tfrac{1}{2} v_1 + \tfrac{1}{3} v_2 + v_3, \qquad \text{or} \qquad a_3 = \sqrt{\tfrac{1}{2}}\, q_1 + \sqrt{\tfrac{1}{6}}\, q_2 + \sqrt{\tfrac{4}{3}}\, q_3.$$

This set of equations practically demands to be written in matrix notation:

$$\begin{bmatrix} & & \\ a_1 & a_2 & a_3 \\ & & \end{bmatrix} = \begin{bmatrix} & & \\ q_1 & q_2 & q_3 \\ & & \end{bmatrix} \begin{bmatrix} \sqrt{2} & \sqrt{\tfrac{1}{2}} & \sqrt{\tfrac{1}{2}} \\ 0 & \sqrt{\tfrac{3}{2}} & \sqrt{\tfrac{1}{6}} \\ 0 & 0 & \sqrt{\tfrac{4}{3}} \end{bmatrix}. \tag{42}$$

The original matrix A is factored into an orthogonal matrix Q times an upper triangular R. The columns of Q are the orthonormal vectors which we wanted, and the entries of R can be seen (almost) in Eq. (41). That equation expresses a_i as a combination of $v_1, \ldots, v_i$, and we have only to replace v_1 by $\| v_1 \| q_1$, v_2 by $\| v_2 \| q_2$, and so on. *Then a_i is a combination of $q_1, \ldots, q_i$, and does not involve the remaining $q_{i+1}, \ldots, q_n$*; that is the reason R is upper triangular. The coefficients in this combination go into the ith column of R, and in particular the diagonal entry is not zero; v_i is replaced in (41) by $\| v_i \| q_i$, so the diagonal entry is the nonzero coefficient $\| v_i \|$.† R has positive entries along its diagonal and is therefore invertible. This produces the main result of the section:

3P Any matrix A with linearly independent columns can be factored into a product $A = QR$: The columns of Q are orthonormal, and R is upper triangular and invertible. If the original matrix A is square, then so are its two factors Q and R, and Q becomes an orthogonal matrix.

Starting from $A = QR$, it is easy to solve the least squares problem $Ax = b$.

† The lengths $\| v_i \|$ can be recognized along the diagonal in (42).

We know from (21) that

$$\bar{x} = (A^{\mathrm{T}}A)^{-1}A^{\mathrm{T}}b = (R^{\mathrm{T}}Q^{\mathrm{T}}QR)^{-1}R^{\mathrm{T}}Q^{\mathrm{T}}b.$$

But $Q^{\mathrm{T}}Q = I$, because the columns of Q are orthonormal. Therefore

$$\bar{x} = (R^{\mathrm{T}}R)^{-1}R^{\mathrm{T}}Q^{\mathrm{T}}b = R^{-1}Q^{\mathrm{T}}b. \tag{43}$$

The computation of $\bar{x}$ requires only the matrix–vector multiplication $Q^{\mathrm{T}}b$, followed by back-substitution in the triangular system $R\bar{x} = Q^{\mathrm{T}}b$. The preliminary work of orthogonalization saves us the task of forming $A^{\mathrm{T}}A$ and solving the normal equations $A^{\mathrm{T}}A\bar{x} = A^{\mathrm{T}}b$.†

EXERCISE 3.3.9   Apply the Gram–Schmidt process to

$$a_1 = \begin{bmatrix} 0 \\ 0 \\ 1 \end{bmatrix}, \qquad a_2 = \begin{bmatrix} 0 \\ 1 \\ 1 \end{bmatrix}, \qquad a_3 = \begin{bmatrix} 1 \\ 1 \\ 1 \end{bmatrix},$$

and write the result in the form $A = QR$.

EXERCISE 3.3.10   Factor $A = \begin{bmatrix} 3 & 0 \\ 4 & 5 \end{bmatrix}$ into QR.

EXERCISE 3.3.11   Express the Gram–Schmidt orthogonalization of

$$a_1 = \begin{bmatrix} 1 \\ 2 \\ 2 \end{bmatrix}, \qquad a_2 = \begin{bmatrix} 1 \\ 3 \\ 1 \end{bmatrix}$$

in the form $A = QR$. Given n vectors a_i, each with m components, what are the shapes of A, Q, and R?

EXERCISE 3.3.12   With the same matrix A, and with $b = (1, 1, 1)^{\mathrm{T}}$, use $A = QR$ to solve the least squares problem $Ax = b$.

EXERCISE 3.3.13   If $A = QR$, find a simple formula for the projection matrix P onto the column space of A.

EXERCISE 3.3.14   Show that the two steps

$$w = c - \frac{v_1^{\mathrm{T}}c}{v_1^{\mathrm{T}}v_1}v_1, \qquad v_3 = w - \frac{v_2^{\mathrm{T}}w}{v_2^{\mathrm{T}}v_2}v_2$$

produce the same third direction v_3 as formula (40). This is an example of modified Gram–Schmidt orthogonalization, which for the sake of numerical stability subtracts off the projections one at a time.

---

† The true saving is not in the operation count, which is actually greater when the preliminary work is included, but in numerical stability—at least if we switch to the modified Gram–Schmidt algorithm of Exercise 3.3.14 or (better) the Householder algorithm of Section 7.3.

Function Spaces and Fourier Series

This is a brief and optional section, but it has a number of good intentions:

(1)   to introduce for the first time an infinite-dimensional vector space;

(2)   to extend the ideas of length and inner product from vectors v to functions $f(x)$;

(3)   to recognize the Fourier series of f as a sum of one-dimensional projections; the orthogonal "columns" which span the space are the sines and cosines;

(4)   to apply Gram–Schmidt orthogonalization to the polynomials 1, x, x^2, . . . ;

(5)   to find the best approximation of $f(x)$ by a straight line.

We will try to follow this outline, which opens up a range of new applications for linear algebra, in a systematic way.

1.   After studying all the finite-dimensional spaces $\mathbf{R}^n$, it is natural to think of the space $\mathbf{R}^\infty$—containing all vectors $v = (v_1, v_2, v_3, \ldots)$ which have an infinite sequence of components. This space is actually too big to be very useful when there is no control on the components v_j. A much better idea is to keep the familiar definition of length, as the square root of Pythagoras' sum of squares, and *to include only those vectors that have a finite length*: The infinite series

$$\| v \|^2 = v_1^2 + v_2^2 + v_3^2 + \cdots \tag{44}$$

must converge to a finite sum. This still leaves an infinite-dimensional set of vectors, including for example the vector $(1, \frac{1}{2}, \frac{1}{3}, \ldots)$ but excluding $(1, 1/\sqrt{2}, 1/\sqrt{3}, \ldots)$. The vectors with finite length can be added together ($\| v + w \| \leq \| v \| + \| w \|$) and multiplied by scalars, so they form a vector space. It is the celebrated **Hilbert space**.

Hilbert space is the most natural way to allow the number of dimensions to become infinite, and at the same time to retain the geometry of ordinary Euclidean space. Ellipses become infinite-dimensional ellipsoids, parabolas become paraboloids, and perpendicular lines are recognized in the same way as before: The vectors $v = (v_1, v_2, \ldots)$ and $w = (w_1, w_2, \ldots)$ are orthogonal when their inner product is zero,

$$v^\mathrm{T} w = v_1 w_1 + v_2 w_2 + v_3 w_3 + \cdots = 0. \tag{45}$$

This sum is guaranteed to converge, and for any two vectors it still obeys the Schwarz inequality $| v^\mathrm{T} w | \leq \| v \| \| w \|$. The cosine, even in Hilbert space, is never larger than one.

There is another remarkable thing about this space: It is found under a great many different disguises. Its "vectors" can turn into functions, and that brings us to the second point.

2. Suppose we think of a function like $f(x) = \sin x$, on the interval $0 \leq x \leq 2\pi$. This f is like a vector with a whole continuum of components, the values of $\sin x$ along the whole interval. To find the length of such a vector, the usual rule of adding the squares of the components becomes impossible. This summation is replaced, in a natural and inevitable way, by *integration*:

$$\| f \|^2 = \int_0^{2\pi} (f(x))^2 \, dx = \int_0^{2\pi} (\sin x)^2 \, dx = \pi. \tag{46}$$

Our Hilbert space has become a function space; the vectors are functions, we have a way to measure their length, and the space contains all those functions that have a finite length—just as in (44) above. It does not contain the function $F(x) = 1/x$, because the integral of $1/x^2$ is infinite.

The same idea of replacing summation by integration produces the *inner product of two functions*: If $f(x) = \sin x$ and $g(x) = \cos x$, then

$$f^{\mathsf{T}} g = \int_0^{2\pi} f(x) g(x) \, dx = \int_0^{2\pi} \sin x \cos x \, dx = 0.$$

This inner product is still related to the length by $f^{\mathsf{T}} f = \| f \|^2$, and again the Schwarz inequality is satisfied: $|f^{\mathsf{T}} g| \leq \| f \| \, \| g \|$. Of course two functions like $\sin x$ and $\cos x$—whose inner product is zero—will be called orthogonal. They are even orthonormal, after division by their length $\sqrt{\pi}$.

3. The **Fourier series** of a function $y(x)$ is an expansion into sines and cosines:

$$y(x) = a_0 + a_1 \cos x + b_1 \sin x + a_2 \cos 2x + b_2 \sin 2x + \cdots$$

To compute a typical coefficient, say b_1, we multiply both sides by its corresponding function $\sin x$, and integrate from 0 to 2π. In other words, we take the inner product of both sides with $\sin x$:

$$\int_0^{2\pi} y(x) \sin x \, dx = a_0 \int_0^{2\pi} \sin x \, dx + a_1 \int_0^{2\pi} \cos x \sin x \, dx$$

$$+ b_1 \int_0^{2\pi} (\sin x)^2 \, dx + \cdots.$$

On the right-hand side, every integral is zero except one—the one in which $\sin x$ multiplies itself. *The sines and cosines are mutually orthogonal.* Therefore the coefficient of $f(x) = \sin x$ is

$$b_1 = \frac{\int_0^{2\pi} y(x) \sin x \, dx}{\int_0^{2\pi} (\sin x)^2 \, dx} = \frac{y^{\mathsf{T}} f}{f^{\mathsf{T}} f}, \quad \text{or} \quad b_1 \sin x = \frac{y^{\mathsf{T}} f}{f^{\mathsf{T}} f} f. \tag{47}$$

The point of this calculation is to see the analogy with projections. The component of the vector b along the line spanned by a was computed at the beginning of the chapter:

$$\bar{x} = \frac{b^{\mathsf{T}}a}{a^{\mathsf{T}}a}, \quad \text{or} \quad p = \bar{x}a = \frac{b^{\mathsf{T}}a}{a^{\mathsf{T}}a}a.$$

In a Fourier series, we are projecting the function y onto the function $\sin x$, and its component p in this direction is exactly the term $b_1 \sin x$. The coefficient b_1 is the least squares solution of the inconsistent equation $b_1 \sin x = y$, in other words, it brings the function $b_1 \sin x$ as close as possible to the function y (see Exercise 3.3.16). The same is true for all the other terms in the series; every one is a projection of y onto a sine or cosine. Since these sines and cosines are orthogonal, *the Fourier series just gives the coordinates of the "vector" y with respect to a set of* (infinitely many) *perpendicular axes.*

4.   There are plenty of useful functions other than sines and cosines, and they are not always orthogonal. The simplest are the polynomials, and unfortunately there is no interval on which even the first three coordinate axes—the functions 1, x, and x^2—are perpendicular. (The inner product of 1 and x^2 is always positive, because it is the integral of x^2.) Therefore the closest parabola to $y(x)$ is not the sum of its projections onto 1, x, and x^2. There will be a coupling term, exactly like $(A^{\mathsf{T}}A)^{-1}$ in the matrix case, and in fact the coupling is given by the ill-conditioned Hilbert matrix of Exercise 1.5.11. On the interval $0 \leq x \leq 1$,

$$A^{\mathsf{T}}A = \begin{bmatrix} 1^{\mathsf{T}}1 & 1^{\mathsf{T}}x & 1^{\mathsf{T}}x^2 \\ x^{\mathsf{T}}1 & x^{\mathsf{T}}x & x^{\mathsf{T}}x^2 \\ (x^2)^{\mathsf{T}}1 & (x^2)^{\mathsf{T}}x & (x^2)^{\mathsf{T}}x^2 \end{bmatrix} = \begin{bmatrix} \int 1 & \int x & \int x^2 \\ \int x & \int x^2 & \int x^3 \\ \int x^2 & \int x^3 & \int x^4 \end{bmatrix} = \begin{bmatrix} 1 & \frac{1}{2} & \frac{1}{3} \\ \frac{1}{2} & \frac{1}{3} & \frac{1}{4} \\ \frac{1}{3} & \frac{1}{4} & \frac{1}{5} \end{bmatrix}.$$

This matrix has a large inverse, because the axes 1, x, x^2 are far from perpendicular. Even for a modern computer, the situation becomes impossible if we add a few more axes; *it is virtually hopeless to solve the normal equations $A^{\mathsf{T}}A\bar{x} = A^{\mathsf{T}}b$ for the closest polynomial of degree ten.*

More precisely, it is hopeless to solve them by Gaussian elimination; every roundoff error would be amplified by more than 10^{13}. On the other hand, we cannot just give up; approximation by polynomials has to be possible. The right idea is to switch to orthogonal axes, and this means a Gram–Schmidt orthogonalization: We look for combinations of 1, x, and x^2 that *are* orthogonal.

It is convenient to work with a symmetrically placed interval like $-1 \leq x \leq 1$, because this makes all the odd powers of x orthogonal to all the even powers:

$$1^{\mathsf{T}}x = \int_{-1}^{1} x\, dx = 0, \quad x^{\mathsf{T}}x^2 = \int_{-1}^{1} x^3\, dx = 0.$$

Therefore the Gram–Schmidt process can begin by accepting $v_1 = 1$ and $v_2 = x$ as the first two perpendicular axes, and it only has to correct the angle between

1 and x^2. By formula (40), the third orthogonal polynomial is

$$v_3 = x^2 - \frac{1^T x^2}{1^T 1} 1 - \frac{x^T x^2}{x^T x} x = x^2 - \frac{\displaystyle\int_{-1}^{1} x^2 \, dx}{\displaystyle\int_{-1}^{1} 1 \, dx} = x^2 - \frac{1}{3}.$$

The polynomials constructed in this way are the Legendre polynomials, orthogonal to each other over the interval $-1 \le x \le 1$.

Check

$$1^T (x^2 - \tfrac{1}{3}) = \int_{-1}^{1} (x^2 - \tfrac{1}{3}) \, dx = \left[\frac{x^3}{3} - \frac{x}{3} \right]_{-1}^{1} = 0.$$

The closest polynomial of degree ten is now computable, without disaster, by projecting onto each of the first 10 (or 11) Legendre polynomials.

5. Suppose we are given a function like $f(x) = \sqrt{x}$, between the limits $x = 0$ and $x = 1$, and we want to fit it by a straight line $y = C + Dx$. There are at least three ways of finding the best line, and if you compare them the whole chapter might become clear!

   (i)   Minimize

$$E^2 = \int_0^1 (\sqrt{x} - C - Dx)^2 \, dx = \frac{1}{2} - \frac{4}{3} C - \frac{4}{5} D + C^2 + CD + \frac{D^2}{3}.$$

We set to zero the derivatives of the error:

$$\frac{\partial E^2}{\partial C} = -\frac{4}{3} + 2C + D = 0, \qquad \frac{\partial E^2}{\partial D} = -\frac{4}{5} + C + \frac{2D}{3} = 0.$$

The solution is $C = \frac{4}{15}$, $D = \frac{4}{5}$.

   (ii)   Solve $[1 \ \ x][\begin{smallmatrix} C \\ D \end{smallmatrix}] = [\sqrt{x}]$, or $Ay = f$, by least squares.

$$A^T A = \begin{bmatrix} 1^T 1 & 1^T x \\ x^T 1 & x^T x \end{bmatrix} = \begin{bmatrix} 1 & \frac{1}{2} \\ \frac{1}{2} & \frac{1}{3} \end{bmatrix}, \qquad (A^T A)^{-1} = \begin{bmatrix} 4 & -6 \\ -6 & 12 \end{bmatrix},$$

$$A^T f = \begin{bmatrix} 1^T \sqrt{x} \\ x^T \sqrt{x} \end{bmatrix} = \begin{bmatrix} \frac{2}{3} \\ \frac{2}{5} \end{bmatrix}.$$

The solution of the normal equations is

$$\bar{y} = (A^T A)^{-1} A^T f, \qquad \text{or} \qquad \begin{bmatrix} C \\ D \end{bmatrix} = \begin{bmatrix} 4 & -6 \\ -6 & 12 \end{bmatrix} \begin{bmatrix} \frac{2}{3} \\ \frac{2}{5} \end{bmatrix} = \begin{bmatrix} \frac{4}{15} \\ \frac{4}{5} \end{bmatrix}.$$

   (iii)   Apply the Gram–Schmidt process to produce a second column $x - \frac{1}{2}$

orthogonal to the first column:

$$1^{\mathrm{T}}(x - \tfrac{1}{2}) = \int (x - \tfrac{1}{2}) \, dx = 0.$$

Then the projection of $\sqrt{x}$ is

$$y = \frac{\sqrt{x}^{\mathrm{T}}1}{1^{\mathrm{T}}1} 1 + \frac{\sqrt{x}^{\mathrm{T}}(x - \tfrac{1}{2})}{(x - \tfrac{1}{2})^{\mathrm{T}}(x - \tfrac{1}{2})}(x - \tfrac{1}{2})$$

$$= \tfrac{2}{3} + \tfrac{4}{5}(x - \tfrac{1}{2}) = \tfrac{4}{15} + \tfrac{4}{5}x.$$

For whatever it is worth, the only mistake in my arithmetic (now corrected!) was in method (i).

EXERCISE 3.3.15　Find the length of the vector $v = (1/\sqrt{2},\ 1/\sqrt{4},\ 1/\sqrt{8}, \ldots)$ and of the function $f(x) = e^x$ (over the interval $0 \le x \le 1$). What is the inner product over this interval of e^x and e^{-x}?

EXERCISE 3.3.16　By setting the derivative to zero, find the value of b_1 that minimizes

$$\| b_1 \sin x - y \|^2 = \int_0^{2\pi} (b_1 \sin x - y(x))^2 \, dx.$$

Compare with the Fourier coefficient (47). If $y(x) = \cos x$, what is b_1 ?

EXERCISE 3.3.17　Find the Fourier coefficients a_0, a_1, b_1 of the step function $y(x)$, which equals 1 on the interval $0 \le x \le \pi$ and 0 on the remaining interval $\pi < x < 2\pi$:

$$a_0 = \frac{y^{\mathrm{T}}1}{1^{\mathrm{T}}1}, \qquad a_1 = \frac{y^{\mathrm{T}} \cos x}{(\cos x)^{\mathrm{T}} \cos x}, \qquad b_1 = \frac{y^{\mathrm{T}} \sin x}{(\sin x)^{\mathrm{T}} \sin x}.$$

EXERCISE 3.3.18　Find the next Legendre polynomial—a cubic orthogonal to 1, x, and $x^2 - \tfrac{1}{3}$.

EXERCISE 3.3.19　What is the closest straight line to the parabola $y = x^2$ over the interval $-1 \le x \le 1$?

3.4　■　THE PSEUDOINVERSE AND THE SINGULAR VALUE DECOMPOSITION

What is the optimal solution $\bar{x}$ to an inconsistent system $Ax = b$? That is the key question in this chapter, and it is not yet completely answered. We now propose to answer it. Our goal is to find a rule that specifies $\bar{x}$, given any coefficient matrix A and any right side b.

This goal is already more than halfway reached because we know the value of $A\bar{x}$. For every x, Ax is necessarily in the column space of A; it is a combination of the columns, weighted by the components of x. Therefore the optimal choice $A\bar{x}$ is the point p in this column space closest to the given b. This choice minimizes the error $E = \| Ax - b \|$. In other words, we cannot do better than to project b onto the column space:

$$A\bar{x} = p = Pb. \tag{48}$$

In a great many cases—I think they include a large majority of practical problems—this equation is enough to determine $\bar{x}$ itself. (It is another form of the normal equation $A^{T}A\bar{x} = A^{T}b$, which appeared in 3I when we minimized E.) Certainly $\bar{x}$ is determined when there is only one combination of the columns of A that will produce p; the weights in this combination will be the components of $\bar{x}$. This is nothing but the "uniqueness case" of Chapter 2, and we know several equivalent conditions for the equation $A\bar{x} = p$ to have only one solution:

(i)   The columns of A are linearly independent.
(ii)   The nullspace of A contains only the zero vector.
(iii) The rank of A is n.
(iv) The square matrix $A^{T}A$ is invertible.

In such a case, the only solution to (48) is the one that has been studied throughout this chapter:

$$\bar{x} = (A^{T}A)^{-1}A^{T}b. \tag{49}$$

This formula, which is comparatively simple, includes the simplest case of all— when A is actually invertible. Then $\bar{x}$ coincides with the one and only solution of the original system $Ax = b$: $\bar{x} = A^{-1}(A^{T})^{-1}A^{T}b = A^{-1}b$.

This suggests another way of describing our goal: We are trying to define the **pseudoinverse** A^{+} of a matrix which may not be invertible.† When the matrix *is* invertible, we are happy with that: $A^{+} = A^{-1}$. When the matrix satisfies the conditions (i)–(iv) listed above, the pseudoinverse is the left-inverse which appears in (49): $A^{+} = (A^{T}A)^{-1}A^{T}$. But when the conditions (i)–(iv) do not hold, and $\bar{x}$ is not uniquely determined by $A\bar{x} = p$, the pseudoinverse remains to be defined. We have to choose one of the many vectors that satisfy $A\bar{x} = p$—and that choice will be, by definition, the optimal solution $\bar{x} = A^{+}b$ to the inconsistent system $Ax = b$.

The choice is made according to the following rule: *The optimal solution, among all solutions of $A\bar{x} = p$, is the one that has the minimum length.* To find it, the key is to remember that the row space and nullspace of A are orthogonal complements in $\mathbf{R}^{n}$. This means that any vector can be split into two perpendicular pieces, its projection onto the row space and its projection onto the nullspace. Suppose we apply this splitting to one of the solutions (call it $\bar{x}_{0}$)

---

† A^{+} is also called the *Moore–Penrose inverse,* after its discoverers, or more commonly a *generalized inverse* of A. But a great many other matrices, sharing some but not all of the properties we intend for A^{+}, have also been described as generalized inverses. With the term *pseudoinverse,* we are on safer ground.

of the equation $A\bar{x} = p$. Then $\bar{x}_0 = \bar{x}_r + w$, where $\bar{x}_r$ is in the row space and w is in the nullspace. Now there are three important points:

(1)   The component $\bar{x}_r$ is itself a solution of $A\bar{x} = p$: Since $Aw = 0$,

$$A\bar{x}_r = A(\bar{x}_r + w) = A\bar{x}_0 = p.$$

(2)   All solutions of $A\bar{x} = p$ share this same component $\bar{x}_r$ in the row space, and differ only in the nullspace component w. "The general solution is the sum of one particular solution (in this case $\bar{x}_r$) and an arbitrary solution w of the homogeneous equation."

(3)   The length of such a solution $\bar{x}_r + w$ obeys Pythagoras' law, since the two components are orthogonal:

$$\| \bar{x}_r + w \|^2 = \| \bar{x}_r \|^2 + \| w \|^2.$$

Our conclusion is this: *The solution that has the minimum length is $\bar{x}_r$.* We should choose the nullspace component to be zero, leaving a solution that is entirely in the row space.

3Q   The optimal least squares solution of any system $Ax = b$ is the vector $\bar{x}_r$ (or $\bar{x}$, if we return to the original notation) which is determined by two conditions:

(1) $A\bar{x}$ equals the projection of b onto the column space of A.
(2) $\bar{x}$ lies in the row space of A.

The matrix that "solves" $Ax = b$ is the pseudoinverse A^+, defined by $\bar{x} = A^+b$.

The best way to understand this pseudoinverse is from the geometry. We look again at an illustration of the four fundamental subspaces (Fig. 3.7).

The matrix A^+ combines the effect of two separate steps: It projects b onto the point p, and then finds the only vector $\bar{x}$ in the row space solving $A\bar{x} = p$. One extreme case occurs when b is perpendicular to the column space—in other words, when b is in the left nullspace. Then $p = 0$ and $\bar{x} = 0$: A^+ sends all of $\Re(A^T)$ into zero. The other extreme finds b right inside the column space; then $p = b$, and $\bar{x}$ is found by "inverting" A. (We already remarked in 2Z that A is invertible if we think of it only as a mapping from its row space to its column space; A^+ is the inverse.)

An arbitrary b is split into these two extremes: The component p is inverted to give $\bar{x}$, and the other component $b - p$ is annihilated. From this description, and from Fig. 3.7, we can list a few basic properties of the pseudoinverse:

(1) A^+ is an n by m matrix. It starts with the vector b in $\mathbf{R}^m$, and produces the vector $\bar{x}$ in $\mathbf{R}^n$.

(2)   The column space of A^+ is the row space of A, and the row space of A^+ is the column space of A. (As an important consequence, rank A^+ = rank A.)

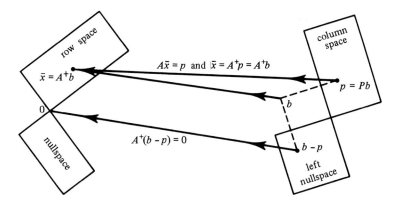

Fig. 3.7.   The matrix A and its pseudoinverse A^+.

(3)   The pseudoinverse of A^+ is A itself.

(4)   In general $AA^+ \neq I$, since A may not have a right-inverse, but AA^+ always equals the projection P onto the column space:

$$AA^+b = A\bar{x} = p = Pb, \quad \text{or} \quad AA^+ = P. \quad (50)$$

EXAMPLE 1

$$A = \begin{bmatrix} 1 & 0 & 0 \\ 0 & 1 & 0 \\ 0 & 0 & 0 \end{bmatrix},$$

a noninvertible matrix. The column space and row space both coincide with the "x-y plane" in $\mathbf{R}^3$, containing all vectors $(x, y, 0)$. The nullspace is the "z axis", and of course it is orthogonal to the row space. In order to find $\bar{x}$, we begin by projecting b onto the column space:

$$\text{If} \quad b = \begin{bmatrix} b_1 \\ b_2 \\ b_3 \end{bmatrix}, \quad \text{then} \quad p = Pb = \begin{bmatrix} b_1 \\ b_2 \\ 0 \end{bmatrix}.$$

Then we solve $A\bar{x} = p$, or

$$\begin{bmatrix} 1 & 0 & 0 \\ 0 & 1 & 0 \\ 0 & 0 & 0 \end{bmatrix} \begin{bmatrix} \bar{x}_1 \\ \bar{x}_2 \\ \bar{x}_3 \end{bmatrix} = \begin{bmatrix} b_1 \\ b_2 \\ 0 \end{bmatrix}. \quad (51)$$

The solutions are easy to write down: $\bar{x}_1 = b_1$, $\bar{x}_2 = b_2$, and $\bar{x}_3$ is arbitrary. Among this infinite family of solutions, we choose the one of minimum length: The third component $\bar{x}_3$ should be zero. This leaves $\bar{x} = (b_1, b_2, 0)^T$, which lies in the row space (the x-y plane) exactly as predicted by the theory. Finally, we identify the pseudoinverse A^+:

$$A^+ = \begin{bmatrix} 1 & 0 & 0 \\ 0 & 1 & 0 \\ 0 & 0 & 0 \end{bmatrix}, \quad \text{and} \quad A^+b = \begin{bmatrix} b_1 \\ b_2 \\ 0 \end{bmatrix} = \bar{x}.$$

For this special matrix, A^+ is identical with A. This is because A acts like the identity matrix in the way it maps its row space to its column space, and the pseudoinverse ignores the rest.

To summarize: The optimal solution of $Ax = b$, which is an inconsistent set

$$x_1 + 0x_2 + 0x_3 = b_1$$

$$0x_1 + x_2 + 0x_3 = b_2$$

$$0x_1 + 0x_2 + 0x_3 = b_3,$$

is to satisfy the first two equations and set $x_3 = 0$.

EXAMPLE 2

$$A = \begin{bmatrix} \mu_1 & 0 & 0 & 0 \\ 0 & \mu_2 & 0 & 0 \\ 0 & 0 & 0 & 0 \end{bmatrix},$$

with $\mu_1 > 0$, $\mu_2 > 0$. The column space is again the x-y plane, so that projecting b again annihilates its z component: $p = Pb = (b_1, b_2, 0)^{\mathrm{T}}$. The equation $A\bar{x} = p$ becomes

$$\begin{bmatrix} \mu_1 & 0 & 0 & 0 \\ 0 & \mu_2 & 0 & 0 \\ 0 & 0 & 0 & 0 \end{bmatrix} \begin{bmatrix} \bar{x}_1 \\ \bar{x}_2 \\ \bar{x}_3 \\ \bar{x}_4 \end{bmatrix} = \begin{bmatrix} b_1 \\ b_2 \\ 0 \end{bmatrix}.$$

The first two components are again determined: $\bar{x}_1 = b_1/\mu_1$, $\bar{x}_2 = b_2/\mu_2$. The other two components must be zero, by the requirement of minimum length, and therefore the optimal solution is

$$\bar{x} = \begin{bmatrix} b_1/\mu_1 \\ b_2/\mu_2 \\ 0 \\ 0 \end{bmatrix}, \quad \text{or} \quad A^+b = \begin{bmatrix} \mu_1^{-1} & 0 & 0 \\ 0 & \mu_2^{-1} & 0 \\ 0 & 0 & 0 \\ 0 & 0 & 0 \end{bmatrix} \begin{bmatrix} b_1 \\ b_2 \\ b_3 \end{bmatrix}.$$

These two examples are typical of a whole family of special matrices, namely those with positive numbers $\mu_1, \ldots, \mu_r$ in the first r entries on the main diagonal, and zeros everywhere else. Denoting this matrix by Σ, its pseudoinverse is computed exactly as in the examples:

$$\Sigma = \begin{bmatrix} \mu_1 & & & \\ & \ddots & & \\ & & \mu_r & \\ & & & \end{bmatrix} \qquad \text{and} \qquad \Sigma^+ = \begin{bmatrix} \mu_1^{-1} & & & \\ & \ddots & & \\ & & \mu_r^{-1} & \\ & & & \end{bmatrix}.$$

If Σ is m by n, then Σ^+ is n by m. It is easy to see that the rank of Σ^+ equals the rank of Σ, which is r, and that the pseudoinverse of Σ^+ brings us back to Σ: $(\Sigma^+)^+ = \Sigma$.

These simple matrices are much more valuable than they look, because of a new way to factor the matrix A. It is called the ***singular value decomposition***, and it is not nearly as famous as it should be.

> **3R** Any m by n matrix can be factored into $A = Q_1 \Sigma Q_2^T$, where Q_1 is an m by m orthogonal matrix, Q_2 is an n by n orthogonal matrix, and Σ has the special diagonal form described above.

The numbers μ_i are called the singular values of A. This factorization leads immediately to an explicit formula for the pseudoinverse of A, something that has so far been lacking. We are entitled in this special case (but not in general, see (60) below) to "pseudoinvert" the three factors separately, and multiply in the usual reverse order; since the inverse of an orthogonal matrix is its transpose, the formula becomes

$$A^+ = Q_2 \Sigma^+ Q_1^T. \tag{52}$$

This formula can be proved directly from the least squares principle. Multiplication by the orthogonal matrix Q_1^T leaves the length unchanged, by 3N, so the error to be minimized is

$$\| Ax - b \| = \| Q_1 \Sigma Q_2^T x - b \| = \| \Sigma Q_2^T x - Q_1^T b \|. \tag{53}$$

Introduce the new unknown $y = Q_2^T x = Q_2^{-1} x$, which has the same length as x. Then we want to minimize $\| \Sigma y - Q_1^T b \|$, and the optimal solution $\bar{y}$—the shortest of all vectors that are minimizing—is $\bar{y} = \Sigma^+ Q_1^T b$. Therefore

$$\bar{x} = Q_2 \bar{y} = Q_2 \Sigma^+ Q_1^T b, \qquad \text{or} \qquad A^+ = Q_2 \Sigma^+ Q_1^T.\dagger \tag{54}$$

The proof of the singular value decomposition requires the main result of Chapter 5: For a symmetric matrix like $A^T A$, there is an orthonormal set of

---

† In practice, it is essential to catch the very small μ_i and set them to zero. Otherwise roundoff or experimental error can easily create an unwanted $\mu_r = 10^{-6}$, and then $\mu_r^{-1} = 10^6$ will overwhelm the pseudoinverse. This reflects the extreme instability of the rank, and of the pseudoinverse itself.

"eigenvectors" $x_1, \ldots, x_n$ (which go into the columns of Q_2):

$$A^T A x_i = \lambda_i x_i, \qquad \text{with} \qquad x_i^T x_i = 1 \quad \text{and} \quad x_j^T x_i = 0 \text{ for } j \neq i. \quad \text{(55a)}$$

Taking the inner product with x_i, we discover that $\lambda_i \geq 0$:

$$x_i^T A^T A x_i = \lambda_i x_i^T x_i, \qquad \text{or} \qquad \| A x_i \|^2 = \lambda_i. \quad \text{(55b)}$$

Suppose that $\lambda_1, \ldots, \lambda_r$ are positive, and the remaining $n - r$ of the $A x_i$ and λ_i are zero. For the positive ones, we set $\mu_i = \sqrt{\lambda_i}$ and $y_i = A x_i / \mu_i$. These y_i are unit vectors from (55b), and they are orthogonal from (55a):

$$y_j^T y_i = \frac{x_j^T A^T A x_i}{\mu_j \mu_i} = \frac{\lambda_i x_j^T x_i}{\mu_j \mu_i} = 0 \qquad \text{for} \quad j \neq i.$$

Therefore we have r orthonormal vectors, and by Gram–Schmidt they can be extended to a full orthonormal basis $y_1, \ldots, y_r, \ldots, y_m$—which goes into the columns of Q_1. Then the entries of $Q_1^T A Q_2$ are the numbers $y_j^T A x_i$, and they are zero whenever $i > r$ (because then $A x_i = 0$). Otherwise they equal $y_j^T \mu_i y_i$, which is zero for $j \neq i$ and μ_i for $j = i$. In other words, $Q_1^T A Q_2$ is exactly the special matrix Σ with the μ_i along its main diagonal, and therefore $A = Q_1 \Sigma Q_2^T$. The rotations Q_1 and Q_2 just swing the column space of A into line with the row space, and A becomes the diagonal matrix Σ.

EXAMPLE 3 If A is the column vector $\left[\begin{smallmatrix}3\\4\end{smallmatrix}\right]$, then $A^T A$ is the 1 by 1 matrix $[25]$, and its unit eigenvector is just $x = [1]$. This is Q_2. Next we find $\lambda = 25$, $\mu = 5$, and $y_1 = A x / \mu = \left[\begin{smallmatrix}3/5\\4/5\end{smallmatrix}\right]$. The other column of Q_1 has to be orthogonal to y_1, so

$$Q_1^T A Q_2 = \begin{bmatrix} \frac{3}{5} & \frac{4}{5} \\ \frac{4}{5} & -\frac{3}{5} \end{bmatrix} \begin{bmatrix} 3 \\ 4 \end{bmatrix} [1] = \begin{bmatrix} 5 \\ 0 \end{bmatrix} = \Sigma,$$

which has the special diagonal form we expected. The pseudoinverse of the column Σ is the row $\Sigma^+ = \left[\begin{smallmatrix}\frac{1}{5} & 0\end{smallmatrix}\right]$, so finally $A^+ = Q_2 \Sigma^+ Q_1^T = \left[\begin{smallmatrix}\frac{3}{25} & \frac{4}{25}\end{smallmatrix}\right]$. Note that $A^+ A = I$; this is an example with rank $r = n$ ($= 1$), so the pseudoinverse is one particular left-inverse.

Of course the singular value decomposition requires us to find the orthogonal matrices Q_1 and Q_2, and that is generally impossible to do by row operations. It would be closer to the spirit of these opening chapters to try to compute $\bar{x}$ directly from the Gauss factors L and U. The result is just like the $U^{-1} L^{-1}$ factorization of A^{-1}, and since it cannot be found elsewhere, we give a complete proof—which it is certainly permissible to omit!†

The key lies in the idea, which was introduced on p. 91, of throwing away the zero rows of U. If A has rank r, there will be $m - r$ such rows. At the

---

† We are not suggesting that the new and more "elementary" formula (56) will be better for numerical calculations than either (52) or Gram–Schmidt. Probably the opposite is the case.

same time, we also throw away the last $m - r$ columns of L—or of $P^{-1}L$, in case some row exchanges were required during elimination. $\bar{U}$ is the r by n matrix that is left in U, and $\bar{L}$ is the m by r matrix left in L.

EXAMPLE 4 (*with rank* $r = 2$)

$$A = \begin{bmatrix} 1 & 2 \\ 2 & 5 \\ 3 & 7 \end{bmatrix} = \begin{bmatrix} 1 & 0 & 0 \\ 2 & 1 & 0 \\ 3 & 1 & 1 \end{bmatrix} \begin{bmatrix} 1 & 2 \\ 0 & 1 \\ 0 & 0 \end{bmatrix} = \begin{bmatrix} 1 & 0 \\ 2 & 1 \\ 3 & 1 \end{bmatrix} \begin{bmatrix} 1 & 2 \\ 0 & 1 \end{bmatrix} = \bar{L}\bar{U}.$$

The last column of L—which only multiplies the last row of U—was erased without changing the product: $LU = \bar{L}\bar{U}$. The important facts are:

(1)　$\bar{U}$ is an r by n matrix of rank r; it has the same row space as U, since only zero rows were removed, and therefore *the same row space as A*.

(2)　$\bar{L}$ is an m by r matrix which also has rank r; L was invertible (and so was P), so its first r columns were certainly independent. As we noted in Section 2.6, $\bar{L}$ has *the same column space as A*.

With the help of this factorization $A = \bar{L}\bar{U}$, we can finally construct A^+ in a completely explicit way:

3S　The pseudoinverse of any m by n matrix is given by

$$A^+ = \bar{U}^T(\bar{U}\bar{U}^T)^{-1}(\bar{L}^T\bar{L})^{-1}\bar{L}^T. \tag{56}$$

The goal is to verify that A^+b is the optimal solution $\bar{x}$ to the least squares problem $Ax = b$. Then the formula (56) must be correct. Remembering that $\bar{x}$ was determined by two requirements—it lies in the row space of A, and $A\bar{x}$ equals the projection p of b onto the column space—the proof depends on matching these requirements to the two properties (one of $\bar{U}$ and one of $\bar{L}$) which were listed above.

(1′)　When we form A^+b according to (56), it will be the product of $\bar{U}^T$ (at the left end of the formula) and some complicated vector y. In other words, it will be a combination of the columns of $\bar{U}^T$. Since these are the rows of $\bar{U}$, A^+b lies in the row space of $\bar{U}$—which is the row space of A. The first requirement is met.

(2′)　Multiplying A^+b by A, we hope to obtain p:

$$AA^+b = (\bar{L}\bar{U})[\bar{U}^T(\bar{U}\bar{U}^T)^{-1}(\bar{L}^T\bar{L})^{-1}\bar{L}^Tb] = \bar{L}(\bar{L}^T\bar{L})^{-1}\bar{L}^Tb. \tag{57}$$

But $\bar{L}(\bar{L}^T\bar{L})^{-1}\bar{L}^T$ is exactly the projection onto the column space of $\bar{L}$, which is the column space of A, so (57) is the vector $p = Pb$. The vector A^+b satisfies both requirements for an optimal solution, and (56) is correct.

EXERCISE 3.4.1   Find the pseudoinverse of the m by n matrix of zeros, $A = 0$, and explain your reasoning.

EXERCISE 3.4.2   For a matrix of rank one, the factorizations into $A = \bar{L}\bar{U}$ and $A = uv^T$ are identical; both are products of a column vector and a row vector. With

$$A = \begin{bmatrix} 1 & 1 & 1 \\ 1 & 1 & 1 \end{bmatrix} \quad \text{and} \quad b = \begin{bmatrix} 0 \\ 1 \end{bmatrix},$$

compute A^+ and $\bar{x}$. Project b onto the column space, and check $A\bar{x} = p$.

EXERCISE 3.4.3   If A has orthonormal columns, what is its pseudoinverse?

EXERCISE 3.4.4   If AA^T is invertible, then $A^+ = A^T(AA^T)^{-1}$ and $\bar{x} = A^+b$.

(a)   Verify that $A\bar{x} = b$.
(b)   With $A = \begin{bmatrix} 1 & 1 \end{bmatrix}$, find the optimal solution to $u + v = 3$.

EXERCISE 3.4.5 A^+ was originally defined by Penrose in a completely different way; he proved that for any A there is one and only one matrix A^+ satisfying the four conditions:

(i) $AA^+A = A$ (ii) $A^+AA^+ = A^+$

(iii) $(AA^+)^T = AA^+$ (iv) $(A^+A)^T = A^+A$.

We preferred a more geometrical and intuitive definition (of the same A^+!), but his algebraic definition makes it simple to check that a proposed matrix meets the requirements for a pseudoinverse.

(a)   Show that if A is a projection ($A^2 = A$ and $A = A^T$), then the choice $A^+ = A$ satisfies requirements (i)–(iv). Thus the pseudoinverse is A itself, as in Example 1 and Exercise 3.4.1.
(b)   Show that the requirements on the pseudoinverse of A^T are satisfied by $(A^+)^T$, so that $(A^T)^+ = (A^+)^T$. (Corollary: If A is symmetric, so is A^+.)
(c)   Show that the pseudoinverse of A^+ is A, as verified earlier for Σ and claimed for every A.

3.5 ■ WEIGHTED LEAST SQUARES

Suppose we return for a moment to the simplest kind of least squares problem, the estimate $\bar{x}$ of a patient's weight on the basis of two observations $x = b_1$ and $x = b_2$. Unless these measurements are identical, we are faced with an inconsistent system of two equations in one unknown:

$$\begin{bmatrix} 1 \\ 1 \end{bmatrix} [x] = \begin{bmatrix} b_1 \\ b_2 \end{bmatrix}.$$

Up to now, we have regarded the two observations as equally reliable, and looked for the value $\bar{x}$ that minimized $E^2 = (x - b_1)^2 + (x - b_2)^2$:

$$\frac{dE^2}{dx} = 0 \quad \text{at} \quad \bar{x} = \frac{b_1 + b_2}{2}.$$

The optimal $\bar{x}$ is the average of the measurements.

Suppose instead that the two observations are not trusted to the same degree. The value $x = b_1$ may be obtained from a more accurate scale—or, in a statistical problem, from a larger sample—than the value $x = b_2$. Nevertheless, if there is some information content in the second observation, we are not willing to rely totally on $x = b_1$. The simplest compromise is to attach different weights w_i^2 to the two observations, and to choose the $\bar{x}$ that minimizes the *weighted sum of squares*

$$E^2 = w_1^2(x - b_1)^2 + w_2^2(x - b_2)^2. \tag{58}$$

If $w_1 > w_2$, then more importance is attached to the first observation, and the minimizing process tries harder to make $(x - b_1)^2$ small. We can easily compute

$$\frac{dE^2}{dx} = 2[w_1^2(x - b_1) + w_2^2(x - b_2)],$$

and setting this to zero gives the new solution $\bar{x}_W$:

$$\bar{x}_W = \frac{w_1^2 b_1 + w_2^2 b_2}{w_1^2 + w_2^2}. \tag{59}$$

Instead of the average of b_1 and b_2, as we had when $w_1 = w_2 = 1$, $\bar{x}_W$ is a weighted average of the data. This average is closer to b_1 than to b_2.

There are two ways to look at weighted least squares. One is to figure out which ordinary least squares problem would lead to the E^2 in (58). It is the equation

$$\begin{bmatrix} w_1 \\ w_2 \end{bmatrix} [x] = \begin{bmatrix} w_1 b_1 \\ w_2 b_2 \end{bmatrix}, \quad \text{with} \quad E^2 = (w_1 x - w_1 b_1)^2 + (w_2 x - w_2 b_2)^2.$$

This is exactly the same as our weighted E^2 for the original problem. Looked at in this way, the weighting has changed the original system $Ax = b$ to a new system $WAx = Wb$, where

$$A = \begin{bmatrix} 1 \\ 1 \end{bmatrix} \quad \text{and} \quad W = \begin{bmatrix} w_1 & 0 \\ 0 & w_2 \end{bmatrix}.$$

The weighted $\bar{x}_W$ is the ordinary least squares solution of $WAx = Wb$.

This way of looking at the problem has one obvious but important consequence: We must expect to change the least squares solution if the system is changed to $WAx = Wb$. We emphasize that if A had been invertible, such a change could not have happened: $x = A^{-1}b$ in one case, and $x = (WA)^{-1}Wb =$

$A^{-1}W^{-1}Wb = A^{-1}b$ in the other. The exact solutions are the same, whereas in the least squares problem $\bar{x} = A^+b$ is different from $\bar{x}_W = (WA)^+Wb$. This is not at all surprising, but it forces us to an unhappy conclusion:

3T The fundamental law for the inverse of a product, $(BA)^{-1} = A^{-1}B^{-1}$, is not valid for pseudoinverses: In general we must expect that

$$(WA)^+ \neq A^+W^+. \tag{60}$$

Proof If the law did hold for the particular matrix W—which is invertible, so that its pseudoinverse is the same as its inverse—then the weighted $\bar{x}_W$ would be

$$\bar{x}_W = (WA)^+Wb = A^+W^{-1}Wb = A^+b = \bar{x}.$$

But $\bar{x}_W$ is different from $\bar{x}$, so the law fails.

We said earlier that there are two ways of looking at weighted least squares. Instead of changing from $Ax = b$ to $WAx = Wb$, we can *keep the original equation and change the definition of length*. The natural length in the weighted problem is a weighted sum of squares:

$$\text{the vector} \quad v = \begin{bmatrix} v_1 \\ v_2 \end{bmatrix} \quad \text{has length} \quad \| v \|_W = (w_1^2 v_1^2 + w_2^2 v_2^2)^{1/2}. \tag{61}$$

With this change, we proceed as usual to minimize the error $E = \| Ax - b \|_W$. Since E^2 is exactly our weighted sum of squares, its minimum will occur at $\bar{x}_W$. With the new definition of length, the closest point to b has moved from $A\bar{x}$ to $A\bar{x}_W$. Perhaps this can best be illustrated by a rescaling of the coordinate axes (Fig. 3.8). The projection has moved down the line, changing

$$\bar{x} = \frac{b_1 + b_2}{2} = \frac{3}{4} \quad \text{to} \quad \bar{x}_W = \frac{w_1^2 b_1 + w_2^2 b_2}{w_1^2 + w_2^2} < \frac{3}{4}.$$

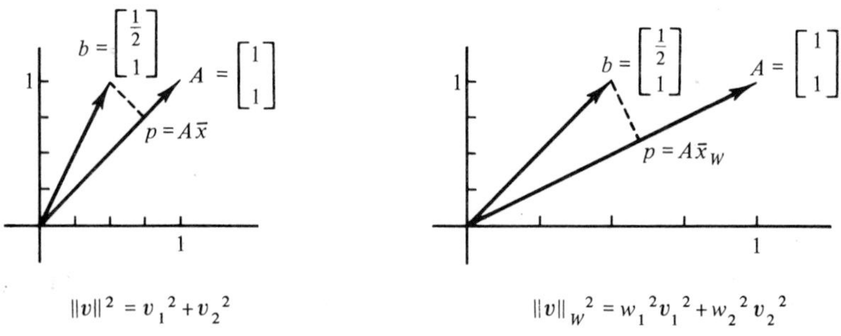

Fig. 3.8. The change in length and change in $\bar{x}$.

Since the weighted average is less than $\frac{3}{4}$, it is closer to the first measurement $b_1 = \frac{1}{2}$ than to the measurement $b_2 = 1$. By stretching one coordinate axis relative to the other, we give it more importance.

Weights, Lengths, and Inner Products

In the next few paragraphs, we want to show how the previous example is typical of all weighted least squares problems. And at the same time, we will find all the possible inner products of two vectors, and all the corresponding definitions of length. Fortunately, this terrific and overwhelming generalization is a very easy task.

We start with a least squares problem $Ax = b$. If the m observations b_1, ..., b_m are not equally reliable, then different weights w_1, ..., w_m are associated to the m equations. This generalizes our original example, but there is another and more subtle generalization. In addition to their unequal reliability, *the observations may not be independent.* In this case we also introduce a coefficient w_{ij} which measures the coupling of observation i to observation j. Then

$$Ax = b \quad \text{is changed to} \quad WAx = Wb. \tag{62}$$

The numbers w_1, ..., w_n lie along the main diagonal of W (from now on we denote them by w_{11}, ..., w_{nn}) and the coupling coefficients, if any, are the off-diagonal entries.

Note   In statistics, this roughly corresponds to the introduction of *regression coefficients*, or after normalization to *correlation coefficients*. The exact relation between our weighting matrix W and the statistician's covariance matrix V is given in the last sentence of this chapter.

To solve this new problem $WAx = Wb$, we have only to look back at the normal equations $A^{T}A\bar{x} = A^{T}b$ for the original problem, and make the appropriate changes—A is replaced by WA, and b is replaced by Wb.

3U   If A has independent columns (its rank is n) and W is invertible, then the least squares solution to $WAx = Wb$ is determined from the weighted normal equations:

$$(A^{T}W^{T}WA)\bar{x}_W = A^{T}W^{T}Wb. \tag{63}$$

If we write H for $W^{T}W$, then $\bar{x}_W = (A^{T}HA)^{-1}A^{T}Hb$.

This completes the first approach to weighted least squares, and we turn to the second approach. You remember that, in the 2 by 1 example with which we started, the key idea was to introduce a change in the definition of length.

Therefore we do exactly the same thing for the more general weighting matrix W:

3V If W is any invertible matrix, we may define a new length and inner product by the following rules: The new length of x equals the old length of Wx, and the new inner product of x and y equals the old inner product of Wx and Wy:

$$\| x \|_W = \| Wx \| \quad \text{and} \quad (x, y)_W = (Wx)^T(Wy) = x^T W^T W y.† \quad (64)$$

The length and inner product are connected by the usual law:

$$\| x \|_W^2 = (x, x)_W > 0 \quad \text{for all nonzero vectors } x. \quad (65)$$

This definition actually describes all possible inner products, in other words, all possible ways of producing a number that depends linearly on x and y, and is positive whenever $x = y \neq 0$. It is this positivity condition that forced W to be invertible, since otherwise there would be a nonzero vector x in its null-space, and the length of that vector would be $\| x \|_W = \| Wx \| = \| 0 \| = 0$. We refuse to allow a nonzero vector to have zero length.

Returning to the least squares problem $Ax = b$, with weighting matrix W, we can recover the geometrical picture by using the new length and inner product. The optimal choice of Ax, minimizing the error $Ax - b$, is again the point p in the column space of A closest to b. But the word *closest* must now be interpreted in terms of the new length; we are minimizing $\| Ax - b \|_W = \| WAx - Wb \|$. This point p (or more properly p_W, since it depends on W) is again the projection of b onto the column space, and the error vector $b - p_W$ is perpendicular to this space. But once more there is a change: Perpendicularity now means that the new inner product is zero. Thus the point $p_W = A\bar{x}_W$ is defined by the property that $b - p_W$ is perpendicular to every vector Ay in the column space:

$$(Ay)^T W^T W (b - A\bar{x}_W) = 0 \quad \text{for all } y.$$

The vector that multiplies y^T must be zero,

$$A^T W^T W (b - A\bar{x}_W) = 0, \quad \text{or} \quad A^T W^T W A \bar{x}_W = A^T W^T W b;$$

and we are back to the weighted normal equations (63).

We shall try, sort of pessimistically, to draw a picture (Fig. 3.9) which illustrates this new kind of perpendicularity. Either the space can be stretched according to W, as in Fig. 3.8, or we can keep the usual axes and suffer from angles that do not look right.‡

---

† This matrix $H = W^T W$, with an invertible W, is exactly what we call a *positive definite matrix* in Chapter 6.

‡ I worked hard on that one.

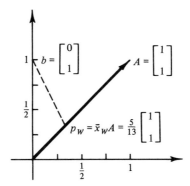

Fig. 3.9. Weighted least squares and W-perpendicularity.

The calculation that goes with the figure is this: Given

$$WAx = Wb, \quad \text{or} \quad \begin{bmatrix} 2 & 1 \\ 1 & 1 \end{bmatrix} \begin{bmatrix} 1 \\ 1 \end{bmatrix} [x] = \begin{bmatrix} 2 & 1 \\ 1 & 1 \end{bmatrix} \begin{bmatrix} 0 \\ 1 \end{bmatrix},$$

the weighted normal equations are

$$(WA)^{\mathrm{T}}WA\bar{x}_W = (WA)^{\mathrm{T}}Wb, \quad \text{or} \quad \begin{bmatrix} 3 & 2 \end{bmatrix} \begin{bmatrix} 3 \\ 2 \end{bmatrix} \bar{x}_W = \begin{bmatrix} 3 & 2 \end{bmatrix} \begin{bmatrix} 1 \\ 1 \end{bmatrix}$$

$$\text{or} \quad \bar{x}_W = \tfrac{5}{13}.$$

As a final remark on least squares, we want to match our approach to the one that is conventional in applications to statistics: The errors in the m equations $Ax = b$ are assumed to have mean zero, and a covariance matrix V. Then the optimal estimate $\bar{x}$ is the solution of the normal equations $A^{\mathrm{T}}V^{-1}A\bar{x} = A^{\mathrm{T}}V^{-1}b$, and these equations are identical with (63) provided that our $H = W^{\mathrm{T}}W$ is chosen to be the inverse of V.

EXERCISE 3.5.1   With $W = \begin{bmatrix} 2 & 0 \\ 0 & 1 \end{bmatrix}$ find the $\bar{x}_W$ that gives twice as much weight to the observation $x = 4$ as it does to $x = 1$: $w_1 = 2$ and $w_2 = 1$.

EXERCISE 3.5.2   With the same W, what vectors are W-perpendicular to $b = \begin{bmatrix} 1 \\ 1 \end{bmatrix}$? What is the weighted length $\| b \|_W$ of b itself?

EXERCISE 3.5.3   With $W = \begin{bmatrix} 2 & 1 \\ 1 & 1 \end{bmatrix}$, what is the optimal average $\bar{x}_W$ of the observations $x = 1$ and $x = 0$?

EXERCISE 3.5.4   What kind of matrix W produces a new length and new inner product that are the same as the old?

REVIEW EXERCISES

3.1 How is $(ABC)^\mathrm{T}$ related to the individual matrices A^T, B^T, and C^T?

3.2 Write the rank one matrix

$$A = \begin{bmatrix} 2 & 3 \\ 6 & 9 \end{bmatrix}$$

in the form uv^T and write A^T in the same form.

3.3 What is the angle between $a = (2, -2, 1)$ and $b = (1, 2, 2)$?

3.4 What is the projection p of $b = (1, 2, 2)$ onto $a = (2, -2, 1)$?

3.5 Find the cosine of the angle between the vectors $(3, 4)$ and $(4, 3)$.

3.6 Where is the projection of $b = (1, 1, 1)$ onto the plane spanned by $(1, 0, 0)$ and $(1, 1, 0)$?

3.7 What are the normal equations for the system $u = 0$, $v = 2$, $u + v = 4$? What are $\bar{u}$ and $\bar{v}$?

3.8 Construct the projection matrix P onto the space spanned by $(1, 1, 1)$ and $(0, 1, 3)$.

3.9 Which straight line gives the best fit to the following data: $y = 0$ at $x = 0$, $y = 0$ at $x = 1$, $y = 12$ at $x = 3$?

3.10 Which line of the restricted form $y = Dt$ best fits the two points $y = 1$, $t = 1$ and $y = 1$, $t = 2$?

3.11 Which constant function is closest to $y = x^4$ in the least squares sense over the interval $0 \le x \le 1$?

3.12 If Q is orthogonal, is the same true of Q^3?

3.13 Find all 3 by 3 orthogonal matrices whose entries are zeros and ones.

3.14 What multiple of

$$a_1 = \begin{bmatrix} 1 \\ 1 \end{bmatrix}$$

should be subtracted from

$$a_2 = \begin{bmatrix} 4 \\ 0 \end{bmatrix},$$

to make the result orthogonal to a_1? Sketch a figure.

3.15 Normalize the orthogonal vectors produced in the previous exercise, thereby completing the Gram–Schmidt process. Factor

$$A = \begin{bmatrix} 1 & 4 \\ 1 & 0 \end{bmatrix}$$

into QR.

3.16 Factor

$$\begin{bmatrix} \cos\theta & \sin\theta \\ \sin\theta & 0 \end{bmatrix}$$

into QR, recognizing that the first column is already a unit vector.

3.17 Find the pseudoinverse of $A = [3 \ \ 0]$.

3.18 With weighting matrix

$$W = \begin{bmatrix} 2 & 1 \\ 1 & 0 \end{bmatrix},$$

what is the inner product of $(1, 0)$ with $(0, 1)$?

3.19 If every entry in an orthogonal matrix is either $\frac{1}{4}$ or $-\frac{1}{4}$, how big is the matrix?

$$\begin{bmatrix} 4 \end{bmatrix}$$

DETERMINANTS

4.1 ■ INTRODUCTION

It is hard to know what to say about determinants. Seventy years ago they seemed more interesting and more important than the matrices they came from, and Muir's "History of Determinants" filled four volumes. Mathematics keeps changing direction, however, and determinants are now far from the center of linear algebra. After all, a single number can tell only so much about a matrix.

One viewpoint is this: The determinant provides an explicit "formula," a concise and definite expression in closed form, for quantities such as A^{-1}. This formula will not change the way we compute A^{-1}, or $A^{-1}b$; even the determinant itself is found by elimination. In fact, elimination can be regarded as the most efficient way to substitute the entries of a given matrix A into the formula. What counts is the fact that the general formula shows how the quantity we want depends on the n^2 entries of the matrix, and how it varies when those entries vary.

We can list some of the main uses of determinants:

(1)   It gives a test for invertibility. *If the determinant of A is zero, then A is singular; and if* det $A \neq 0$, *then A is invertible.* The most important application of this test, and the reason this chapter is essential to the book, is to the family of matrices $A - \lambda I$. The parameter λ is subtracted all along the main diagonal, and the problem is to find those values of λ (the *eigenvalues*) for

which $A - \lambda I$ is singular. The test is to see if the determinant of this matrix is zero. We shall see that $\det(A - \lambda I)$ is a polynomial of degree n in λ, and therefore, counting multiplicities, it has exactly n roots; the matrix has n eigenvalues. This is a fact which follows from the determinant formula, and not from a computer.

(2)   The determinant of A equals the volume of a parallelepiped P in n-dimensional space, provided the edges of P come from the rows of A (Fig. 4.1).† This volume may seem an odd thing to want to compute. In practice, P is often the infinitesimal volume element in a multiple integral. The simplest element is a little cube $dV = dx\,dy\,dz$, as in $\int\int\int f(x, y, z)\,dV$. Suppose, in order to simplify the integral, we decide to change variables to x', y', z'; say $x = x'\cos y'$, $y = x'\sin y'$, $z = z'$. Then, just as we have to remember for one variable that the differential dx is stretched into $(dx/dx')\,dx'$, so the volume element $dx\,dy\,dz$ is modified to $J\,dx'\,dy'\,dz'$. The *Jacobian determinant*, which is the three-dimensional analog of the stretching factor dx/dx', is

$$J = \begin{vmatrix} \partial x/\partial x' & \partial x/\partial y' & \partial x/\partial z' \\ \partial y/\partial x' & \partial y/\partial y' & \partial y/\partial z' \\ \partial z/\partial x' & \partial z/\partial y' & \partial z/\partial z' \end{vmatrix} = \begin{vmatrix} \cos y' & -x'\sin y' & 0 \\ \sin y' & x'\cos y' & 0 \\ 0 & 0 & 1 \end{vmatrix}.$$

The value of this determinant turns out to be $J = x'$. It is the r in the polar element $r\,dr\,d\theta$, or more accurately in the cylindrical element $r\,dr\,d\theta\,dz$; this element is our little parallelepiped. (It looks curved if we try to draw it, but probably it gets straighter as the edges become infinitesimal.)

(3)   The determinant gives an explicit formula for the pivots. Theoretically, we could use it to predict when a pivot will be zero, and a row exchange will be necessary. More importantly, from the formula

$$\det A = \pm(\text{product of the pivots}),$$

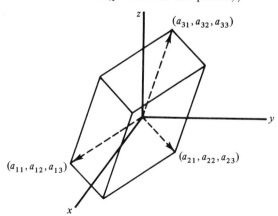

Fig. 4.1.   The parallelepiped formed from the rows of A.

† Or the edges could come from the columns of A, giving an entirely different parallelepiped with the same volume.

it follows that *regardless of the order of elimination, the product of the pivots remains the same apart from sign*. Years ago, this led to the belief that it was useless to escape a very small pivot by exchanging rows, since eventually the small pivot would catch up with us. But what usually happens in practice, if an abnormally small pivot is not avoided, is that it is very soon followed by an abnormally large one; this brings the product back to normal but leaves the numerical solution in ruins.

(4)   The determinant measures the dependence of $A^{-1}b$ on each element of b. If one parameter is changed in an experiment, or one observation is corrected, the "influence coefficient" on $x = A^{-1}b$ is a ratio of determinants.

There is one more problem about the determinant. It is difficult not only to decide on its importance, and its proper place in the theory of linear algebra, but also to decide on its definition. Obviously, det A will not be some extremely simple function of n^2 variables, otherwise A^{-1} would be much easier to find than it actually is. The explicit formula given in Section 4.3 will require a good deal of explanation, and its connection with the inverse is far from evident.

The simple things about the determinant are not the explicit formulas by which it can be expressed, but the properties it possesses. This suggests the natural place to begin. The determinant can be (and will be) defined by its three most basic properties. The problem is then to show how, by systematically using these properties to simplify the formulas, the value of the determinant can be computed. This will bring us back to Gaussian elimination, and to the product of the pivots. And the more difficult theoretical problem is to show that whatever the order in which the properties are used, the result is always the same—the defining properties are self-consistent.

The next section lists the defining properties of the determinant, and their most important consequences. Then Section 4.3 gives several possible formulas for the determinant—one is an explicit formula with $n!$ terms, another is a formula "by induction," and the third is the one involving pivots from which the determinant of a large matrix is actually computed. In Section 4.4 the determinant is applied to find A^{-1} and then to solve for $x = A^{-1}b$; the latter is *Cramer's rule*. And finally, in an optional remark on permutations, we prove that the properties are self-consistent, so that there is no ambiguity in the definition.

4.2 ■ THE PROPERTIES OF THE DETERMINANT

This will be a pretty long list. Fortunately each of the properties is easy to understand, and even easier to illustrate, for a 2 by 2 example. Therefore we shall verify that the familiar definition in the 2 by 2 case,

$$\det \begin{bmatrix} a & b \\ c & d \end{bmatrix} = \begin{vmatrix} a & b \\ c & d \end{vmatrix} = ad - bc,$$

possesses every property in the list. (There are two accepted notations for the determinant of A: either det A or $|A|$.) From the fourth property on, we also indicate how it can be deduced from the previous ones; then every property is a consequence of the first three.

1. *The determinant is a linear function of the first row.* This means that if the matrices A, B, and C are identical from the second row onward, and the first row of A is a combination of the first rows of B and C, then det A is the same combination of det B and det C. Since linear combinations include scalar multiplication as well as addition of vectors, this amounts to two rules in one:

$$\begin{vmatrix} a + a' & b + b' \\ c & d \end{vmatrix} = (a + a')d - (b + b')c = \begin{vmatrix} a & b \\ c & d \end{vmatrix} + \begin{vmatrix} a' & b' \\ c & d \end{vmatrix}$$

$$\begin{vmatrix} ta & tb \\ c & d \end{vmatrix} = tad - tbc = t \begin{vmatrix} a & b \\ c & d \end{vmatrix}.$$

We emphasize that this rule is completely different from $\det(B + C) = \det B + \det C$, which is false. Only one row is allowed to vary.

2. *The determinant changes sign when two rows are exchanged.*

$$\begin{vmatrix} c & d \\ a & b \end{vmatrix} = cb - ad = - \begin{vmatrix} a & b \\ c & d \end{vmatrix}.$$

It follows that in rule 1, there is nothing special about the first row. It can be exchanged for any other, and exchanged back, so the determinant is a linear function of each row separately.

3. *The determinant of the identity matrix is* 1.

$$\begin{vmatrix} 1 & 0 \\ 0 & 1 \end{vmatrix} = 1.$$

This condition just normalizes the value; rules 1 and 2 leave a "one-dimensional space" of possible determinants, and this rule picks out one of them.

4. *If two rows of A are equal, then* det $A = 0$.

$$\begin{vmatrix} a & b \\ a & b \end{vmatrix} = ab - ba = 0.$$

This follows from rule 2, since if the equal rows are exchanged, the determinant is supposed to change sign. But it also has to stay the same, because the matrix stays the same. The only number which can do that is zero, so det $A = 0$. (The reasoning fails if $1 = -1$, which is the case in Boolean algebra. Then 4 should replace 2 as one of the defining properties.)

5. *The elementary operation of subtracting a multiple of one row from another leaves the determinant unchanged.*

$$\begin{vmatrix} a - lc & b - ld \\ c & d \end{vmatrix} = \begin{vmatrix} a & b \\ c & d \end{vmatrix}.$$

Rule 1 would say that there is a further term $l \begin{vmatrix} c & d \\ c & d \end{vmatrix}$, but rule 4 ensures that this extra term is zero.

6. *If A has a zero row, then* $\det A = 0$.

$$\begin{vmatrix} 0 & 0 \\ c & d \end{vmatrix} = 0.$$

One proof is to add some other row to the zero row. The determinant is unchanged, by 5, and because the matrix will now have two identical rows, $\det A = 0$ by 4.

7. *If A is triangular, then* $\det A$ *is the product* $a_{11}a_{22} \cdots a_{nn}$ *of the entries on the main diagonal. In particular, if A has* 1*'s along the diagonal,* $\det A = 1$.

$$\begin{vmatrix} a & b \\ 0 & d \end{vmatrix} = ad, \qquad \begin{vmatrix} a & 0 \\ c & d \end{vmatrix} = ad.$$

Proof Suppose the diagonal entries are nonzero. Then elementary operations will eliminate all the off-diagonal entries, without changing the determinant (by rule 5). These operations leave us with the diagonal matrix

$$D = \begin{bmatrix} a_{11} & & & \\ & a_{22} & & \\ & & \ddots & \\ & & & a_{nn} \end{bmatrix}.$$

(This is the usual elimination if A is lower triangular; if A is upper triangular, then multiples of each row are subtracted from rows above it.) To find the determinant of D, we patiently apply rule 1; factoring out each a_{ii} leaves the identity matrix whose determinant we know:

$$\det D = a_{11}a_{22} \cdots a_{nn} \det I = a_{11}a_{22} \cdots a_{nn}.$$

If any diagonal entry is zero, then A is singular, and by the following rule its determinant is zero.

8. *If A is singular, then* $\det A = 0$. *If A is nonsingular, then* $\det A \neq 0$.

$$\begin{bmatrix} a & b \\ c & d \end{bmatrix} \quad \text{is singular if and only if} \quad ad - bc = 0.$$

If A is singular, then elementary operations lead to a matrix U with a zero row. By 5 and 6, $\det A = \det U = 0$. Suppose on the other hand that A is nonsingular. Then elementary operations and row exchanges lead to an upper triangular U with nonzero entries along the main diagonal. These entries are the pivots $d_1, \ldots, d_n$, and by rule 7, $\det A = \pm \det U = \pm d_1 d_2 \cdots d_n$. The plus or minus sign depends, by 2, on whether the number of row exchanges was even or odd.

9. *For any two n by n matrices, the determinant of the product is the product of the determinants*: $\det AB = (\det A)(\det B)$.

$$\begin{vmatrix} a & b \\ c & d \end{vmatrix} \begin{vmatrix} e & f \\ g & h \end{vmatrix} = \begin{vmatrix} ae + bg & af + bh \\ ce + dg & cf + dh \end{vmatrix}.$$

This is the same as the identity

$$(ad - bc)(eh - fg) = (ae + bg)(cf + dh) - (af + bh)(ce + dg).$$

In particular, if A is invertible, then $\det A \det A^{-1} = \det AA^{-1} = \det I = 1$, *or* $\det A^{-1} = 1/\det A$.

For this rule, which is surely the least obvious so far, we suggest two possible proofs. In both, we assume that A and B are nonsingular; otherwise AB is singular, and the equation $\det AB = \det A \det B$ is easily verified. By 8, it becomes $0 = 0$.

(i) We may consider the quantity $d(A) = \det AB/\det B$, and prove that it has properties 1–3; then, because these properties define the determinant, $d(A)$ must equal $\det A$. For example, if A is the identity, then certainly $d(I) = \det B/\det B = 1$; thus rule 3 is satisfied by $d(A)$. If two rows of A are exchanged, so are the same two rows of AB, and the sign of d changes as required by rule 2. And a linear combination appearing in the first row of A gives the same linear combination in the first row of AB. Therefore rule 1 for the determinant of AB, divided by the fixed quantity $\det B$, leads to rule 1 for the ratio $d(A)$. Thus $d(A)$ coincides with the determinant of A, and $\det AB/\det B = \det A$.

(ii) This second proof is less elegant than the first. It starts by supposing that A is a diagonal matrix D. Then $\det DB = \det D \det B$ follows from rule 1, by factoring out each of the diagonal elements d_i from its row. For a general matrix A, we reduce it to D by a "Gauss–Jordan" series of elimination steps— first from A to U by the usual sequence, and then from U to D by using each row to operate on the rows above it. The determinant does not change, except for a sign reversal whenever two rows are exchanged. The same steps reduce AB to DB, with precisely the same effect on the determinant. But for the product DB it is already confirmed that the rule 9 is correct.

10.   *The transpose of A has the same determinant as A itself:* $\det A^{\mathsf{T}} = \det A$.

$$\begin{vmatrix} a & b \\ c & d \end{vmatrix} = \begin{vmatrix} a & c \\ b & d \end{vmatrix}.$$

Again the singular case is separate; A is singular if and only if A^{T} is, and we have $0 = 0$. If A is not singular, then it allows the factorization $PA = LDU$, and we apply the previous rule 9 for the determinant of a product:

$$\det P \det A = \det L \det D \det U. \tag{1}$$

Transposing $PA = LDU$ gives $A^{\mathsf{T}}P^{\mathsf{T}} = U^{\mathsf{T}}D^{\mathsf{T}}L^{\mathsf{T}}$, and again by rule 9,

$$\det A^{\mathsf{T}} \det P^{\mathsf{T}} = \det U^{\mathsf{T}} \det D^{\mathsf{T}} \det L^{\mathsf{T}}. \tag{2}$$

This is simpler than it looks, because L, U, L^{T}, and U^{T} are triangular with unit diagonal. By rule 7, their determinants all equal one. Also, any diagonal matrix is the same as its transpose: $D = D^{\mathsf{T}}$. This leaves only the permutation matrices.

Certainly the determinant of P is either 1 or -1, because it comes from the identity matrix by a sequence of row exchanges. Observe that for any permutation matrix, we have $PP^{\mathsf{T}} = I$. (Multiplying P and P^{T}, the 1 in the first row of P matches the 1 in the first column of P^{T}, and misses the 1's in the other columns.) Therefore $\det P \det P^{\mathsf{T}} = \det I = 1$, and P and P^{T} must have the same determinant; both equal 1 or both equal -1.

We conclude that the products (1) and (2) are the same, and $\det A = \det A^{\mathsf{T}}$. This fact practically doubles our list of properties, because every rule which applied to the rows can now be applied to the columns: *The determinant changes sign when two columns are exchanged, two equal columns* (or a column of zeros) *produce a zero determinant, and the determinant depends linearly on each individual column.* The proof is just to transpose the matrix and work with the rows.

I think it is time to stop and call the list complete. It only remains to find a definite formula for the determinant, and to put that formula to use.

EXERCISE 4.2.1   How are $\det(2A)$, $\det(-A)$, and $\det(A^2)$ related to $\det A$?

EXERCISE 4.2.2   Show—by carrying out each step on a 2 by 2 example—that an exchange of rows i and j can be produced by adding row i to row j, then subtracting the new row j from row i, then adding the new row i to row j, and finally multiplying row i by -1. Which rules could we then use to deduce rule 2?

EXERCISE 4.2.3   By applying row operations to produce an upper triangular U, compute

$$\det \begin{bmatrix} 1 & 2 & -2 & 0 \\ 2 & 3 & -4 & 1 \\ -1 & -2 & 0 & 2 \\ 0 & 2 & 5 & 3 \end{bmatrix} \quad \text{and} \quad \det \begin{bmatrix} 2 & -1 & 0 & 0 \\ -1 & 2 & -1 & 0 \\ 0 & -1 & 2 & -1 \\ 0 & 0 & -1 & 2 \end{bmatrix}.$$

Note Many readers will already know the formula for 3 by 3 determinants, which contains six terms; see Eq. (5) below. It is natural to hope for a similar rule for 4 by 4 determinants, and it is true that there is one; *but it contains* 4! = 24 *terms, and not just the eight terms which might seem to correspond to the* 3 *by* 3 *case.* It is for that reason that we recommend row operations and a triangular U.

EXERCISE 4.2.4   A skew-symmetric matrix satisfies $A^T = -A$, as in

$$A = \begin{bmatrix} 0 & a & b \\ -a & 0 & c \\ -b & -c & 0 \end{bmatrix}.$$

Show that det A is zero, by comparing it to det A^T and det$(-A)$. Show also that det A can be nonzero in the 2 by 2 skew-symmetric case; the proof of det $A = 0$ requires the order of the matrix to be odd.

EXERCISE 4.2.5   Find the determinants of:

(a)   the matrix product

$$A = \begin{bmatrix} 1 \\ 4 \\ 2 \end{bmatrix} \begin{bmatrix} 2 & -1 & 2 \end{bmatrix}$$

of rank one;
(b)   the upper triangular matrix

$$U = \begin{bmatrix} 4 & 4 & 8 & 8 \\ 0 & 1 & 2 & 2 \\ 0 & 0 & 2 & 6 \\ 0 & 0 & 0 & 2 \end{bmatrix};$$

(c)   the lower triangular matrix U^T;
(d)   the inverse matrix U^{-1};
(e)   the "reverse-triangular" matrix that results from row exchanges,

$$M = \begin{bmatrix} 0 & 0 & 0 & 2 \\ 0 & 0 & 2 & 6 \\ 0 & 1 & 2 & 2 \\ 4 & 4 & 8 & 8 \end{bmatrix}.$$

EXERCISE 4.2.6   Find an alternative way of deducing rule 6 from the preceding rules.

EXERCISE 4.2.7 If Q is an orthogonal matrix, so that $Q^TQ = I$, prove that det Q equals $+1$ or -1. What kind of parallelepiped is formed from the rows (or columns) of Q?

EXERCISE 4.2.8 Use row operations to verify that the 4 by 4 Vandermonde determinant is

$$V_4 = \begin{vmatrix} 1 & x_1 & x_1^2 & x_1^3 \\ 1 & x_2 & x_2^2 & x_2^3 \\ 1 & x_3 & x_3^2 & x_3^3 \\ 1 & x_4 & x_4^2 & x_4^3 \end{vmatrix}$$

$$= (x_2 - x_1)(x_3 - x_1)(x_4 - x_1)(x_3 - x_2)(x_4 - x_2)(x_4 - x_3).$$

4.3 ■ FORMULAS FOR THE DETERMINANT

The first formula has already appeared:

4A If A is nonsingular, then $A = P^{-1}LDU$, and

$$\det A = \det P^{-1} \det L \det D \det U$$

$$= \pm(\text{product of the pivots}). \tag{3}$$

The sign ±1 is the determinant of P^{-1} (or of P), and depends on whether the number of row exchanges is even or odd. The triangular factors have $\det L = \det U = 1$ and $\det D = d_1 \cdots d_n$.

In the 2 by 2 case, the standard LDU factorization is

$$\begin{bmatrix} a & b \\ c & d \end{bmatrix} = \begin{bmatrix} 1 & 0 \\ c/a & 1 \end{bmatrix} \begin{bmatrix} a & 0 \\ 0 & (ad - bc)/a \end{bmatrix} \begin{bmatrix} 1 & b/a \\ 0 & 1 \end{bmatrix} \tag{4}$$

and the product of the pivots is $ad - bc$. If the first step had been a row exchange, then

$$PA = \begin{bmatrix} c & d \\ a & b \end{bmatrix} = \begin{bmatrix} 1 & 0 \\ a/c & 1 \end{bmatrix} \begin{bmatrix} c & 0 \\ 0 & (cb - da)/c \end{bmatrix} \begin{bmatrix} 1 & d/c \\ 0 & 1 \end{bmatrix}$$

and the product of the pivots is now $-\det A$.

EXAMPLE The finite difference matrix in Section 1.6 had the $A = LDU$

factorization

$$
\begin{bmatrix}
2 & -1 & & & \\
-1 & 2 & -1 & & \\
 & -1 & 2 & \cdot & \\
 & & \cdot & \cdot & -1 \\
 & & & -1 & 2
\end{bmatrix}
$$

$$
=
\begin{bmatrix}
1 & & & & \\
-\dfrac{1}{2} & 1 & & & \\
 & -\dfrac{2}{3} & 1 & & \\
 & & \cdot & \cdot & \\
 & & & \dfrac{1-n}{n} & 1
\end{bmatrix}
\begin{bmatrix}
2 & & & & \\
 & \dfrac{3}{2} & & & \\
 & & \dfrac{4}{3} & & \\
 & & & \cdot & \\
 & & & & \dfrac{n+1}{n}
\end{bmatrix}
\begin{bmatrix}
1 & -\dfrac{1}{2} & & & \\
 & 1 & -\dfrac{2}{3} & & \\
 & & 1 & \cdot & \\
 & & & \cdot & \dfrac{1-n}{n} \\
 & & & & 1
\end{bmatrix}
$$

Its determinant is the product of its pivots:

$$
\det A = 2 \left(\frac{3}{2}\right)\left(\frac{4}{3}\right) \cdots \left(\frac{n+1}{n}\right) = n + 1.
$$

This is the way determinants are calculated, except for very special matrices. In fact, the pivots are the result of systematically condensing the information that was originally spread over all n^2 entries of the matrix. From a theoretical point of view, however, this concentration of information into the pivots has a disadvantage: It is impossible to figure out how a change in one particular entry would affect the determinant. Therefore we now propose to find an explicit expression for the determinant in terms of the n^2 entries. For $n = 2$, we will be proving that the formula $ad - bc$ is correct. For $n = 3$, the corresponding formula is again pretty well known:

$$
\begin{vmatrix}
a_{11} & a_{12} & a_{13} \\
a_{21} & a_{22} & a_{23} \\
a_{31} & a_{32} & a_{33}
\end{vmatrix}
= +a_{11}a_{22}a_{33} + a_{12}a_{23}a_{31} + a_{13}a_{21}a_{32}
$$

$$
- a_{11}a_{23}a_{32} - a_{12}a_{21}a_{33} - a_{13}a_{22}a_{31} . \tag{5}
$$

Our goal is to derive these formulas directly from the defining properties 1–3 of the previous section. If we can manage to handle $n = 2$ and $n = 3$ in a sufficiently organized way, you will have some grasp of the determinant of a larger matrix.

To start with, each row of A can be broken down into vectors in the coordinate directions: With $n = 2$ this splitting is simply

$$[a \quad b] = [a \quad 0] + [0 \quad b] \qquad \text{and} \qquad [c \quad d] = [c \quad 0] + [0 \quad d]$$

Then we apply the key property of linearity in each row separately—first in row 1 and then in row 2:

$$\begin{vmatrix} a & b \\ c & d \end{vmatrix} = \begin{vmatrix} a & 0 \\ c & d \end{vmatrix} + \begin{vmatrix} 0 & b \\ c & d \end{vmatrix}$$

$$= \begin{vmatrix} a & 0 \\ c & 0 \end{vmatrix} + \begin{vmatrix} 0 & b \\ c & 0 \end{vmatrix} + \begin{vmatrix} a & 0 \\ 0 & d \end{vmatrix} + \begin{vmatrix} 0 & b \\ 0 & d \end{vmatrix}. \tag{6}$$

For an n by n matrix, every row will be split into n coordinate directions. Then in the expansion analogous to (6), there will be n^n terms: In our case $2^2 = 4$. Fortunately, most of them (like the first and last terms above) will be automatically zero. Whenever two rows are in the same coordinate direction, one will be a multiple of the other, and

$$\begin{vmatrix} a & 0 \\ c & 0 \end{vmatrix} = 0, \qquad \begin{vmatrix} 0 & b \\ 0 & d \end{vmatrix} = 0.$$

There is a column of zeros, and a zero determinant. Therefore, we pay attention *only when the rows point in different directions*; the nonzero terms come in *different columns*. Suppose the first row has a nonzero entry in column α, the second row is nonzero in column β, and finally the nth row is nonzero in column ν. The column numbers $\alpha, \beta, \ldots, \nu$ are all different; they are just a reordering, or permutation, of the numbers $1, 2, \ldots, n$. For the 3 by 3 case, there will be six terms of this kind:

$$\begin{vmatrix} a_{11} & a_{12} & a_{13} \\ a_{21} & a_{22} & a_{23} \\ a_{31} & a_{32} & a_{33} \end{vmatrix} = \begin{vmatrix} a_{11} & & \\ & a_{22} & \\ & & a_{33} \end{vmatrix} + \begin{vmatrix} a_{12} & \\ & a_{23} \\ a_{31} & \end{vmatrix} + \begin{vmatrix} & & a_{13} \\ a_{21} & & \\ & a_{32} & \end{vmatrix}$$

$$+ \begin{vmatrix} a_{11} & & \\ & & a_{23} \\ & a_{32} & \end{vmatrix} + \begin{vmatrix} & a_{12} & \\ a_{21} & & \\ & & a_{33} \end{vmatrix} + \begin{vmatrix} & & a_{13} \\ & a_{22} & \\ a_{31} & & \end{vmatrix}. \tag{7}$$

To repeat, the expansion (6) would have $3^3 = 27$ terms; all but $3! = 6$ are zero, because of the duplication of a coordinate direction. In general, since there are n choices for the first column α, $n - 1$ remaining choices for β, and finally only one choice for the last column ν—all but one column will be used by that time, when we "snake" down the matrix as illustrated in (7)—there are $n!$ terms left in the expansion. In other words, there are $n!$ ways to permute the numbers $1, 2, \ldots, n$. We look at the sequence of column numbers to find the associated permutations; the six terms in (7) come from the following column sequences:

$$(\alpha, \beta, \nu) = (1, 2, 3), (2, 3, 1), (3, 1, 2), (1, 3, 2), (2, 1, 3), (3, 2, 1).$$

These are all of the $3! = 6$ permutations of $(1, 2, 3)$; the first permutation in the list is the identity.

The determinant of A is now reduced to six separate and much simpler determinants. Factoring out the a_{ij}, there is a term for every one of the six permutations:

$$\det A = a_{11}a_{22}a_{33} \begin{vmatrix} 1 & & \\ & 1 & \\ & & 1 \end{vmatrix} + a_{12}a_{23}a_{31} \begin{vmatrix} & 1 & \\ & & 1 \\ 1 & & \end{vmatrix} + a_{13}a_{21}a_{32} \begin{vmatrix} & & 1 \\ 1 & & \\ & 1 & \end{vmatrix}$$

$$+ a_{11}a_{23}a_{32} \begin{vmatrix} 1 & & \\ & & 1 \\ & 1 & \end{vmatrix} + a_{12}a_{21}a_{33} \begin{vmatrix} & 1 & \\ 1 & & \\ & & 1 \end{vmatrix} + a_{13}a_{22}a_{31} \begin{vmatrix} & & 1 \\ & 1 & \\ 1 & & \end{vmatrix}. \quad (7')$$

Every term is a product of $n = 3$ entries a_{ij}, with each row and column represented once. In other words, there is a term corresponding to every path that goes down through the matrix and uses each column once; if the columns are used in the order $(\alpha, \ldots, \nu)$, then that term is the product $a_{1\alpha} \cdots a_{n\nu}$ multiplied by the determinant of a permutation matrix P_σ. The determinant of the whole matrix is the sum of these terms, and **that sum is the explicit formula we are after**:

$$\det A = \sum_\sigma (a_{1\alpha}a_{2\beta} \cdots a_{n\nu}) \det P_\sigma. \quad (8)$$

For an n by n matrix, this sum is taken over all $n!$ permutations $\sigma = (\alpha, \ldots, \nu)$ of the numbers $(1, \ldots, n)$. The permutation gives the sequence of column numbers as we go down the rows of the matrix, and it also specifies the permutation matrix P_σ: The 1's appear in P_σ at the same places where the a's appeared in A.

It remains only to find the determinant of P_σ. Since row exchanges transform P_σ into the identity matrix, and each exchange reverses the sign of the determinant, $\det P_\sigma$ is either $+1$ or -1. The sign depends on whether the number of exchanges is even or odd, and we consider two examples:

$$P_\sigma = \begin{bmatrix} 1 & & \\ & & 1 \\ & 1 & \end{bmatrix}, \quad \text{with column sequence} \quad (\alpha, \beta, \nu) = (1, 3, 2);$$

$$P_\sigma = \begin{bmatrix} & & 1 \\ 1 & & \\ & 1 & \end{bmatrix}, \quad \text{with column sequence} \quad \sigma = (3, 1, 2).$$

The first requires one exchange (the second and third rows), so that det $P_\sigma = -1$. The second requires two exchanges to recover the identity (the first and second rows, followed by the second and third), so that det $P_\sigma = (-1)^2 = 1$. These are two of the six $\pm$ signs that appear in (5).

Equation (8) is an explicit expression for the determinant, and it is easy enough to check the 2 by 2 case: The $2! = 2$ permutations are $\sigma = (1, 2)$ and $\sigma = (2, 1)$, and therefore

$$\det A = a_{11}a_{22} \det \begin{bmatrix} 1 & 0 \\ 0 & 1 \end{bmatrix} + a_{12}a_{21} \det \begin{bmatrix} 0 & 1 \\ 1 & 0 \end{bmatrix} = a_{11}a_{22} - a_{12}a_{21} = ad - bc.$$

No one can claim that the explicit formula (8) is particularly simple. Nevertheless, it is possible to see why it has properties 1–3. Property 3, the fact that det $I = 1$, is of course the simplest; the products of the a_{ij} will always be zero, except for the special column sequence $\sigma = (1, 2, \ldots, n)$, in other words the identity permutation. This term gives det $I = 1$. Property 2 will be checked in the next section, because here we are most interested in property 1: The determinant should depend linearly on the row $a_{11}, a_{12}, \ldots, a_{1n}$. To see this dependence, look at the terms in formula (8) involving a_{11}. They occur when the choice of the first column is $\alpha = 1$, leaving some permutation $\sigma' = (\beta, \ldots, \nu)$ of the remaining column numbers $(2, \ldots, n)$. We collect all these terms together as $a_{11}A_{11}$, where the coefficient of a_{11} is

$$A_{11} = \sum_{\sigma'} (a_{2\beta} \cdots a_{n\nu}) \det P_{\sigma'}. \tag{9}$$

Similarly, the entry a_{12} is multiplied by some messy expression A_{12}. Grouping all the terms which start with the same a_{1j}, the formula (8) becomes

$$\det A = a_{11}A_{11} + a_{12}A_{12} + \cdots + a_{1n}A_{1n}. \tag{10}$$

Obviously det A depends linearly on the entries $a_{11}, \ldots, a_{1n}$ of the first row. Therefore property 1 is verified, whatever the coefficients A_{1j} may be.

EXAMPLE   For a 3 by 3 matrix, this way of collecting terms gives

$$\det A = a_{11}(a_{22}a_{33} - a_{23}a_{32}) + a_{12}(a_{23}a_{31} - a_{21}a_{33}) + a_{13}(a_{21}a_{32} - a_{22}a_{31}).$$

The "cofactors" A_{11}, A_{12}, A_{13} are written out in the three parentheses.

Expansion of det A in Cofactors

We want one more formula for the determinant. If this meant starting again from scratch, it would be too much. But *the formula is already discovered—it is* (10), *and the only point is to identify the cofactors A_{1j}.*

We know that this number A_{1j} depends on rows $2, \ldots, n$; row 1 is already accounted for by the factor a_{1j}. Furthermore, a_{1j} also accounts for the jth column, so its cofactor A_{1j} must depend entirely on the other columns. No row or column can be used twice in the same term, and what we are really doing is to split the determinant into the following sum:

$$
\begin{vmatrix} a_{11} & a_{12} & a_{13} \\ a_{21} & a_{22} & a_{23} \\ a_{31} & a_{32} & a_{33} \end{vmatrix} = \begin{vmatrix} a_{11} & & \\ & a_{22} & a_{23} \\ & a_{32} & a_{33} \end{vmatrix} + \begin{vmatrix} & a_{12} & \\ a_{21} & & a_{23} \\ a_{31} & & a_{33} \end{vmatrix} + \begin{vmatrix} & & a_{13} \\ a_{21} & a_{22} & \\ a_{31} & a_{32} & \end{vmatrix} .
$$

This splitting reduces a determinant of order n to several smaller determinants (called *minors*) of order $n - 1$; you can see the 2 by 2 submatrices that appear on the right-hand side. The submatrix M_{1j} is formed by throwing away row 1 and column j, and each term on the right is a product of a_{1j} and the determinant of M_{1j}—with the correct plus or minus sign. These signs alternate as we go along the row, and the cofactors are finally identified as

$$ A_{1j} = (-1)^{1+j} \det M_{1j} . \tag{11} $$

For example, the second cofactor A_{12} is $a_{23}a_{31} - a_{21}a_{33}$, which matches $(-1)^{1+2} \det M_{12}$. This same technique works on square matrices of any size, and a close look at Eq. (9) confirms that A_{11} is the determinant of the lower right corner M_{11} of the original matrix A.

There is a similar expansion on any other row, say row i, which could be proved by exchanging row i with row 1:

4B The determinant of A can be computed by expanding it in the cofactors of the ith row:

$$ \det A = a_{i1}A_{i1} + a_{i2}A_{i2} + \cdots + a_{in}A_{in} . \tag{12} $$

The **cofactor** A_{ij} is the determinant of M_{ij} with the correct sign:

$$ A_{ij} = (-1)^{i+j} \det M_{ij} . $$

The minor M_{ij} is formed by deleting row i and column j of A.

These formulas express $\det A$ as a combination of determinants of order $n - 1$. Therefore *we could have defined the determinant by induction on n.* For 1 by 1 matrices, we would set $\det A = a_{11}$—and then use (12) to define successively the determinants of 2 by 2 matrices, 3 by 3 matrices, and so on indefinitely. We have preferred to define the determinant by its properties, which are much simpler to explain, and then to deduce the explicit formula (8) and the induction formula (12) from these properties.

Finally, there is one more consequence of the fact that $\det A = \det A^{\text{T}}$. This property allows us to expand in the cofactors of a column instead of a row; for every j,

$$ \det A = \sum_{i=1}^{n} a_{ij}A_{ij} . \tag{13} $$

The proof is simply to expand $\det A^{\text{T}}$ in the cofactors of its jth row, which is the jth column of A.

EXAMPLE 1 Expansions in cofactors are most useful when some rows are almost entirely zero, as for the triangular

$$A = \begin{bmatrix} 4 & 0 & 0 & 0 \\ 2 & 3 & 0 & 0 \\ 1 & 2 & -1 & 0 \\ 6 & 6 & 7 & 2 \end{bmatrix}.$$

The only term produced by the first row is $a_{11}A_{11}$, since $a_{12} = a_{13} = a_{14} = 0$:

$$a_{11} = 4, \qquad A_{11} = \det \begin{bmatrix} 3 & 0 & 0 \\ 2 & -1 & 0 \\ 6 & 7 & 2 \end{bmatrix}.$$

Expanding A_{11} itself on the first row, we are down to

$$\det A = 4A_{11} = 4 \cdot 3 \cdot \det \begin{bmatrix} -1 & 0 \\ 7 & 2 \end{bmatrix}.$$

Repeating this reduction once more,

$$\det A = 4 \cdot 3 \cdot (-1) \det[2] = -24.$$

The same result would be produced by expanding in the cofactors of the last column—and it confirms the fact that the determinant of any triangular matrix is the product of its diagonal entries.

EXAMPLE 2 Consider again the finite difference matrix

$$A = \begin{bmatrix} 2 & -1 & & & \\ -1 & 2 & -1 & & \\ & -1 & 2 & \cdot & \\ & & \cdot & \cdot & -1 \\ & & & -1 & 2 \end{bmatrix}.$$

Suppose $\det A$ is computed by an expansion on the first row. The only nonzero entries are $a_{11} = 2$ and $a_{12} = -1$, so that

$$\det A = a_{11}A_{11} + a_{12}A_{12} + \cdots + a_{1n}A_{1n} = 2A_{11} - A_{12}.$$

A_{11} is computed by removing the first row and column, which leaves a sub-matrix exactly like A itself—except that it is of order $n - 1$. The other cofactor is formed by excluding row 1 and column 2:

$$A_{12} = (-1)^{1+2} \det \begin{bmatrix} -1 & -1 & & \\ & 2 & \cdot & \\ & \cdot & \cdot & -1 \\ & & -1 & 2 \end{bmatrix}.$$

This time we expand on the first column, which has only one nonzero entry:

$$A_{12} = (-1)^{1+2}(-1) \det \begin{bmatrix} 2 & \cdot & \\ \cdot & \cdot & -1 \\ & -1 & 2 \end{bmatrix}.$$

Again the same kind of difference matrix appears, but now it is a submatrix of order $n-2$. If D_n is the determinant of A when there are n rows and columns, then we have found a recurrence relation

$$D_n = 2A_{11} - A_{12} = 2D_{n-1} - D_{n-2}.$$

For the small matrices with $n = 1$ and $n = 2$, it is easy to compute

$$D_1 = |\, 2 \,| = 2, \qquad D_2 = \begin{vmatrix} 2 & -1 \\ -1 & 2 \end{vmatrix} = 3.$$

The recurrence relation gives successively $D_3 = 2D_2 - D_1 = 4$, $D_4 = 5$, and in general $D_n = n + 1$:

$$D_n = 2D_{n-1} - D_{n-2} \qquad \text{becomes} \qquad n + 1 = 2(n) - (n - 1).$$

This agrees with the product of pivots computed at the start of this section, and the determinant of A is $n + 1$.

EXERCISE 4.3.1   For the matrix

$$A = \begin{bmatrix} 0 & 1 & 0 & 0 \\ 1 & 0 & 1 & 0 \\ 0 & 1 & 0 & 1 \\ 0 & 0 & 1 & 0 \end{bmatrix}$$

find the only nonzero term to appear in formula (8)—the only way of choosing four entries which come from different rows and different columns, without choosing any zeros. By deciding whether this permutation is even or odd, compute det A.

EXERCISE 4.3.2   Carry out the expansion in cofactors for the first row of the preceding matrix A, and reduce det A to a 3 by 3 determinant. Do the same for that determinant (still watching the sign $(-1)^{i+j}$) and again for the resulting 2 by 2 determinant. Finally compute det A.

EXERCISE 4.3.3   Find a 4 by 4 example in which

$$\det \begin{bmatrix} A & B \\ C & D \end{bmatrix} \neq \det A \det D - \det B \det C.$$

Here A, B, C, and D represent 2 by 2 submatrices—and equality does hold if B or C is zero.

EXERCISE 4.3.4   Compute the determinant of

$$A_4 = \begin{bmatrix} 0 & 1 & 1 & 1 \\ 1 & 0 & 1 & 1 \\ 1 & 1 & 0 & 1 \\ 1 & 1 & 1 & 0 \end{bmatrix}$$

either by using row operations to produce zeros, or by expanding in cofactors of the first row, or otherwise. Find also the determinants of the smaller matrices A_3 and A_2, with the same pattern of zeros on the diagonal and ones elsewhere. Can you predict $\det A_n$?

EXERCISE 4.3.5   Find the determinant and all nine cofactors of

$$A = \begin{bmatrix} 1 & 2 & 3 \\ 0 & 4 & 0 \\ 0 & 0 & 5 \end{bmatrix}.$$

Then form a new matrix B, whose i, j entry is the cofactor A_{ji}. Verify that AB is the identity matrix times the determinant. What is A^{-1}?

4.4 ■ APPLICATIONS OF DETERMINANTS

We shall try to go systematically through each of the applications described in the introduction to the chapter.

1.   *The computation of A^{-1}.*   For this we use the expansion in cofactors of the ith row,

$$\det A = \sum_{j=1}^{n} a_{ij} A_{ij}. \tag{14}$$

We also need to combine the entries from one row with the cofactors from another, and see that the result is zero:

$$0 = \sum_{j=1}^{n} a_{kj} A_{ij} \qquad \text{for} \quad i \neq k. \tag{15}$$

This is because we are really computing the determinant of a different matrix B, formed by throwing away the ith row of A and putting the kth row in its place. This kth row appears twice in B, and a matrix with two equal rows has $\det B = 0$. Then (15) is just the usual expansion along the ith row of B: The cofactors A_{ij} are not changed, since they are independent of row i, but they are multiplied by the entries a_{kj} which went into the ith row of B.

From (14) and (15) we can read off the remarkable multiplication

$$\begin{bmatrix} a_{11} & a_{12} & \cdots & a_{1n} \\ a_{21} & a_{22} & \cdots & a_{2n} \\ \vdots & \vdots & & \vdots \\ a_{n1} & a_{n2} & \cdots & a_{nn} \end{bmatrix} \begin{bmatrix} A_{11} & A_{21} & \cdots & A_{n1} \\ A_{12} & A_{22} & \cdots & A_{n2} \\ \vdots & \vdots & & \vdots \\ A_{1n} & A_{2n} & \cdots & A_{nn} \end{bmatrix}$$

$$= \begin{bmatrix} \det A & 0 & \cdots & 0 \\ 0 & \det A & \cdots & 0 \\ \vdots & \vdots & & \vdots \\ 0 & 0 & \cdots & \det A \end{bmatrix}. \qquad (16)$$

EXAMPLE The cofactors of $A = \begin{bmatrix} a & b \\ c & d \end{bmatrix}$ are $A_{11} = d$, $A_{12} = -c$, $A_{21} = -b$, $A_{22} = a$. These cofactors go into the "adjugate matrix" which is in the middle of Eq. (16), and which we denote by adj A: In the 2 by 2 case

$$\text{adj } A = \begin{bmatrix} A_{11} & A_{21} \\ A_{12} & A_{22} \end{bmatrix} = \begin{bmatrix} d & -b \\ -c & a \end{bmatrix}.$$

Then our multiplication (16) becomes

$$A \text{ adj } A = \begin{bmatrix} a & b \\ c & d \end{bmatrix} \begin{bmatrix} d & -b \\ -c & a \end{bmatrix} = \begin{bmatrix} ad - bc & 0 \\ 0 & ad - bc \end{bmatrix} = (\det A)I.$$

Now we can divide by the number $\det A$ (as long as it is not zero!) and we have the desired inverse:

$$A \frac{\text{adj } A}{\det A} = I, \quad \text{or} \quad A^{-1} = \frac{\text{adj } A}{\det A}. \qquad (17)$$

Every entry in the inverse is *a cofactor of A divided by the determinant of A.* This confirms that A is invertible if and only if $\det A \neq 0$, and gives a precise formula for A^{-1}.

2. *The solution of $Ax = b$.* This second application just amounts to

$$x = A^{-1}b = \frac{(\text{adj } A)b}{\det A}.$$

This is simply the product of a matrix and a vector, divided by the number $\det A$, but there is a famous way in which to write the answer:

4C *Cramer's rule*: The jth component of $x = A^{-1}b$ is

$$x_j = \frac{\det B_j}{\det A}, \quad \text{where} \quad B_j = \begin{bmatrix} a_{11} & a_{12} & b_1 & a_{1n} \\ \vdots & \vdots & \vdots & \vdots \\ a_{n1} & a_{n2} & b_n & a_{nn} \end{bmatrix}. \qquad (18)$$

In B_j, the vector b replaces the jth column of A.

Proof Expanding det B_j in cofactors of the jth column (which is b), formula (13) gives

$$\det B_j = b_1 A_{1j} + b_2 A_{2j} + \cdots + b_n A_{nj}.$$

This is exactly the jth component in the matrix–vector product $(\text{adj } A)b$. Dividing by det A, the result is the jth component of x.

Thus each component of x is a ratio of two determinants, a polynomial of degree n divided by another polynomial of degree n. This fact might have been recognized from Gaussian elimination, but it never was.

EXAMPLE The solution of

$$x_1 + 3x_2 = 0$$

$$2x_1 + 4x_2 = 6$$

is

$$x_1 = \frac{\begin{vmatrix} 0 & 3 \\ 6 & 4 \end{vmatrix}}{\begin{vmatrix} 1 & 3 \\ 2 & 4 \end{vmatrix}} = \frac{-18}{-2} = 9, \qquad x_2 = \frac{\begin{vmatrix} 1 & 0 \\ 2 & 6 \end{vmatrix}}{\begin{vmatrix} 1 & 3 \\ 2 & 4 \end{vmatrix}} = \frac{6}{-2} = -3.$$

3. *The volume of a parallelepiped.* The connection between the determinant and the volume is not at all obvious, but we can suppose first that all angles are right angles—the edges are mutually perpendicular, and we have a rectangular parallelepiped. Then the volume is just the product of the lengths of the edges: volume $= l_1 l_2 \cdots l_n$.

We want to obtain the same formula from the determinant. Recall that the edges of the parallelepiped were the rows of A. In our right-angled case, these rows are mutually orthogonal, and so

$$AA^{\mathrm{T}} = \begin{bmatrix} a_{11} & \cdots & a_{1n} \\ & & \\ \vdots & & \vdots \\ & & \\ a_{n1} & \cdots & a_{nn} \end{bmatrix} \begin{bmatrix} a_{11} & & a_{n1} \\ & & \\ \vdots & & \vdots \\ & & \\ a_{1n} & & a_{nn} \end{bmatrix} = \begin{bmatrix} l_1{}^2 & & 0 & 0 \\ & \cdot & & 0 \\ 0 & & \cdot & \\ 0 & 0 & & l_n{}^2 \end{bmatrix}$$

The $l_i{}^2$ are the lengths of the rows, that is, the edges, and the zeros off the diagonal come because the rows are orthogonal. Taking determinants, and using rules 9 and 10,

$$l_1{}^2 l_2{}^2 \cdots l_n{}^2 = \det(AA^{\mathrm{T}}) = (\det A)(\det A^{\mathrm{T}}) = (\det A)^2.$$

The square root of this equation is the required result: *The determinant equals the volume.* The *sign* of det A will indicate whether the edges form a "right-handed" set of coordinates, as in the usual x-y-z system, or a left-handed system like y-x-z.

If the region is not rectangular, then the volume is no longer the product of the edge lengths. In the plane (Fig. 4.2), the volume of a parallelogram equals the base l_1 times the height h. The vector pb of length h is the second row vector $0b = (a_{21}, a_{22})$, minus its projection $0p$ onto the first row. The key point is this: By property 5 of determinants, $\det A$ is unchanged if a multiple of the first row is subtracted from the second row. At the same time, the volume is unchanged if we switch to a rectangle of base l_1 and height h. So we can return the problem to the rectangular case, where it is already proved that volume = determinant.

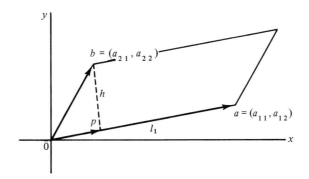

Fig. 4.2. Volume of a parallelogram = $\det A$.

In n dimensions, it takes longer to make each parallelepiped into a rectangular one, but the idea is the same. Neither the volume nor the determinant will be changed if, systematically for rows 2, 3, ..., n, we subtract from each row its projection onto the space spanned by the preceding rows—leaving a "height" vector like pb which is perpendicular to the base. The result of this Gram–Schmidt process is a set of mutually orthogonal rows—but with the same determinant and the same volume as the original set. Since volume = determinant in the rectangular case, the same equality must have held for the original rows.

This completes the link between volumes and determinants, but it is worth coming back one more time to the simplest case. We know that

$$\det \begin{bmatrix} 1 & 0 \\ 0 & 1 \end{bmatrix} = 1, \qquad \det \begin{bmatrix} 1 & 0 \\ c & 1 \end{bmatrix} = 1.$$

These determinants give the volumes—or areas, since we are in two dimensions—of the "parallelepipeds" drawn in Fig. 4.3. The first is the unit square, whose area is certainly 1. The second is a parallelogram with unit base and unit height; independent of the "shearing" produced by the coefficient c, its area is also equal to 1.

4. *A formula for the pivots.* The last application (in this very utilitarian treatment of determinants) is to the question of zero pivots; we can finally

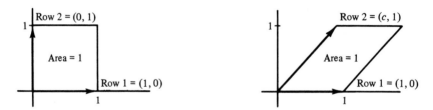

Fig. 4.3. The areas of a square and of a parallelogram.

discover when Gaussian elimination is possible without row exchanges. The key observation is that the first k pivots are completely determined by the submatrix A_k in the upper left corner of A. *The remaining rows and columns of A have no effect on this corner of the problem.*

EXAMPLE

$$A = \begin{bmatrix} a & b & e \\ c & d & f \\ g & h & i \end{bmatrix} \rightarrow \begin{bmatrix} a & b & e \\ 0 & (ad - bc)/a & (af - ec)/a \\ g & h & i \end{bmatrix}.$$

Certainly the first pivot depended only on the first row and column; it was $d_1 = a$. And the second pivot has become visible after a single elimination step; it is $d_2 = (ad - bc)/a$, and it depends only on the entries a, b, c, and d. The rest of A does not enter until the third pivot. Actually it is not just the pivots, but the entire upper left corners of L, D, and U, which are determined by the upper left corner of A:

$$A = LDU = \begin{bmatrix} 1 & & \\ c/a & 1 & \\ * & * & 1 \end{bmatrix} \begin{bmatrix} a & & \\ & (ad - bc)/a & \\ & & * \end{bmatrix} \begin{bmatrix} 1 & b/a & * \\ & 1 & * \\ & & 1 \end{bmatrix}.$$

What we see in the first two rows and columns is exactly the factorization (written out earlier in (4)) of the submatrix $A_2 = \begin{bmatrix} a & b \\ c & d \end{bmatrix}$. This is a general rule:

4D   If A is factored into LDU, then the upper left corners satisfy

$$A_k = L_k D_k U_k. \tag{19}$$

For every k, the submatrix A_k is going through a Gaussian elimination of its own.

The proof is either just to see that this corner can be settled first, before even looking at the eliminations elsewhere, or to use the laws for *block multiplication of matrices*. These laws are the same as the ordinary element by element rule:

$LDU = A$ becomes

$$\begin{bmatrix} L_k & 0 \\ B & C \end{bmatrix}\begin{bmatrix} D_k & 0 \\ 0 & E \end{bmatrix}\begin{bmatrix} U_k & F \\ 0 & G \end{bmatrix} = \begin{bmatrix} L_kD_k & 0 \\ BD_k & CE \end{bmatrix}\begin{bmatrix} U_k & F \\ 0 & G \end{bmatrix}$$

$$= \begin{bmatrix} L_kD_kU_k & L_kD_kF \\ BD_kU_k & BD_kF + CEG \end{bmatrix}.$$

As long as matrices are partitioned to make each multiplication permissible—the square or rectangular submatrices are of the right sizes for multiplication—they can be multiplied this way in blocks.† Comparing the last matrix with A, obviously the corner $L_kD_kU_k$ coincides with A_k, and (19) is correct.

The formulas for the pivots follow immediately. Taking determinants in (19),

$$\det A_k = \det L_k \det D_k \det U_k = \det D_k = d_1d_2 \cdots d_k. \tag{20}$$

The product of the first k pivots is the determinant of A_k; this is the same rule for A_k that we know already for the whole matrix $A = A_n$. Since the determinant of A_{k-1} will be given similarly by $d_1d_2\cdots d_{k-1}$, we can isolate the pivot d_k as a *ratio of determinants*:

$$\frac{\det A_k}{\det A_{k-1}} = \frac{d_1d_2 \cdots d_k}{d_1d_2 \cdots d_{k-1}} = d_k. \tag{21}$$

In our example above, the second pivot was exactly this ratio $(ad - bc)/a$—the determinant of A_2 divided by the determinant of A_1. (By convention $\det A_0 = 1$, so that the first pivot is $a/1 = a$.) Multiplying together all the individual pivots, we recover

$$d_1d_2\cdots d_n = \frac{\det A_1}{\det A_0}\frac{\det A_2}{\det A_1} \cdots \frac{\det A_n}{\det A_{n-1}} = \frac{\det A_n}{\det A_0} = \det A.$$

From (21), we can finally read off the answer to our original question: *The pivots are all nonzero whenever the numbers* $\det A_k$ *are all nonzero*:

4E Gaussian elimination can be carried out on A, without row exchanges or a permutation matrix or zero pivots, if and only if the leading submatrices $A_1, A_2, \ldots, A_n$ are all nonsingular.

That does it for determinants, except for the optional remark promised at the beginning of the chapter. That remark concerns the self-consistency of the defining properties 1–3. The key is property 2, the sign reversal on row exchanges, which led to the rule for the determinant of a permutation matrix P_σ. This was the only questionable point in the explicit formula (8): Is it true that, independent of the particular sequence of row exchanges linking P_σ to

† This is a very useful rule, even though we meet it rather late in the book.

the identity, the number of exchanges is either always even or always odd? If so, we are justified in calling the permutation "even" or "odd," and its determinant is well defined by rule 2 as either $+1$ or -1.

Starting from the permutation $(3, 2, 1)$, a single exchange of 3 and 1 would achieve the natural order $(1, 2, 3)$. So would an exchange of 3 and 2, then 3 and 1, and then 2 and 1. In both sequences, the number of exchanges is odd. The assertion is that *an even number of exchanges can never produce the natural order, beginning with* $(3, 2, 1)$.

Here is a proof. Look at each pair of numbers in the permutation, and let N count the pairs in which the larger number comes first. Certainly $N = 0$ for the natural order $(1, 2, 3)$, the identity permutation; and $N = 3$ for the order $(3, 2, 1)$, since all the pairs $(3, 2)$, $(3, 1)$, and $(2, 1)$ are wrong. Now the point is to show that the permutation is odd or even according as N is odd or even. In other words, starting with any permutation, every exchange will alter N by an odd number. Then to arrive at $N = 0$ (the natural order) takes a number of exchanges having the same parity—evenness or oddness—as the initial N.

If the pair being exchanged lie next to one another, obviously N changes by $+1$ or -1, both of which are odd numbers. Therefore, once it is observed that *any exchange can be achieved by an odd number of exchanges of neighbors*, the proof is complete; an odd number of odd numbers is odd. This observation is easy to confirm by an example; to exchange the first and fourth entries below, which happen to be 2 and 3, we use five exchanges (an odd number) of neighbors:

$$(2, 1, 4, 3) \rightarrow (1, 2, 4, 3) \rightarrow (1, 4, 2, 3)$$
$$\rightarrow (1, 4, 3, 2) \rightarrow (1, 3, 4, 2) \rightarrow (3, 1, 4, 2).$$

In general, to exchange the entries in places k and l, we need $l - k$ exchanges to move the entry in place k to place l, and $l - k - 1$ to move the one originally in place l (and now to be found in place $l - 1$) back down to place k. Since $(l - k) + (l - k - 1)$ is odd, the proof is complete. The determinant not only has all the properties found earlier, it even exists.

EXERCISE 4.4.1 Use (17) to invert

$$A = \begin{bmatrix} 1 & 1 & 1 \\ 0 & 1 & 1 \\ 0 & 0 & 1 \end{bmatrix} \quad \text{and} \quad B = \begin{bmatrix} a & b \\ c & d \end{bmatrix}.$$

EXERCISE 4.4.2 Find x, y, and z by Cramer's rule:

$$x + 4y - z = 1$$
$$x + y + z = 0$$
$$2x + 3z = 0.$$

EXERCISE 4.4.3 Find the Jacobian determinant J for the change from rectangular coordinates x, y, z to spherical coordinates r, θ, ϕ: $x = r \cos \theta \cos \phi$, $y = r \sin \theta \cos \phi$, $z = r \sin \phi$.

EXERCISE 4.4.4 (a) Draw the triangle whose vertices are $A = (2, 2)$, $B = (-1, 3)$, and $C = (0, 0)$. By regarding it as half of a parallelogram, explain why its area equals

$$\text{area } (ABC) = \frac{1}{2} \det \begin{bmatrix} 2 & 2 \\ -1 & 3 \end{bmatrix}.$$

(b) Suppose the third vertex, instead of being at the origin, is at $C = (1, -4)$. Evaluate and then justify the formula

$$\text{area } (ABC) = \frac{1}{2} \det \begin{bmatrix} x_1 & y_1 & 1 \\ x_2 & y_2 & 1 \\ x_3 & y_3 & 1 \end{bmatrix} = \frac{1}{2} \det \begin{bmatrix} 2 & 2 & 1 \\ -1 & 3 & 1 \\ 1 & -4 & 1 \end{bmatrix}.$$

Hint: Subtracting the last row from each of the others leaves

$$\det \begin{bmatrix} 2 & 2 & 1 \\ -1 & 3 & 1 \\ 1 & -4 & 1 \end{bmatrix} = \det \begin{bmatrix} 1 & 6 & 0 \\ -2 & 7 & 0 \\ 1 & -4 & 1 \end{bmatrix} = \det \begin{bmatrix} 1 & 6 \\ -2 & 7 \end{bmatrix}.$$

How are the new vertices $A' = (1, 6)$, $B' = (-2, 7)$, and $C' = (0, 0)$ related to the given A, B, and C? Sketch a figure.

EXERCISE 4.4.5 Explain in terms of volumes why $\det 3A = 3^n \det A$ for an n by n matrix A.

EXERCISE 4.4.6 Predict in advance the pivots of

$$A = \begin{bmatrix} 2 & 1 & 1 \\ 4 & 5 & 0 \\ 0 & 6 & -4 \end{bmatrix},$$

and confirm them by elimination. Is there a row exchange that will avoid any zero pivots?

EXERCISE 4.4.7 Is the permutation of $(1, 2, \ldots, n)$ into $(n, n - 1, \ldots, 1)$ even or odd?

EXERCISE 4.4.8 Find a series of exchanges taking $\sigma = (5, 3, 1, 2, 4)$ into $(1, 2, 3, 4, 5)$. Is σ even or odd?

EXERCISE 4.4.9 Decide whether or not $(1, 3, 2)$, $(2, 4, 4)$, and $(1, 5, 2)$ are linearly independent, from the determinant they form.

REVIEW EXERCISES

4.1   Where can you put zeros into a 4 by 4 matrix, using as few as possible but enough to guarantee that the determinant is zero?

4.2   Where can you put zeros and ones into a 4 by 4 matrix, using as few as possible but enough to guarantee that the determinant is one?

4.3   Find the determinant of

$$\begin{bmatrix} 1 & 1 & 1 & 1 \\ 1 & 1 & 1 & 2 \\ 1 & 1 & 3 & 1 \\ 1 & 4 & 1 & 1 \end{bmatrix}.$$

4.4   If $B = M^{-1}AM$, why is $\det B = \det A$?

4.5   Give a counterexample to $\det(A + B) = \det A + \det B$.

4.6   Starting with a matrix A, multiply its first row by 3 to produce B, and then subtract the first row of B from the second to produce C. How is $\det C$ related to $\det A$?

4.7   Solve $3u + 2v = 7$, $4u + 3v = 11$ by Cramer's rule.

4.8   If the entries of A are integers, and $\det A$ is 1 or -1, how do you know that the entries of A^{-1} are integers?

4.9   If the entries of A and A^{-1} are all integers, how do you know that both determinants are 1 or -1? Hint: What is $\det A$ times $\det A^{-1}$?

4.10   Find all the cofactors, and the inverse, of

$$\begin{bmatrix} 3 & 5 \\ 6 & 9 \end{bmatrix}.$$

4.11   What is the volume of the parallelepiped with four of its vertices at $(0, 0, 0)$, $(-1, 2, 2)$, $(2, -1, 2)$, and $(2, 2, -1)$? Where are the other four vertices?

4.12   How many terms are there in the expansion of a 5 by 5 matrix, and how many are sure to be zero if $a_{21} = 0$?

4.13   If every row of A adds up to zero, and x is a column vector of ones, what is Ax? How do you know that $\det A = 0$?

4.14   Why are there an even number of permutations of $(1, 2, \ldots, 9)$, and why are exactly half of them odd permutations?

EIGENVALUES AND EIGENVECTORS

This chapter begins the "second half" of matrix theory. The first part was almost completely involved with linear systems $Ax = b$, and the fundamental technique was elimination. From now on that technique will play only a minor role. The new problems will still be solved by simplifying a matrix—making it diagonal or upper triangular—but *the basic step is no longer to subtract a multiple of one row from another*. We are not interested any more in preserving the row space of a matrix, but in preserving its eigenvalues. The elementary row operations will not do that.

The chapter on determinants was really a transition from the old problem $Ax = b$ to the new problem of eigenvalues. In both cases the determinant leads to a "formal solution": to Cramer's rule in the case of $Ax = b$ and to the polynomial $\det(A - \lambda I)$ whose roots will be the eigenvalues. (We emphasize that all matrices are now square; the eigenvalues of a rectangular matrix make no more sense than its determinant.) As always, the determinant can actually be used to solve the problem, if $n = 2$ or 3; for large n the computation of eigenvalues is a much longer and more difficult task than solving $Ax = b$, and even Gauss himself did not help much. But that can wait.

The first step is to understand what eigenvalues are and how they can be useful. One of their applications, the one by which we want to introduce them, is to the solution of a system of ordinary differential equations. We shall not assume that the reader is an expert on differential equations; if he can differentiate the usual functions like x^n, $\sin x$, and e^x, he knows more than enough. As a specific example, consider the coupled pair of equations

$$\frac{dv}{dt} = 4v - 5w, \qquad v = 8 \quad \text{at} \quad t = 0,$$

$$\frac{dw}{dt} = 2v - 3w, \qquad w = 5 \quad \text{at} \quad t = 0. \tag{1}$$

This is an *initial-value problem*, as distinguished from the boundary-value problem of Section 1.6. The unknown is specified only at time $t = 0$, and not at both endpoints of an interval; we are interested in a transient rather than a steady state. The system evolves in time from the given initial values, and the problem is to follow this evolution.

It is easy to write the system in matrix form. Let the unknown vector be u, its initial value be u_0, and the coefficient matrix be A:

$$u(t) = \begin{bmatrix} v(t) \\ w(t) \end{bmatrix}, \qquad u_0 = \begin{bmatrix} 8 \\ 5 \end{bmatrix}, \qquad A = \begin{bmatrix} 4 & -5 \\ 2 & -3 \end{bmatrix}.$$

In this notation, the system becomes

$$\frac{du}{dt} = Au, \qquad u = u_0 \quad \text{at} \quad t = 0. \tag{2}$$

This is the basic statement of the problem. Note that it is a first-order equation—no higher derivatives appear—and it is linear in the unknowns. The most general linear first-order initial-value problem would be

$$\frac{du}{dt} = A(t)u + b(t), \qquad u = u_0 \quad \text{at} \quad t = 0. \tag{3}$$

By comparison, our example is homogeneous $(b = 0)$, and it also has constant coefficients; the matrix A is independent of time.

How do we find the solution? If there were only one unknown instead of two, that question would be easy to answer. We would have a scalar instead of a vector differential equation, and if it is again homogeneous with constant coefficients, it can only have the simple form

$$\frac{du}{dt} = au, \qquad u = u_0 \quad \text{at} \quad t = 0. \tag{4}$$

The solution is probably known to the reader:

$$u(t) = e^{at}u_0. \tag{5}$$

At the initial time $t = 0$, u assumes the required value u_0. The equation is unstable if $a > 0$, neutrally stable if $a = 0$, or stable if $a < 0$; the solution either approaches infinity, remains bounded, or goes to zero. If a were a complex number, $a = \alpha + i\beta$, then the same tests would be applied to the *real part* α. The solution is $e^{(\alpha+i\beta)t}u_0$, and the complex part produces oscillations $e^{i\beta t} = \cos \beta t + i \sin \beta t$; but stability is governed by the factor $e^{\alpha t}$.

So much for a single equation. We shall take a direct approach to systems, and look for solutions with the same kind of exponential dependence on t just found in the scalar case. In other words, we look for solutions of the form

$$v(t) = e^{\lambda t}y$$
$$w(t) = e^{\lambda t}z, \tag{6a}$$

or in vector notation

$$u(t) = e^{\lambda t}x. \tag{6b}$$

Substituting this hoped-for solution into the system of differential equations, we find

$$\lambda e^{\lambda t}y = 4e^{\lambda t}y - 5e^{\lambda t}z$$
$$\lambda e^{\lambda t}z = 2e^{\lambda t}y - 3e^{\lambda t}z.$$

The factor $e^{\lambda t}$ is common to every term, and can be removed. This cancellation is the reason for assuming the same exponent λ for both unknowns; if v and w were proportional to different exponentials $e^{\lambda t}$ and $e^{\rho t}$, these factors would appear in both equations and their cancellation would be impossible. As it is, we are left with

$$4y - 5z = \lambda y$$
$$2y - 3z = \lambda z. \tag{7}$$

The substitution of $u = e^{\lambda t}x$ **into** $du/dt = Au$ **gave** $\lambda e^{\lambda t}x = Ae^{\lambda t}x$, **and the cancellation produced**

$$Ax = \lambda x. \tag{8}$$

This is the fundamental equation for the eigenvalue λ and the eigenvector x. Notice that it is nonlinear, because it involves the product of both unknowns λ and x. If we could discover the number λ, then the equation for the vector x alone becomes linear; in fact, we could write λIx in place of λx,† and bring this term over to the left side:

$$(A - \lambda I)x = 0. \tag{9}$$

Evidently the eigenvector x lies in the nullspace of the matrix

$$A - \lambda I = \begin{bmatrix} 4 - \lambda & -5 \\ 2 & -3 - \lambda \end{bmatrix}.$$

† The introduction of the identity matrix is simply to keep matrices and vectors and scalars straight; the equation $(A - \lambda)x = 0$ is shorter, but mixed up.

Here is one key point: For every value of λ, the vector $x = 0$ will always satisfy $Ax = \lambda x$; the nullspace of any matrix contains $x = 0$. But the zero vector is useless in our problem of building the solution $u(t)$ out of exponentials $e^{\lambda t}x$. Therefore *we are interested only in those particular values λ for which there is a nonzero eigenvector x.* To be of any use, the nullspace of $A - \lambda I$ must contain some nonzero vector, and therefore the rank must be something less than the order of the matrix. In short, $A - \lambda I$ *must be singular.*

For this, the determinant gives a conclusive test.

5A   The number λ will be an *eigenvalue* of A, with a corresponding nonzero eigenvector, if and only if

$$\det(A - \lambda I) = 0. \tag{10}$$

This is the *characteristic equation* for the matrix A.

In our example,

$$\det \begin{bmatrix} 4 - \lambda & -5 \\ 2 & -3 - \lambda \end{bmatrix} = (4 - \lambda)(-3 - \lambda) + 10 = \lambda^2 - \lambda - 2.$$

The *characteristic polynomial* $\lambda^2 - \lambda - 2$ factors into $(\lambda + 1)(\lambda - 2)$, and the matrix A has two distinct eigenvalues: $\lambda_1 = -1$ and $\lambda_2 = 2$. To each of these special values there corresponds a space of eigenvectors, satisfying $Ax = \lambda x$ or $(A - \lambda I)x = 0$. The computations for λ_1 and λ_2 are done separately:

$$\lambda_1 = -1: \qquad (A - \lambda_1 I)x = \begin{bmatrix} 5 & -5 \\ 2 & -2 \end{bmatrix}\begin{bmatrix} y \\ z \end{bmatrix} = \begin{bmatrix} 0 \\ 0 \end{bmatrix},$$

and the solution is any multiple of

$$x_1 = \begin{bmatrix} 1 \\ 1 \end{bmatrix};$$

$$\lambda_2 = 2: \qquad (A - \lambda_2 I)x = \begin{bmatrix} 2 & -5 \\ 2 & -5 \end{bmatrix}\begin{bmatrix} y \\ z \end{bmatrix} = \begin{bmatrix} 0 \\ 0 \end{bmatrix},$$

and the solution is any multiple of

$$x_2 = \begin{bmatrix} 5 \\ 2 \end{bmatrix}.$$

The eigenvectors are *not uniquely determined*; any vector in the nullspace of $A - \lambda I$ (which we call the eigenspace corresponding to λ) is an eigenvector, and what we want is a basis for the space. In this example, both eigenspaces are one-dimensional, and they are spanned by x_1 and x_2, respectively.

Returning to the differential equation, we have found two pure exponential solutions:

$$u = e^{\lambda_1 t} x_1 = e^{-t} \begin{bmatrix} 1 \\ 1 \end{bmatrix} \quad \text{and} \quad u = e^{\lambda_2 t} x_2 = e^{2t} \begin{bmatrix} 5 \\ 2 \end{bmatrix}.$$

Because the equation is linear and homogeneous, superposition is permitted; any combination of these two special solutions,

$$u = c_1 e^{\lambda_1 t} x_1 + c_2 e^{\lambda_2 t} x_2 , \tag{11}$$

is again a solution. Now we have two free parameters c_1 and c_2 , and it is reasonable to hope that they can be chosen to satisfy the initial condition $u = u_0$ at $t = 0$:

$$c_1 x_1 + c_2 x_2 = u_0 , \tag{12}$$

or

$$\begin{bmatrix} 1 & 5 \\ 1 & 2 \end{bmatrix} \begin{bmatrix} c_1 \\ c_2 \end{bmatrix} = \begin{bmatrix} 8 \\ 5 \end{bmatrix}. \tag{13}$$

The constants are $c_1 = 3$ and $c_2 = 1$, and *the required solution to the original equation* (1) *is*

$$u(t) = 3e^{-t} \begin{bmatrix} 1 \\ 1 \end{bmatrix} + e^{2t} \begin{bmatrix} 5 \\ 2 \end{bmatrix}. \tag{14}$$

Writing the two components separately, this means that

$$v(t) = 3e^{-t} + 5e^{2t}, \qquad w(t) = 3e^{-t} + 2e^{2t}.$$

The initial conditions $v_0 = 8$ and $w_0 = 5$ are easily checked.

The message seems to be that the key to an equation is in its eigenvalues and eigenvectors. But what the example does not show is their physical significance; they are important in themselves, and not just part of a trick for finding u. Probably the homeliest example† is that of soldiers going over a bridge. Traditionally, they stop marching and just walk across. The reason is that they might happen to march at a frequency equal to one of the eigenvalues of the bridge, and it would begin to oscillate. (Just as a child's swing does; you soon notice the natural frequency of a swing, and by matching it you make the swing go higher.) An engineer tries to keep the natural frequencies of his bridge or his rocket away from those of the wind or the sloshing of fuel. And at the other extreme, a stockbroker spends his life trying to get in line with the natural frequencies of the market. The eigenvalues are the most important feature of practically any dynamical system.

---

† One which I never really believed. But a bridge did crash this way in 1831

We stop now to summarize both what has been done, and what there remains to do. This introduction has shown how the eigenvalues and eigenvectors of A appear naturally and automatically when solving $du/dt = Au$. Such an equation has "pure exponential solutions" $u = e^{\lambda t}x$; the eigenvalue gives the rate of growth or decay, and the eigenvector x develops at this rate. The other solutions will be mixtures of these pure solutions, and the mixture is adjusted to fit the initial conditions.

The key equation was $Ax = \lambda x$. Most vectors x will not satisfy such an equation, whether λ is an eigenvalue or not. A typical x changes direction when it is multiplied by A, so that Ax is not a multiple of x. This means that *only certain special numbers λ are eigenvalues, and only certain special vectors are eigenvectors*. Of course if A were a multiple of the identity matrix, then no vector would change direction, and all vectors would be eigenvectors. But in the usual case, eigenvectors are few and far between.

EXAMPLE   Imagine any rotation of the earth. Then there will always be one direction that stays unchanged, namely the axis of rotation. This is not necessarily the one that the real earth rotates around, but there must be *some* north pole and south pole that stay fixed.† These poles are eigenvectors—with eigenvalue equal to 1. In general, all other points are moved and no more eigenvectors are visible. The only exceptions occur when the rotation is through 360° (with eigenvectors everywhere) or through 180°. In the 180° case, the plane of the equator is filled with eigenvectors; every direction in this plane is exactly reversed, and the eigenvalue is -1. This equatorial plane is a case of a "two-dimensional eigenspace"; the single eigenvalue $\lambda = -1$ has two independent eigenvectors, and therefore a plane of eigenvectors.

In some ways this is a good example, in other ways not. For one thing, eigenvalues are not generally ± 1; vectors are usually stretched or shrunk. More important, this was a rotation of three-dimensional space $\mathbf{R}^3$, but except for 180° and 360°, *there was only one line of eigenvectors and we expect to find three*. In an application to differential equations, we would need three different special solutions to match the initial conditions; one eigenvalue and one eigenvector are not enough. The way to find the other two is to admit the imaginary number i. If there are too few solutions in the real world $\mathbf{R}^3$, we look in the space $\mathbf{C}^3$ of vectors with complex components. Allowing complex numbers, any n by n matrix will have n eigenvalues. The three eigenvalues and eigenvectors for this rotated earth example are found in Section 5.2, and in Section 5.5 we make a complete conversion from real vectors and matrices to the complex case.

The real substance of this chapter, however, lies somewhere else. The most important thing to be done is to explain how a system of equations is decoupled by finding the eigenvectors. These eigenvectors are the "normal modes" of the

---

† I guess this is not so obvious, but it is true—you cannot turn the earth in a way that moves every point.

system, and they act independently. We can watch the behavior of each eigenvector separately, and then combine these normal modes to find the solution. To say the same thing in another way, *the underlying matrix has been diagonalized.*

We plan to devote Section 5.2 to the theory of diagonalization, and the following sections to its applications: first to difference equations and Fibonacci numbers and Markov processes, and afterward to differential equations. In every example, we have to start by computing the eigenvalues and eigenvectors; there is no shortcut to avoid that. But then the examples go in so many directions that a quick summary is impossible, except to emphasize that symmetric matrices are especially easy and that certain other "defective matrices" are especially hard. They lack a full set of eigenvectors, they are not diagonalizable, and they produce a breakdown in the technique of normal modes. Certainly they have to be discussed, but we do not intend to allow them to take over the book.

To conclude this introduction, we review some of the most basic facts about eigenvalues and eigenvectors: $Ax = \lambda x$. The n by n matrix A is given, and the problem is to find those special vectors x on which A acts like a simple multiplication; Ax points in the same direction as x. The trick is to find the eigenvalues first, by rewriting λx as $\lambda I x$ and moving this term to the left side: $(A - \lambda I)x = 0$. An eigenvector of A is the same as a nullvector of $A - \lambda I$. In other words, the key question is this: If A is shifted by various multiples of the identity matrix, *which shifts make it singular?* These shifts will reveal the eigenvalues, and then we can compute the eigenvectors.

If A is already singular, then one possibility is not to shift it at all. One of the eigenvalues of a singular matrix is $\lambda = 0$, and the nullspace contains the corresponding eigenvectors. But unlike the first half of this book, there is no longer anything special about A being singular; all it means is that $\lambda = 0$ is an eigenvalue. Whether A itself is singular or not, all of its eigenvalues are on the same footing: The combination $A - \lambda I$ is singular, and the nullspace of that combination is the space of eigenvectors corresponding to λ.

To decide when $A - \lambda I$ is singular, we compute its determinant. This determinant is a polynomial in λ of degree n, known as the *characteristic polynomial* of A. The equation $\det(A - \lambda I) = 0$ is the *characteristic equation*, and its roots λ_1, λ_2, ..., λ_n (which may or may not be real numbers, and may or may not include some repetitions of the same λ) are the eigenvalues of A. To summarize:

5B   Each of the following conditions is necessary and sufficient for the number λ to be an eigenvalue of A:

(1)   There is a nonzero vector x such that $Ax = \lambda x$.
(2)   The matrix $A - \lambda I$ is singular.
(3) $\det(A - \lambda I) = 0$.

EXAMPLE   Consider the symmetric matrix

$$A = \begin{bmatrix} 1 & -1 & 0 \\ -1 & 2 & -1 \\ 0 & -1 & 1 \end{bmatrix}.$$

Its characteristic polynomial is

$$\det (A - \lambda I) = \begin{vmatrix} 1 - \lambda & -1 & 0 \\ -1 & 2 - \lambda & -1 \\ 0 & -1 & 1 - \lambda \end{vmatrix} = -\lambda^3 + 4\lambda^2 - 3\lambda.$$

Therefore the characteristic equation is

$$-\lambda^3 + 4\lambda^2 - 3\lambda = -\lambda(\lambda - 1)(\lambda - 3) = 0,$$

and the eigenvalues are real and distinct: $\lambda_1 = 0$, $\lambda_2 = 1$, $\lambda_3 = 3$. Separately for each eigenvalue λ_i, we look for a corresponding eigenvector x_i.

$$\lambda_1 = 0: \quad (A - 0I)x_1 = \begin{bmatrix} 1 & -1 & 0 \\ -1 & 2 & -1 \\ 0 & -1 & 1 \end{bmatrix} x_1 = \begin{bmatrix} 0 \\ 0 \\ 0 \end{bmatrix} \quad \text{and} \quad x_1 = \begin{bmatrix} 1 \\ 1 \\ 1 \end{bmatrix}.$$

$$\lambda_2 = 1: \quad (A - I)x_2 = \begin{bmatrix} 0 & -1 & 0 \\ -1 & 1 & -1 \\ 0 & -1 & 0 \end{bmatrix} x_2 = \begin{bmatrix} 0 \\ 0 \\ 0 \end{bmatrix}, \quad \text{and} \quad x_2 = \begin{bmatrix} 1 \\ 0 \\ -1 \end{bmatrix}.$$

$$\lambda_3 = 3: \quad (A - 3I)x_3 = \begin{bmatrix} -2 & -1 & 0 \\ -1 & -1 & -1 \\ 0 & -1 & -2 \end{bmatrix} x_3 = \begin{bmatrix} 0 \\ 0 \\ 0 \end{bmatrix}, \quad \text{and} \quad x_3 = \begin{bmatrix} 1 \\ -2 \\ 1 \end{bmatrix}.$$

This amount of calculation is unavoidable, and of course we had to arrange for a nice example in the first place; most cubic polynomials will not factor as easily as this one did. There is no doubt that the eigenvalue problem is algebraically and computationally much more difficult than $Ax = b$. For a linear system, a finite number of elimination steps produced the exact answer in a finite time. (Or equivalently, Cramer's rule gave an exact formula for the solution.) In the case of eigenvalues, no such steps and no such formula can exist, or Galois would turn in his grave: The characteristic polynomial of a 5 by 5 matrix is a quintic, and he proved that there can be no algebraic formula for the roots of a quintic polynomial. All he will allow is a few simple checks on the eigenvalues, *after* they have been computed, and we mention two of them:

5C   The sum of the n eigenvalues equals the sum of the n diagonal entries of A:

$$\lambda_1 + \cdots + \lambda_n = a_{11} + \cdots + a_{nn}.$$

This sum is known as the *trace* of A. Furthermore, the product of the n eigenvalues equals the determinant of A.

This is confirmed by the example above, where the trace of A is $0 + 1 + 3 = 1 + 2 + 1 = 4$, and the determinant is $0 \cdot 1 \cdot 3 = 0$.

EXERCISE 5.1.1   Find the eigenvalues and eigenvectors of the matrix $A = \begin{bmatrix} 1 & -1 \\ 2 & 4 \end{bmatrix}$. Verify that the trace equals the sum of the eigenvalues, and that the determinant equals their product.

EXERCISE 5.1.2   With the same matrix A, solve the differential equation $du/dt = Au$, $u_0 = \begin{bmatrix} 0 \\ 6 \end{bmatrix}$.

EXERCISE 5.1.3   Suppose we shift the preceding A by subtracting $7I$:

$$B = A - 7I = \begin{bmatrix} -6 & -1 \\ 2 & -3 \end{bmatrix}.$$

What are the eigenvalues and eigenvectors of B, and how are they related to those of A?

EXERCISE 5.1.4   Using the 3 by 3 example whose eigenvalues and eigenvectors were computed above, choose constants c_i so that the solution

$$u = c_1 e^{\lambda_1 t} x_1 + c_2 e^{\lambda_2 t} x_2 + c_3 e^{\lambda_3 t} x_3$$

will match the initial condition $u_0 = \begin{bmatrix} 1 & 3 & 1 \end{bmatrix}^{\mathrm{T}}$.

There should be no confusion between the eigenvalues of a matrix and its diagonal entries. Normally they are completely different. Nevertheless, at the risk of introducing a little confusion where there was none, we point out one situation where these numbers coincide.

5D   If the matrix A is triangular—it may be either upper or lower triangular, and in particular it may be diagonal—then the eigenvalues $\lambda_1, \ldots, \lambda_n$ are exactly the same as the diagonal entries $a_{11}, \ldots, a_{nn}$.

The reason will be obvious from an example. If

$$A = \begin{bmatrix} 1 & \frac{1}{4} & 0 \\ 0 & \frac{3}{4} & \frac{1}{2} \\ 0 & 0 & \frac{1}{2} \end{bmatrix},$$

then its characteristic polynomial is

$$
\det \begin{bmatrix} 1 - \lambda & \frac{1}{4} & 0 \\ 0 & \frac{3}{4} - \lambda & \frac{1}{2} \\ 0 & 0 & \frac{1}{2} - \lambda \end{bmatrix} = (1 - \lambda)\,(\tfrac{3}{4} - \lambda)\,(\tfrac{1}{2} - \lambda).
$$

The determinant is just the product of the diagonal entries. Obviously the roots are $\lambda = 1$, $\lambda = \frac{3}{4}$, and $\lambda = \frac{1}{2}$; the eigenvalues were already sitting along the main diagonal.

EXERCISE 5.1.5   Find the eigenvectors of this triangular matrix, and the eigenvalues as well as the eigenvectors of the diagonal matrix

$$
A = \begin{bmatrix} 3 & 0 & 0 \\ 0 & 1 & 0 \\ 0 & 0 & 2 \end{bmatrix}.
$$

These examples, in which the eigenvalues can be found by inspection, point to the main theme of the whole chapter: To transform A into a diagonal or triangular matrix *without changing its eigenvalues.* We emphasize once more that the Gaussian factorization $A = LU$ is not suited to this purpose. The eigenvalues of U may be visible on the diagonal, but they are not the eigenvalues of A.

There is one more situation in which the calculations are easy. Suppose we have already found the eigenvalues and eigenvectors of a matrix A. Then **the eigenvalues of A^2 are exactly $\lambda_1^2, \ldots, \lambda_n^2$, and every eigenvector of A is also an eigenvector of A^2.** The proof is typical of mathematics: If we try to study $\det(A^2 - \lambda I)$ we are sunk; but if we start from $Ax = \lambda x$, the whole thing is obvious. Multiplying again by A,

$$
A^2x = A\lambda x = \lambda A x = \lambda^2 x.
$$

Thus λ^2 is an eigenvalue of A^2, with the same eigenvector x. If the first multiplication by A leaves the direction of x unchanged, then so does the second.

This reasoning does not apply when two different matrices are involved. Suppose λ is an eigenvalue of A, and μ is an eigenvalue of B. Then *in general $\lambda\mu$ is not an eigenvalue of AB.* An attempted proof would be this: If $Ax = \lambda x$ and $Bx = \mu x$, then $ABx = A\mu x = \mu Ax = \mu\lambda x$. The fallacy lies in assuming that A and B share the same eigenvector x. In general, they do not.

EXERCISE 5.1.6   Give an example to show that the eigenvalues can be changed when a multiple of one row is subtracted from another.

EXERCISE 5.1.7   Suppose that λ is an eigenvalue of A, and x is an associated eigenvector: $Ax = \lambda x$.

(a) Show that this same x is an eigenvector of $B = A - 7I$, and find the eigenvalue. This should confirm Exercise 5.1.3.

(b) Assuming $\lambda \neq 0$, show that x is also an eigenvector of A^{-1}—and again find the eigenvalue.

EXERCISE 5.1.8 Show that the determinant equals the product of the eigenvalues by imagining that the characteristic polynomial is factored into

$$\det(A - \lambda I) = (\lambda_1 - \lambda)(\lambda_2 - \lambda) \cdots (\lambda_n - \lambda), \tag{15}$$

and making a clever choice of λ.

EXERCISE 5.1.9 Show that the trace equals the sum of the eigenvalues, in two steps. First, find the coefficient of $(-\lambda)^{n-1}$ on the right side of (15). Next, look for all the terms in

$$\det (A - \lambda I) = \det \begin{bmatrix} a_{11} - \lambda & a_{12} & \cdots & a_{1n} \\ a_{21} & a_{22} - \lambda & \cdots & a_{2n} \\ \vdots & \vdots & & \vdots \\ a_{n1} & a_{n2} & \cdots & a_{nn} - \lambda \end{bmatrix}$$

which involve $(-\lambda)^{n-1}$. Explain why they all come from the product down the main diagonal, and find the coefficient of $(-\lambda)^{n-1}$ on the left side of (15). Compare.

EXERCISE 5.1.10 (a) Construct 2 by 2 matrices such that the eigenvalues of AB are not the products of the eigenvalues of A and B, and the eigenvalues of $A + B$ are not the sums of the individual eigenvalues.

(b) Verify however that the sum of the eigenvalues of $A + B$ equals the sum of all the individual eigenvalues of A and B, and similarly for products. Why is this true?

EXERCISE 5.1.11 Prove that A and A^{T} have the same eigenvalues, by comparing their characteristic polynomials.

EXERCISE 5.1.12 Find the eigenvalues and eigenvectors of $A = \begin{bmatrix} 3 & 4 \\ 4 & -3 \end{bmatrix}$.

EXERCISE 5.1.13 Find the eigenvalues, and check that their sum is the trace, for

$$A = \begin{bmatrix} 0 & 0 & 2 \\ 0 & 0 & 0 \\ 2 & 0 & 0 \end{bmatrix}.$$

THE DIAGONAL FORM OF A MATRIX ■ 5.2

We start right off with the one essential computation. It is perfectly simple and will be used in every section of this chapter.

5E Suppose the n by n matrix A has n linearly independent eigenvectors. Then if these vectors are chosen to be the columns of a matrix S, it follows that $S^{-1}AS$ is a diagonal matrix Λ, with the eigenvalues of A along its diagonal:

$$S^{-1}AS = \Lambda = \begin{bmatrix} \lambda_1 & & & \\ & \lambda_2 & & \\ & & \cdot & \\ & & & \cdot \\ & & & & \lambda_n \end{bmatrix}. \tag{16}$$

Proof Put the eigenvectors x_i in the columns of S, and compute the product AS one column at a time:

$$AS = A \begin{bmatrix} | & | & & | \\ x_1 & x_2 & \cdots & x_n \\ | & | & & | \end{bmatrix} = \begin{bmatrix} | & | & & | \\ \lambda_1 x_1 & \lambda_2 x_2 & \cdots & \lambda_n x_n \\ | & | & & | \end{bmatrix}.$$

Then the trick is to split this last matrix into a quite different product:

$$\begin{bmatrix} | & | & & | \\ \lambda_1 x_1 & \lambda_2 x_2 & \cdots & \lambda_n x_n \\ | & | & & | \end{bmatrix} = \begin{bmatrix} | & | & & | \\ x_1 & x_2 & \cdots & x_n \\ | & | & & | \end{bmatrix} \begin{bmatrix} \lambda_1 & & & \\ & \lambda_2 & & \\ & & \cdot & \\ & & & \cdot \\ & & & & \lambda_n \end{bmatrix}.$$

Regarded simply as an exercise in matrix multiplication, it is crucial to keep these matrices in the right order. If Λ came before S instead of after, then λ_1 would multiply the entries in the first row, whereas we want it to appear in the first column. As it is, we have the correct product $S\Lambda$. Therefore

$$AS = S\Lambda, \quad \text{or} \quad S^{-1}AS = \Lambda, \quad \text{or even} \quad A = S\Lambda S^{-1}. \tag{17}$$

The matrix S is invertible, because its columns (the eigenvectors) were assumed to be linearly independent.

We add four remarks before giving any examples or applications.

Remark 1 If the matrix A has no repeated eigenvalues—the numbers $\lambda_1, \ldots,$ λ_n are distinct—then the n eigenvectors are automatically independent (see 5F below). Therefore *any matrix with distinct eigenvalues can be diagonalized.*

Remark 2 The diagonalizing matrix S is *not unique*. In the first place, an eigenvector x can be multiplied by a constant, and will remain an eigenvector.

Therefore we can multiply the columns of S by any nonzero constants, and produce a new diagonalizing S. Repeated eigenvalues leave even more freedom, and for the trivial example $A = I$, any invertible S will do: $S^{-1}IS$ is always diagonal (and the diagonal matrix Λ is just I). This reflects the fact that all vectors are eigenvectors of the identity.

Remark 3   The equation $AS = S\Lambda$ holds if the columns of S are the eigenvectors of A, and not otherwise. Other matrices S will not produce a diagonal Λ. The reason lies in the rules for matrix multiplication. Suppose the first column of S is some vector y. Then the first column of $S\Lambda$ is $\lambda_1 y$. If this is to agree with the first column of AS, which by matrix multiplication is Ay, then y must be an eigenvector: $Ay = \lambda_1 y$. In fact, the order of appearance of the eigenvectors in S and the eigenvalues in Λ is automatically the same.

Remark 4   Not all matrices possess n linearly independent eigenvectors, and therefore *not all matrices are diagonalizable*. The standard example of a "defective matrix" is

$$A = \begin{bmatrix} 0 & 1 \\ 0 & 0 \end{bmatrix}.$$

Its eigenvalues are $\lambda_1 = \lambda_2 = 0$, since it is triangular:

$$\det(A - \lambda I) = \det \begin{bmatrix} -\lambda & 1 \\ 0 & -\lambda \end{bmatrix} = \lambda^2.$$

If x is an eigenvector, then it must satisfy

$$\begin{bmatrix} 0 & 1 \\ 0 & 0 \end{bmatrix} x = \begin{bmatrix} 0 \\ 0 \end{bmatrix}, \quad \text{or} \quad x = \begin{bmatrix} x_1 \\ 0 \end{bmatrix}.$$

Although $\lambda = 0$ is a double eigenvalue—its *algebraic multiplicity* is 2—it has only a one-dimensional space of eigenvectors. The *geometric multiplicity* of this eigenvalue is 1, and we cannot construct S.

Here is a more direct proof that A is not diagonalizable. Since $\lambda_1 = \lambda_2 = 0$, Λ would have to be the zero matrix. But if $S^{-1}AS = 0$, then we premultiply by S and postmultiply by S^{-1}, to deduce that $A = 0$. Since A is not 0, the contradiction proves that no S can achieve $S^{-1}AS = \Lambda$.

Some matrices with repeated eigenvalues can be diagonalized (for example, $A = I$); others are defective. The only test is to compute all the eigenvectors, and look to see whether there are enough. There is no difficulty when the eigenvalues are distinct; every algebraic and geometric multiplicity equals 1. But if an eigenvalue λ is repeated m times, then everything hinges on the null-space of $A - \lambda I$; the test is passed only if there are m corresponding eigenvectors. When all the eigenvalues pass this test, there is a full set of eigenvectors and A can be diagonalized.

To complete this circle of ideas, we have to prove a useful but not very exciting theorem:

5F   If the nonzero eigenvectors $x_1, \ldots, x_k$ correspond to different eigenvalues $\lambda_1, \ldots, \lambda_k$, then those eigenvectors are linearly independent.

Suppose first that $k = 2$, and that some combination of x_1 and x_2 produces zero: $c_1x_1 + c_2x_2 = 0$. Multiplying by A, we find $c_1\lambda_1x_1 + c_2\lambda_2x_2 = 0$. Subtracting λ_2 times the previous equation, the vector x_2 disappears:

$$c_1(\lambda_1 - \lambda_2)x_1 = 0.$$

Since $\lambda_1 \neq \lambda_2$ and $x_1 \neq 0$, we are forced into $c_1 = 0$. Similarly $c_2 = 0$, and the two vectors are independent; only the trivial combination gives zero.

This same argument extends to any number of eigenvectors: We assume some combination produces zero, multiply by A, subtract λ_k times the original combination, and the vector x_k disappears—leaving a combination of $x_1, \ldots,$ x_{k-1} which produces zero. By repeating the same steps (or by saying the words *mathematical induction*) we end up with a multiple of x_1 that produces zero. This forces $c_1 = 0$, and ultimately every $c_i = 0$. Therefore eigenvectors that come from distinct eigenvalues are automatically independent.

EXERCISE 5.2.1   For the matrix

$$A = \begin{bmatrix} 1 & -1 & 0 \\ -1 & 2 & -1 \\ 0 & -1 & 1 \end{bmatrix}$$

on p. 178, find S and compute $S^{-1}AS$. Also, diagonalize $B = \begin{bmatrix} 0 & 1 \\ 1 & 0 \end{bmatrix}$.

EXERCISE 5.2.2   Find the matrix A whose eigenvalues are 1 and 4, and whose eigenvectors are $\begin{bmatrix} 3 \\ 1 \end{bmatrix}$ and $\begin{bmatrix} 2 \\ 1 \end{bmatrix}$, respectively. (Hint: $A = S\Lambda S^{-1}$.)

EXERCISE 5.2.3   Find all the eigenvalues and eigenvectors of

$$A = \begin{bmatrix} 1 & 1 & 1 \\ 1 & 1 & 1 \\ 1 & 1 & 1 \end{bmatrix}$$

and write down two different diagonalizing matrices S.

EXERCISE 5.2.4   By transposing $S^{-1}AS = \Lambda$, find the matrix that will diagonalize A^{T}, and the diagonal matrix it produces.

EXERCISE 5.2.5   If A and B share the same eigenvector matrix S, so that $A = S\Lambda_1 S^{-1}$ and $B = S\Lambda_2 S^{-1}$, prove that $AB = BA$. (Remember that $\Lambda_1\Lambda_2 = \Lambda_2\Lambda_1$ because they are diagonal.) If $A = \begin{bmatrix} 1 & 1 \\ 1 & 1 \end{bmatrix}$, find an example of such a B.

The previous exercise is important in quantum mechanics: Matrices with the same eigenvectors must commute. Its converse is also true, and even more important: *If $AB = BA$, then these matrices share the same eigenvectors.* The key step is to notice that $Ax = \lambda x$ implies $ABx = BAx = B\lambda x = \lambda Bx$. Thus x and Bx are eigenvectors sharing the same λ, and if we assume for convenience that the eigenvalues of A are distinct—the eigenspaces are all one dimensional— then Bx must be a multiple of x. In other words x is an eigenvector of B as well as A, which completes the proof.

In the introduction to this chapter, we spoke about rotations of the earth. A rotation through 90° leaves only one direction unchanged; the axis that connects the poles was the only eigenvector we could find. In the plane of the equator, the x axis is rotated into the y axis, and the y direction goes into what was originally the negative x direction. If the coordinates of a point on the earth were originally (x_0, y_0, z_0), then after rotation they are $(-y_0, x_0, z_0)$. It is easy to see that this change of coordinates could be carried out by a matrix multiplication:

$$\begin{bmatrix} -y_0 \\ x_0 \\ z_0 \end{bmatrix} = \begin{bmatrix} 0 & -1 & 0 \\ 1 & 0 & 0 \\ 0 & 0 & 1 \end{bmatrix} \begin{bmatrix} x_0 \\ y_0 \\ z_0 \end{bmatrix}.$$

We propose to diagonalize this rotation matrix A. The first step is to find its eigenvalues and eigenvectors, starting with the characteristic polynomial

$$\det(A - \lambda I) = \begin{vmatrix} -\lambda & -1 & 0 \\ 1 & -\lambda & 0 \\ 0 & 0 & 1 - \lambda \end{vmatrix} = (1 - \lambda)(\lambda^2 + 1).$$

The roots of this polynomial are the eigenvalues, and *two of them are imaginary even though the matrix is real*:

$$\lambda_1 = 1, \qquad \lambda_2 = i, \qquad \lambda_3 = -i.$$

Since they are distinct (even though imaginary!), the matrix is certain to be diagonalizable. There must be a full set of eigenvectors, and they are computed as usual:

$$\lambda_1 = 1: \qquad (A - I)x_1 = \begin{bmatrix} -1 & -1 & 0 \\ 1 & -1 & 0 \\ 0 & 0 & 0 \end{bmatrix} x_1 = 0, \qquad \text{or} \qquad x_1 = \begin{bmatrix} 0 \\ 0 \\ 1 \end{bmatrix}.$$

This eigenvector is just the north-south axis which stays fixed.

$$\lambda_2 = i: \quad (A - iI)x_2 = \begin{bmatrix} -i & -1 & 0 \\ 1 & -i & 0 \\ 0 & 0 & 1-i \end{bmatrix} x_2 = 0, \quad \text{or} \quad x_2 = \begin{bmatrix} 1 \\ -i \\ 0 \end{bmatrix}.$$

$$\lambda_3 = -i: \quad (A + iI)x_3 = \begin{bmatrix} i & -1 & 0 \\ 1 & i & 0 \\ 0 & 0 & 1+i \end{bmatrix} x_3 = 0, \quad \text{or} \quad x_3 = \begin{bmatrix} 1 \\ i \\ 0 \end{bmatrix}.$$

Therefore the diagonalizing matrix S and the diagonal Λ are

$$S = \begin{bmatrix} 0 & 1 & 1 \\ 0 & -i & i \\ 1 & 0 & 0 \end{bmatrix}, \quad \Lambda = \begin{bmatrix} 1 & 0 & 0 \\ 0 & i & 0 \\ 0 & 0 & -i \end{bmatrix}.$$

The theory guarantees that $S^{-1}AS$ coincides with Λ.

This is an example in which we can make sense of A^2. Geometrically, it is a rotation through 180°; that is the effect of applying A twice. Algebraically, $A = S\Lambda S^{-1}$ leads to $A^2 = S\Lambda S^{-1}S\Lambda S^{-1} = S\Lambda^2 S^{-1}$. This confirms what we already know, that the eigenvector matrix S is the same for A and A^2, and that each of the eigenvalues is squared. The eigenvalues of a 180° rotation are $1^2 = 1$, $i^2 = -1$, and $(-i)^2 = -1$; the repetition of -1 confirms that the whole equatorial plane is reversed. We can go on to compute A^4, which is a complete rotation through 360°:

$$A^4 = (S\Lambda S^{-1})(S\Lambda S^{-1})(S\Lambda S^{-1})(S\Lambda S^{-1}) = S\Lambda^4 S^{-1}.$$

Remembering that $i^4 = 1$, the eigenvalue matrix is $\Lambda^4 = I$, and $A^4 = SIS^{-1} = I$. A 360° rotation is the identity.

EXERCISE 5.2.6 Can you find a square root R of the 90° rotation matrix A, and verify that $R^2 = A$? Remember that R and R^2 share the same eigenvectors, in other words the same S. There are actually eight possible square roots R, because $\sqrt{1} = \pm 1$, $\sqrt{i} = \pm(1+i)/\sqrt{2}$, $\sqrt{-i} = \pm(1-i)/\sqrt{2}$.

EXERCISE 5.2.7 Decide whether $A = \begin{bmatrix} 1 & 1 \\ -1 & -1 \end{bmatrix}$ has two eigenvectors and can be diagonalized, or has one eigenvector and cannot.

5.3 ■ DIFFERENCE EQUATIONS AND THE POWERS A^k

Difference equations are not as well known as differential equations, but they should be. They move forward in a finite number of finite steps, while a differential equation takes an infinite number of infinitesimal steps—but the two theories stay absolutely in parallel. It is the same analogy between the discrete and the continuous that appears over and over in mathematics. Perhaps the best illustration is one which really does not involve n-dimensional linear algebra, because money in a bank is only a scalar.

Suppose you invest \$1000 for five years at 6% interest. If it is compounded once a year, then the principal is multiplied by 1.06, and $P_{k+1} = 1.06P_k$. *This is a difference equation with a time step of one year.* It relates the principal after $k + 1$ years to the principal the year before, and it is easy to solve: After 5 years, the original principal $P_0 = 1000$ has been multiplied 5 times, and

$$P_5 = (1.06)^5 P_0 = (1.06)^5 1000 = \$1338.$$

Now suppose the time step is reduced to a month. The new difference equation is $p_{k+1} = (1 + .06/12)p_k$. After 5 years, or 60 months,

$$p_{60} = \left(1 + \frac{.06}{12}\right)^{60} p_0 = (1.005)^{60} 1000 = \$1349.$$

The next step is to compound the interest daily:

$$\left(1 + \frac{.06}{365}\right)^{5 \cdot 365} 1000 = \$1349.83.$$

Finally, to keep their employees really moving, banks now offer *continuous compounding*. The interest is added on at every instant, and the difference equation breaks down. In fact you can hope that the treasurer does not know calculus, and cannot figure out what he owes you. But he has two different possibilities: Either he can compound the interest more and more frequently, and see that the limit is

$$\left(1 + \frac{.06}{N}\right)^{5N} 1000 \rightarrow e^{.30} 1000 = \$1349.87.$$

Or he can switch to a differential equation, which will be the limit of the difference equation $p_{k+1} = (1 + .06\,\Delta t)p_k$. Moving p_k to the left side and dividing by the time step Δt,

$$\frac{p_{k+1} - p_k}{\Delta t} = .06p_k \qquad \text{approaches} \qquad \frac{dp}{dt} = .06p.$$

The solution is $p(t) = e^{.06t} p_0$, and after 5 years this again amounts to \$1349.87. The principal stays finite, even when it is compounded every instant—and the difference is only four cents.

This example included both difference equations and differential equations, with one approaching the other as the time step disappeared. But there are plenty of difference equations that stand by themselves, and our second example comes from the famous *Fibonacci sequence*:

$$0, 1, 1, 2, 3, 5, 8, 13, \ldots..$$

Probably you see the pattern: Every Fibonacci number is the sum of its two predecessors,

$$F_{k+2} = F_{k+1} + F_k . \tag{18}$$

That is the difference equation. It turns up in a most fantastic variety of applications, and deserves a book of its own. Thorns and leaves grow in a spiral pattern, and on the hawthorn or apple or oak you find five growths for every two turns around the stem. The pear tree has eight for every three turns, and the willow is even more complicated, 13 growths for every five spirals. The champion seems to be a sunflower of Daniel T. O'Connell (*Scientific American*, November 1951) whose seeds chose an almost unbelievable ratio of $F_{12}/F_{13} = 144/233$.†

How could we find the 1000th Fibonacci number, other than by starting with $F_0 = 0$, $F_1 = 1$, and working all the way out to F_{1000}? The goal is to solve the difference equation $F_{k+2} = F_{k+1} + F_k$, and as a first step it can be reduced to a "one-step equation" $u_{k+1} = Au_k$. This is just like compound interest, $P_{k+1} = 1.06P_k$, except that now the unknown has to be a vector and the multiplier A has to be a matrix: if

$$u_k = \begin{bmatrix} F_{k+1} \\ F_k \end{bmatrix},$$

then

$$\begin{matrix} F_{k+2} = F_{k+1} + F_k \\ F_{k+1} = F_{k+1} \end{matrix} \quad \text{becomes} \quad u_{k+1} = \begin{bmatrix} 1 & 1 \\ 1 & 0 \end{bmatrix} u_k .$$

This conversion is a standard device for any equation of order s; $s - 1$ trivial equations like $F_{k+1} = F_{k+1}$ combine with the given equation to produce a one-step system. For Fibonacci, $s = 2$.

Formally, the difference equation $u_{k+1} = Au_k$ is easy to solve. Since every step brings a multiplication by A, **the solution u_k is related to the initial value u_0 by $u_k = A^k u_0$**. The real problem is to find some quick way to compute the powers A^k, and thereby find the 1000th Fibonacci number. The key lies in the eigenvalues and eigenvectors of A:

---

† For these botanical applications, see D'Arcy Thompson's book *On Growth and Form* (Cambridge Univ. Press, London and New York, 1942) or Peter Stevens' beautiful *Patterns in Nature* (Little, Brown, 1974). Hundreds of other properties of the F_n have been published in the *Fibonacci Quarterly*. Apparently Fibonacci himself was also the first to bring Arabic numerals into Europe, about 1200 A.D.

5G If A can be diagonalized, $A = S\Lambda S^{-1}$, then automatically

$$u_k = A^k u_0 = (S\Lambda S^{-1})(S\Lambda S^{-1}) \cdots (S\Lambda S^{-1}) u_0 = S\Lambda^k S^{-1} u_0 . \tag{19}$$

Each S^{-1} cancels an S, except for the first S and the last S^{-1}. The columns of S are the eigenvectors x_i of the matrix A, and matrix multiplication leads to

$$u_k = \begin{bmatrix} x_1 & \cdots & x_n \end{bmatrix} \begin{bmatrix} \lambda_1^k & & \\ & \ddots & \\ & & \lambda_n^k \end{bmatrix} S^{-1} u_0 = c_1 \lambda_1^k x_1 + \cdots + c_n \lambda_n^k x_n . \tag{20}$$

The general solution is a combination of the special solutions $\lambda_i^k x_i$, and the coefficients c_i that match the initial condition u_0 are

$$c_1 \lambda_1^0 x_1 + \cdots + c_n \lambda_n^0 x_n = u_0 , \quad \text{or} \quad Sc = u_0 , \quad \text{or} \quad c = S^{-1} u_0 . \tag{21}$$

These formulas are really giving two different approaches to the same solution $u_k = S\Lambda^k S^{-1} u_0$. The first formula (19) recognized that A^k is identical with $S\Lambda^k S^{-1}$, and we could have stopped there. But the second approach brings out more clearly the analogy with solving a differential equation: *Instead of the pure exponential solutions $e^{\lambda_i t} x_i$, we now have the pure powers $\lambda_i^k x_i$.* The normal modes are again the eigenvectors x_i, and at each step they are amplified by the eigenvalues λ_i. By combining these special solutions in such a way as to match u_0, we recover the correct solution $u_k = S\Lambda^k S^{-1} u_0$.

In any specific example, like Fibonacci's equation, the first step is to diagonalize the matrix A:

$$A = \begin{bmatrix} 1 & 1 \\ 1 & 0 \end{bmatrix}, \quad \det(A - \lambda I) = \lambda^2 - \lambda - 1,$$

$$\lambda_1 = \frac{1 + \sqrt{5}}{2} , \quad \lambda_2 = \frac{1 - \sqrt{5}}{2} .$$

$$A = S\Lambda S^{-1} = \begin{bmatrix} \lambda_1 & \lambda_2 \\ 1 & 1 \end{bmatrix} \begin{bmatrix} \lambda_1 & \\ & \lambda_2 \end{bmatrix} \begin{bmatrix} 1 & -\lambda_2 \\ -1 & \lambda_1 \end{bmatrix} \frac{1}{\lambda_1 - \lambda_2} .$$

Once these eigenvalues and eigenvectors are known, the formula (19) takes over. The initial $F_0 = 0$ and $F_1 = 1$ give $u_0 = \begin{bmatrix} 1 \\ 0 \end{bmatrix}$, and then

$$\begin{bmatrix} F_{k+1} \\ F_k \end{bmatrix} = u_k = A^k u_0 = S\Lambda^k S^{-1} u_0 = \begin{bmatrix} \lambda_1 & \lambda_2 \\ 1 & 1 \end{bmatrix} \begin{bmatrix} \lambda_1^k & \\ & \lambda_2^k \end{bmatrix} \begin{bmatrix} 1 \\ -1 \end{bmatrix} \frac{1}{\lambda_1 - \lambda_2} .$$

The Fibonacci number F_k is the second component of this product:

$$F_k = \frac{\lambda_1^k}{\lambda_1 - \lambda_2} - \frac{\lambda_2^k}{\lambda_1 - \lambda_2} = \frac{1}{\sqrt{5}} \left[\left(\frac{1 + \sqrt{5}}{2} \right)^k - \left(\frac{1 - \sqrt{5}}{2} \right)^k \right] .$$

This is the answer we wanted. In one way it is rather surprising, because Fibonacci's rule $F_{k+2} = F_{k+1} + F_k$ must always produce whole numbers, and we have ended up with fractions and square roots. Somehow these must cancel out, and leave an integer. In fact, since the second term $[(1 - \sqrt{5})/2]^k/\sqrt{5}$ is always less than $\frac{1}{2}$, it must just move the first term to the nearest integer. Subtraction leaves only the integer part, and

$$F_{1000} = \text{nearest integer to } \frac{1}{\sqrt{5}} \left(\frac{1 + \sqrt{5}}{2} \right)^{1000}.$$

Of course this is an enormous number, and F_{1001} will be even bigger. It is pretty clear that the fractional parts are becoming completely insignificant compared to the integers; the ratio F_{1001}/F_{1000} must be very close to the quantity $(1 + \sqrt{5})/2 \approx 1.618$, which the Greeks called the "golden mean."† In other words λ_2^k is becoming insignificant compared to λ_1^k, and the ratio F_{k+1}/F_k approaches $\lambda_1^{k+1}/\lambda_1^k = \lambda_1$.

EXERCISE 5.3.1 Suppose Fibonacci had started his sequence with $F_0 = 1$ and $F_1 = 3$, and then followed the same rule $F_{k+2} = F_{k+1} + F_k$. Find the new initial vector u_0, the new coefficients $c = S^{-1}u_0$, and the new Fibonacci numbers. Show that the ratios F_{k+1}/F_k still approach the golden mean.

EXERCISE 5.3.2 If each number is the *average* of the two previous numbers, $G_{k+2} = \frac{1}{2}(G_{k+1} + G_k)$, set up the matrix A and diagonalize it. Starting from $G_0 = 0$ and $G_1 = \frac{1}{2}$, find a formula for G_k and compute its limit as $k \to \infty$.

EXERCISE 5.3.3 Bernadelli considered a species of beetle "which lives three years only, and propagates in its third year." If the first age group survives with probability $\frac{1}{2}$, and then the second with probability $\frac{1}{3}$, and then the third produces six females on the way out, the associated matrix is

$$A = \begin{bmatrix} 0 & 0 & 6 \\ \frac{1}{2} & 0 & 0 \\ 0 & \frac{1}{3} & 0 \end{bmatrix}.$$

Show that $A^3 = I$, and follow the distribution of beetles for six years starting with 3000 beetles in each age group.

A Markov Process

There was an exercise in Chapter 1, about moving in and out of California, which is worth another look. These were the rules:

Each year $\frac{1}{10}$ of the people outside California move in, and $\frac{2}{10}$ of the people inside California move out.

---

† The most elegant rectangles have their sides in the ratio of 1.618 to 1.

This suggests a difference equation; we start with y_0 people outside and z_0 inside, and at the end of the first year there are

$$\tfrac{9}{10}y_0 + \tfrac{2}{10}z_0 = y_1 \text{ people outside}$$

$$\tfrac{1}{10}y_0 + \tfrac{8}{10}z_0 = z_1 \text{ people inside}$$

or

$$\begin{bmatrix} y_1 \\ z_1 \end{bmatrix} = \begin{bmatrix} .9 & .2 \\ .1 & .8 \end{bmatrix} \begin{bmatrix} y_0 \\ z_0 \end{bmatrix}.$$

Of course this problem was produced out of thin air, but it has the two essential properties of what is called a **Markov process**: The total number of people stays fixed, and the numbers outside and inside can never become negative.† The first property is reflected in the fact that *each column of the matrix adds up to 1*; everybody is accounted for, and nobody is gained or lost. The second property is reflected in the fact that *the matrix has no negative entries*; as long as the initial y_0 and z_0 are nonnegative, the same will be true of y_1 and z_1, y_2 and z_2, and so on forever. The powers A^k are all nonnegative.

We propose first to solve this specific difference equation (using the formula $S\Lambda^k S^{-1}u_0$), then to see whether the population eventually approaches a "steady state," and finally to discuss Markov processes in general. To start the computations, A has to be diagonalized:

$$A = \begin{bmatrix} .9 & .2 \\ .1 & .8 \end{bmatrix}, \qquad \det(A - \lambda I) = \lambda^2 - 1.7\lambda + .7,$$

$$\lambda_1 = 1 \quad \text{and} \quad \lambda_2 = .7.$$

$$A = S\Lambda S^{-1} = \begin{bmatrix} \tfrac{2}{3} & \tfrac{1}{3} \\ \tfrac{1}{3} & -\tfrac{1}{3} \end{bmatrix} \begin{bmatrix} 1 & \\ & .7 \end{bmatrix} \begin{bmatrix} 1 & 1 \\ 1 & -2 \end{bmatrix}.$$

Now we can find A^k, and the distribution after k years:

$$\begin{bmatrix} y_k \\ z_k \end{bmatrix} = A^k \begin{bmatrix} y_0 \\ z_0 \end{bmatrix} = \begin{bmatrix} \tfrac{2}{3} & \tfrac{1}{3} \\ \tfrac{1}{3} & -\tfrac{1}{3} \end{bmatrix} \begin{bmatrix} 1^k & \\ & .7^k \end{bmatrix} \begin{bmatrix} 1 & 1 \\ 1 & -2 \end{bmatrix} \begin{bmatrix} y_0 \\ z_0 \end{bmatrix}$$

$$= (y_0 + z_0) \begin{bmatrix} \tfrac{2}{3} \\ \tfrac{1}{3} \end{bmatrix} + (y_0 - 2z_0)(.7)^k \begin{bmatrix} \tfrac{1}{3} \\ -\tfrac{1}{3} \end{bmatrix}.$$

This is the solution we want, and it is easy to see what happens in the long run: The factor $(.7)^k$ becomes extremely small, and the solution approaches a limiting state

$$\begin{bmatrix} y_\infty \\ z_\infty \end{bmatrix} = (y_0 + z_0) \begin{bmatrix} \tfrac{2}{3} \\ \tfrac{1}{3} \end{bmatrix}.$$

The total population is still $y_0 + z_0$, just as it was initially, but in the limit $\tfrac{2}{3}$ of this population is outside California and $\tfrac{1}{3}$ is inside. This is true no matter

---

† Furthermore, history is completely disregarded; each new situation u_{k+1} depends only on the current u_k, and the record of $u_0, \ldots, u_{k-1}$ can be thrown away. Perhaps even our lives are examples of Markov processes, but I hope not.

what the initial distribution may have been. You may recognize that this steady state is exactly the distribution that was asked for in Exercise 1.3.3; if the year starts with $\frac{2}{3}$ outside and $\frac{1}{3}$ inside, then it ends the same way:

$$\begin{bmatrix} .9 & .2 \\ .1 & .8 \end{bmatrix} \begin{bmatrix} \frac{2}{3} \\ \frac{1}{3} \end{bmatrix} = \begin{bmatrix} \frac{2}{3} \\ \frac{1}{3} \end{bmatrix}, \qquad \text{or} \qquad A u_\infty = u_\infty .$$

The steady state is the eigenvector of A corresponding to $\lambda = 1$. Multiplication by A, which takes us from one time step to the next, leaves u_∞ unchanged.

The above description of a Markov process was completely deterministic; populations moved in fixed proportions. But if we look instead at a single individual, the rules for moving can be given a probabilistic interpretation. If the individual is outside California, then with probability $\frac{1}{10}$ he moves in; if he is inside, then with probability $\frac{2}{10}$ he moves out. His movement becomes a random process, and the matrix A that governs it is called a *transition matrix*. We no longer know exactly where he is, but every year the components of $u_k = A^k u_0$ specify the probability that he is outside the state, and the probability that he is inside. These probabilities add up to 1—he has to be somewhere—and they are never negative, which brings us back to the two fundamental properties of a transition matrix: Each column adds up to 1, and every entry satisfies $a_{ij} \geq 0$.

The key step in the theory is to understand why $\lambda = 1$ is always an eigenvalue, and why its eigenvector is the steady state. The first point is easy to explain: Each column of $A - I$ adds up to $1 - 1 = 0$. Therefore the rows of $A - I$ add up to the zero row, they are linearly dependent, $A - I$ is singular, and $\lambda_1 = 1$ is an eigenvalue. Except for very special cases,† u_k will eventually approach the corresponding eigenvector. This is suggested by the formula $u_k = c_1 \lambda_1^k x_1 + \cdots + c_n \lambda_n^k x_n$, in which no eigenvalue can be larger than 1; otherwise the probabilities u_k would blow up like Fibonacci numbers, which is impossible. If all the other eigenvalues are strictly smaller than $\lambda_1 = 1$, then the first term in the formula will be completely dominant; the other λ_i^k go rapidly to zero, and $u_k \rightarrow c_1 x_1 = u_\infty$. This steady state is assured if the matrix A is not only nonnegative but actually positive: $a_{ij} > 0$. Then the vector $c_1 x_1$ has only positive components, they add up to 1, and they are the limiting probabilities for the Markov process.

EXERCISE 5.3.4 Suppose there are three major centers for Move-It-Yourself trucks. Every month half of those in Boston and in Los Angeles go to Chicago, the other half stay where they are, and the trucks in Chicago are split equally between Boston and Los Angeles. Set up the 3 by 3 transition matrix A, and find the steady state u_∞ corresponding to the eigenvalue $\lambda = 1$.

EXERCISE 5.3.5 Suppose there is an epidemic in which every month half of those who are well become sick, and a quarter of those who are sick become dead. Find the steady

---

† If everybody outside moves in and everybody inside moves out, then the populations are reversed every year and a steady state is impossible. The transition matrix is $A = \begin{bmatrix} 0 & 1 \\ 1 & 0 \end{bmatrix}$, and -1 is an eigenvalue as well as $+1$.

state for the corresponding Markov process

$$
\begin{bmatrix} d_{k+1} \\ s_{k+1} \\ w_{k+1} \end{bmatrix} = \begin{bmatrix} 1 & \frac{1}{4} & 0 \\ 0 & \frac{3}{4} & \frac{1}{2} \\ 0 & 0 & \frac{1}{2} \end{bmatrix} \begin{bmatrix} d_k \\ s_k \\ w_k \end{bmatrix}.
$$

EXERCISE 5.3.6 Suppose that the men in a given generation are either dominant, hybrid, or recessive—the genes that determine eye color are either both brown, one brown and one blue, or both blue—with probabilities d_0, h_0, r_0. (I think in the hybrid case both eyes will actually look brown, but nevertheless the two genes in the father are equally likely to be inherited.) If all women are assumed hybrid,† then the gene inherited from the mother is brown or blue with equal probability. Find the matrix A that gives the probabilities d_1, h_1, and r_1 for a son in the next generation:

$$
\begin{bmatrix} d_1 \\ h_1 \\ r_1 \end{bmatrix} = A \begin{bmatrix} d_0 \\ h_0 \\ r_0 \end{bmatrix};
$$

$a_{11} = a_{21} = \frac{1}{2}$ because the gene from the father must be brown. After infinitely many generations, what will be the limiting distribution u_∞ ?

Stability

There is an obvious difference between Fibonacci numbers and Markov processes; the numbers F_k become larger and larger, while by definition any "probability" is between 0 and 1. The Fibonacci equation is unstable, and so is the compound interest equation $P_{k+1} = 1.06 P_k$; the principal keeps growing forever. If the Markov probabilities decreased to zero, then that equation would be stable; but they do not, since at every stage they must add up to 1. Therefore a Markov process is neutrally stable.

Now suppose we are given any difference equation $u_{k+1} = A u_k$, and we want to study its behavior as $k \to \infty$. Assuming A can be diagonalized, the solution u_k will be a combination of pure solutions,

$$
u_k = S \Lambda^k S^{-1} u_0 = c_1 \lambda_1^k x_1 + \cdots + c_n \lambda_n^k x_n .
$$

The growth of u_k is governed by the factors λ_i^k, and therefore **stability depends on the eigenvalues of** A.

5H The difference equation $u_{k+1} = A u_k$ is **stable** and $u_k \to 0$ whenever all the eigenvalues satisfy $|\lambda_i| < 1$, it is **neutrally stable** and u_k is bounded whenever all $|\lambda_i| \le 1$, and it is **unstable** (u_k is unbounded) when at least one eigenvalue of A has $|\lambda_i| > 1$.

† Which is not so, and I hope that such an assumption will be forgiven.

EXAMPLE 1   The matrix

$$A = \begin{bmatrix} 0 & 4 \\ 0 & \frac{1}{2} \end{bmatrix}$$

is certainly stable; its eigenvalues are 0 and $\frac{1}{2}$, lying on the main diagonal because A is triangular. Starting from any initial vector u_0, and following the rule $u_{k+1} = Au_k$, the solution must eventually approach zero:

$$u_0 = \begin{bmatrix} 0 \\ 1 \end{bmatrix}, \quad u_1 = \begin{bmatrix} 4 \\ \frac{1}{2} \end{bmatrix}, \quad u_2 = \begin{bmatrix} 2 \\ \frac{1}{4} \end{bmatrix}, \quad u_3 = \begin{bmatrix} 1 \\ \frac{1}{8} \end{bmatrix}, \quad u_4 = \begin{bmatrix} \frac{1}{2} \\ \frac{1}{16} \end{bmatrix}, \quad \dots$$

You can see how the larger eigenvalue $\lambda = \frac{1}{2}$ governs the decay; after the first step every vector u_k is half of the preceding one. The real effect of the first step is to split u_0 into the two eigenvectors of A,

$$\begin{bmatrix} 0 \\ 1 \end{bmatrix} = \begin{bmatrix} 8 \\ 1 \end{bmatrix} + \begin{bmatrix} -8 \\ 0 \end{bmatrix},$$

and to annihilate the second eigenvector (corresponding to $\lambda = 0$). The first eigenvector is multiplied by $\lambda = \frac{1}{2}$ at every step.

Remark   Even though A was very stable, with small eigenvalues, the vectors u_k actually grew in length at the beginning. It was not until u_4 that they became shorter than the original u_0, and then decay continued at a geometric rate. To guarantee *immediate decay*, so that $\| u_1 \| < \| u_0 \|$ or $\| Au \| < \| u \|$ for all nonzero starting vectors u, we need

$$\| Au \|^2 = u^{\mathrm{T}} A^{\mathrm{T}} A u < \| u \|^2 = u^{\mathrm{T}} u, \quad \text{or} \quad u^{\mathrm{T}} (A^{\mathrm{T}} A - I) u < 0. \quad (22)$$

This is a stronger requirement than the condition $| \lambda_i | \leq 1$ on the eigenvalues. In the language of Chapter 6 it means that $A^{\mathrm{T}} A - I$ is *negative definite* (its eigenvalues are all negative) and in the language of Chapter 7 it means that the "norm" of A is less than 1.

EXERCISE 5.3.7   Show that there is immediate decay—$A^{\mathrm{T}} A - I$ has negative eigenvalues—if the off-diagonal entry $a_{12} = 4$ is reduced below $a_{12} = \sqrt{3}/2$.

EXERCISE 5.3.8   For the system $v_{n+1} = \alpha (v_n + w_n)$, $w_{n+1} = \alpha (v_n + w_n)$, what values of α produce instability?

EXAMPLE 2   *von Neumann's model of an expanding economy.*   Suppose we accept a simple model in which there are three "goods"; steel, food, and labor. The production of each good consumes a part of what was produced the year before, and the economist's question is whether (and at what rate) the economy can expand. Suppose that a new unit of steel requires .4 unit of existing steel and .5 unit of labor, a unit of food requires .1 unit of food and .7 unit of labor, and producing (or maintaining) a unit of labor needs .8 unit of food and .1

unit of steel and labor. Then the inputs s_0 , f_0 , and l_0 are related to the outputs s_1 , f_1 , and l_1 by

$$u_0 = \begin{bmatrix} s_0 \\ f_0 \\ l_0 \end{bmatrix} = \begin{bmatrix} .4 & 0 & .1 \\ 0 & .1 & .8 \\ .5 & .7 & .1 \end{bmatrix} \begin{bmatrix} s_1 \\ f_1 \\ l_1 \end{bmatrix} = A u_1 .$$

Notice that the difference equation is backward! Instead of $u_1 = A u_0$ we have $u_0 = A u_1$, so it must be the eigenvalues of A^{-1} rather than A that govern the expansion. But there is also a second twist to this problem, because steel, food, and labor cannot come in negative amounts: von Neumann asked for the maximum rate α at which the economy could expand and *still stay nonnegative*, meaning that $u_1 \geq \alpha u_0 \geq 0$.

If these inequalities hold—starting with a vector u_0 of steel, food, and labor, we end up with at least αu_0—then starting the next year with u_1 produces at least αu_1 , or at least $\alpha^2 u_0$. Therefore the economy can continue to expand at the rate α. Of course if the largest possible α is less than 1, it must contract rather than expand—and if A is a Markov matrix there will be an equilibrium at which $\alpha = 1$. In fact von Neumann's theory is very close to Markov's, because in both cases the matrix A is nonnegative—which guarantees that its largest eigenvalue λ_1 is nonnegative and so are the components of the corresponding eigenvector. In the Markov case $\alpha = 1$ because $\lambda_1 = 1$; in every case we have $\lambda_1 = 1/\alpha$, since von Neumann proved that it is always the nonnegative eigenvector that gives the fastest possible expansion. The example has $\lambda_1 = \frac{9}{10}$, so its expansion factor is $\frac{10}{9}$; the eigenvector is

$$x = \begin{bmatrix} .2 \\ 1 \\ 1 \end{bmatrix}, \quad \text{and} \quad Ax = \begin{bmatrix} .4 & 0 & .1 \\ 0 & .1 & .8 \\ .5 & .7 & .1 \end{bmatrix} \begin{bmatrix} .2 \\ 1 \\ 1 \end{bmatrix} = \begin{bmatrix} .18 \\ .9 \\ .9 \end{bmatrix} = .9x.$$

EXERCISE 5.3.9   Explain in economic terms why increasing any entry of the "consumption matrix" A must increase its largest eigenvalue λ_1 (and slow down the expansion).

EXAMPLE 3   *Leontief's input–output matrix.* Unlike von Neumann, Leontief was concerned first of all with consumption and production within a single year; his input–output system was one of the first great successes of mathematical economics. To illustrate it, we keep the same consumption matrix A and ask whether a given production vector y can be achieved: Can we end up with y_1 units of steel, y_2 units of food, and y_3 units of labor? To do so, these goods must be produced in larger amounts x_1 , x_2 , and x_3 , because some of the production is consumed by the production process itself. In fact the amount consumed is exactly Ax, and leaves a net production of $x - Ax$.

Problem   To find x such that

$$x - Ax = y, \quad \text{or} \quad x = (I - A)^{-1} y.$$

On the surface, we are only asking whether $I - A$ is invertible. But there is again a nonnegative twist to the problem: We assume a demand vector $y \geq 0$, and require a production vector $x \geq 0$. Therefore the real question is whether $(I - A)^{-1}$ has nonnegative entries; then the product $(I - A)^{-1}y$ will be ≥ 0. Mathematically, given a consumption matrix A with largest eigenvalue λ_1, the key result of the theory is this: $(I - A)^{-1}$ *is nonnegative if and only if* $\lambda_1 < 1$. In this case the economy can produce any combination of goods; production is ahead of consumption, if the eigenvalues are right.

EXAMPLE   Suppose $A = \begin{bmatrix} 0 & 2 \\ 2 & 0 \end{bmatrix}$, so that producing steel consumes two units of food and vice versa. Then we cannot produce anything! The largest eigenvalue is $\lambda_1 = 2$, and

$$(I - A)^{-1} = -\frac{1}{3}\begin{bmatrix} 1 & 2 \\ 2 & 1 \end{bmatrix} \quad \text{is entirely negative.}$$

EXERCISE 5.3.10   Multiplying term by term, check that $(I - A)(I + A + A^2 + \cdots) = I$. This infinite series represents $(I - A)^{-1}$, and is nonnegative whenever A is nonnegative, provided it has a finite sum; the condition for that is $\lambda_1 < 1$. Add up the infinite series, and confirm that it equals $(I - A)^{-1}$, for the consumption matrix

$$A = \begin{bmatrix} 0 & 1 & 1 \\ 0 & 0 & 1 \\ 0 & 0 & 0 \end{bmatrix}.$$

Leontief's inspiration was to find a model which uses genuine data from the real economy; the official U.S. table for 1958 contained 83 industries, each of whom sent in a complete "transactions table" of consumption and production. The theory also reaches beyond $(I - A)^{-1}$, to decide on the natural prices and on questions of optimization; normally labor is separated out as a primary good, which is in limited supply and ought to be minimized. And, of course, the economy is not always linear.

5.4 ■ DIFFERENTIAL EQUATIONS AND THE EXPONENTIAL e^{At}

Wherever you find a system of equations, rather than a single equation, matrix theory has a part to play. This was true for difference equations, where the solution $u_k = A^k u_0$ depended on the powers of A. It is equally true for differential equations, where the solution $u(t) = e^{At}u_0$ depends on the exponential of A. To define this exponential, and to understand it, we turn right away to an example:

$$\frac{du}{dt} = Au = \begin{bmatrix} -2 & 1 \\ 1 & -2 \end{bmatrix}u. \tag{23}$$

The first step is always to find the eigenvalues and eigenvectors:

$$A \begin{bmatrix} 1 \\ 1 \end{bmatrix} = (-1) \begin{bmatrix} 1 \\ 1 \end{bmatrix}, \qquad A \begin{bmatrix} 1 \\ -1 \end{bmatrix} = (-3) \begin{bmatrix} 1 \\ -1 \end{bmatrix}.$$

Then there are several possibilities, all leading to the same answer. Probably the best way is to write down the general solution, and match it to the initial vector u_0 at $t = 0$:

$$u(t) = c_1 e^{\lambda_1 t} x_1 + c_2 e^{\lambda_2 t} x_2 = c_1 e^{-t} \begin{bmatrix} 1 \\ 1 \end{bmatrix} + c_2 e^{-3t} \begin{bmatrix} 1 \\ -1 \end{bmatrix} \tag{24}$$

$$u_0 = c_1 \begin{bmatrix} 1 \\ 1 \end{bmatrix} + c_2 \begin{bmatrix} 1 \\ -1 \end{bmatrix} = \begin{bmatrix} 1 & 1 \\ 1 & -1 \end{bmatrix} \begin{bmatrix} c_1 \\ c_2 \end{bmatrix}.$$

You recognize S, the matrix of eigenvectors. And the coefficients $\begin{bmatrix} c_1 \\ c_2 \end{bmatrix} = S^{-1} u_0$ are the same as they were for difference equations. Substituting them back into (24), the problem is solved. In matrix form, the solution is

$$u(t) = \begin{bmatrix} 1 & 1 \\ 1 & -1 \end{bmatrix} \begin{bmatrix} e^{-t} & \\ & e^{-3t} \end{bmatrix} \begin{bmatrix} c_1 \\ c_2 \end{bmatrix} = S \begin{bmatrix} e^{-t} & \\ & e^{-3t} \end{bmatrix} S^{-1} u_0.$$

Here is the fundamental formula of this section: $Se^{\Lambda t} S^{-1} u_0$ is the solution to a differential equation, just as $S \Lambda^k S^{-1} u_0$ was the solution to a difference equation. The key matrices are

$$\Lambda = \begin{bmatrix} -1 & \\ & -3 \end{bmatrix} \quad \text{and} \quad e^{\Lambda t} = \begin{bmatrix} e^{-t} & \\ & e^{-3t} \end{bmatrix}.$$

Now we can afford to slow down.

There are two more things to be done with this example. One is to complete the mathematical part of the theory, by giving a direct definition of the exponential e^{At} and matching it with the formula $Se^{\Lambda t} S^{-1}$. The other is to give a physical interpretation of the equation itself and of its solution. It is the kind of differential equation that has useful applications.

First, we take up the exponential. The most natural way to define it is to imitate the power series for e^x:

$$e^x = 1 + x + \frac{x^2}{2!} + \frac{x^3}{3!} + \cdots,$$

and

$$e^{At} = I + At + \frac{(At)^2}{2!} + \frac{(At)^3}{3!} + \cdots.$$

e^{At} is an n by n matrix, and it is easy to differentiate:

$$\frac{d}{dt}(e^{At}) = A + \frac{A^2(2t)}{2!} + \frac{A^3(3t^2)}{3!} + \cdots$$

$$= A\left(I + At + \frac{(At)^2}{2!} + \cdots\right) = Ae^{At}.$$

So it is certainly true that $e^{At}u_0$ is the right solution to the differential equation: At $t = 0$ it reduces to the initial vector u_0, and it satisfies the equation $d(e^{At}u_0)/dt = Ae^{At}u_0$. Therefore it must be identical with our other solution $Se^{\Lambda t}S^{-1}u_0$. To prove that identity in a more direct way, remember that the powers of $A = S\Lambda S^{-1}$ telescope into $A^k = (S\Lambda S^{-1})\cdots(S\Lambda S^{-1}) = S\Lambda^k S^{-1}$. Therefore the infinite series for the exponential becomes

$$e^{At} = I + S\Lambda S^{-1}t + \frac{S\Lambda^2 S^{-1}t^2}{2!} + \frac{S\Lambda^3 S^{-1}t^3}{3!} + \cdots$$

$$= S\left(I + \Lambda t + \frac{(\Lambda t)^2}{2!} + \frac{(\Lambda t)^3}{3!} + \cdots\right)S^{-1} = Se^{\Lambda t}S^{-1}.$$

This finishes the mathematical part, and the results can be summarized as follows:

5I If A can be diagonalized, $A = S\Lambda S^{-1}$, then the differential equation $du/dt = Au$ has the solution

$$u(t) = e^{At}u_0 = Se^{\Lambda t}S^{-1}u_0. \tag{25}$$

The columns of S are the eigenvectors of A, so that

$$u(t) = \begin{bmatrix} x_1 & \cdots & x_n \end{bmatrix}\begin{bmatrix} e^{\lambda_1 t} & & \\ & \ddots & \\ & & e^{\lambda_n t} \end{bmatrix} S^{-1}u_0$$

$$= c_1 e^{\lambda_1 t}x_1 + \cdots + c_n e^{\lambda_n t}x_n. \tag{26}$$

The general solution is a linear combination of the special solutions $e^{\lambda_i t}x_i$, and the coefficients c_i that match the initial condition u_0 are $c = S^{-1}u_0$.

This gives a complete analogy with difference equations—you could compare it with 5G on p. 189. In both cases we assumed that A could be diagonalized, since otherwise it has fewer than n eigenvectors and we have not found enough special solutions. The missing solutions do exist, but they are more complicated than our pure exponentials $e^{\lambda t}x$; they involve "generalized eigenvectors" and factors like $te^{\lambda t}$. Nevertheless the formula $u(t) = e^{At}u_0$ remains completely correct.†

Now we turn to the physical significance of this example, which is easy to explain and at the same time genuinely important. The differential equation describes a process of diffusion, which can be visualized by dividing an infinite pipe into four segments, two in the middle which are finite, and two at the ends which are semi-infinite (Fig. 5.1). At time $t = 0$, the two finite segments

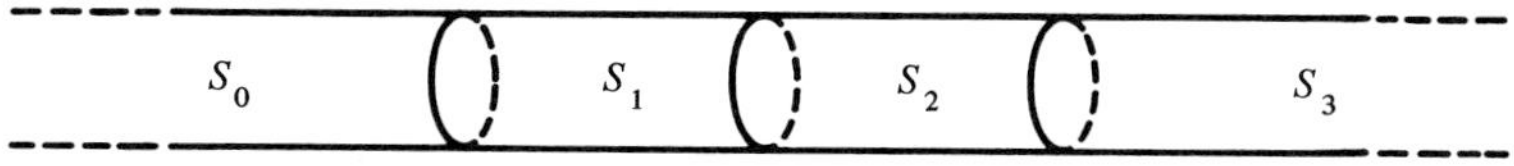

Fig. 5.1. A model of diffusion.

contain concentrations v_0 and w_0 of some chemical solution. At the same time, and for all times, the concentration in the two infinite segments is zero; with an infinite volume, this will be a correct picture of the average concentration in these infinite segments even after the chemical has started to diffuse. Diffusion starts at time $t = 0$, and is governed by the following law: *At each time t, the diffusion rate between two adjacent segments equals the difference in concentrations.* We are imagining that, within each segment, the concentration remains uniform. The process is continuous in time but discrete in space; the only two unknowns are $v(t)$ and $w(t)$ in the two inner segments S_1 and S_2.

The concentration v is changing in two ways, by diffusion into the far left segment S_0 and by diffusion into or out of S_2. The net rate of change is therefore

$$\frac{dv}{dt} = (w - v) + (0 - v),$$

because the concentration in S_0 is identically zero. Similarly,

$$\frac{dw}{dt} = (0 - w) + (v - w).$$

Therefore the system exactly matches our example (23):

$$u = \begin{bmatrix} v \\ w \end{bmatrix}, \qquad \frac{du}{dt} = \begin{bmatrix} -2v + w \\ v - 2w \end{bmatrix} = \begin{bmatrix} -2 & 1 \\ 1 & -2 \end{bmatrix} u.$$

The eigenvalues -1 and -3 will govern the behavior of the solution. They

---

† This defective case is postponed until Appendix B, but an example is easy to give: if $y' = z$ and $z' = 0$, the system is

$$\frac{du}{dt} = Au = \begin{bmatrix} 0 & 1 \\ 0 & 0 \end{bmatrix} u,$$

so

$$A^2 = 0 \qquad \text{and} \qquad e^{At} = I + At = \begin{bmatrix} 1 & t \\ 0 & 1 \end{bmatrix}.$$

give the rate at which the concentrations decay, and λ_1 is the more important because only an exceptional set of starting conditions can lead to "superdecay" at the rate e^{-3t}. In fact, those conditions must come from the eigenvector $(1, -1)$ whose components are of opposite sign. If the experiment admits only nonnegative concentrations, superdecay is impossible and the limiting rate must be e^{-t}. The solution that decays at this rate corresponds to the eigenvector $(1, 1)$, and therefore the two concentrations will become nearly equal as $t \rightarrow \infty$.

One more comment on this example: It is a discrete approximation, with only two unknowns, to the continuous diffusion process described by the partial differential equation

$$\frac{\partial u}{\partial t} = \frac{\partial^2 u}{\partial x^2}, \qquad u(0) = u(1) = 0.$$

This is approached by keeping the two infinite segments at zero concentration, and dividing the middle of the pipe into smaller and smaller segments, of length $h = 1/N$. The discrete system with N unknowns is governed by

$$\frac{d}{dt}\begin{bmatrix} u_1 \\ \cdot \\ \cdot \\ \cdot \\ u_N \end{bmatrix} = \begin{bmatrix} -2 & 1 & & \\ 1 & -2 & \cdot & \\ & \cdot & \cdot & 1 \\ & & 1 & -2 \end{bmatrix}\begin{bmatrix} u_1 \\ \cdot \\ \cdot \\ \cdot \\ u_N \end{bmatrix} = Au.$$

This is exactly the finite difference matrix constructed in Section 1.6 as an approximation to d^2/dx^2 (with the signs reversed for $-d^2/dx^2$). A more careful look at the chemistry will introduce the scale factor $1/h^2$, so we do arrive, in the limit as $h \rightarrow 0$ and $N \rightarrow \infty$, at exactly the partial differential equation $\partial u/\partial t = \partial^2 u/\partial x^2$. It is known as the *heat equation*. Its solutions can still be expanded into the normal modes of the problem; but these are no longer eigenvectors with N discrete components, they are *eigenfunctions*. In fact, they are exactly the functions $\sin n\pi x$, and the general solution to the heat equation is

$$u(t) = \sum_{n=1}^{\infty} c_n e^{-n^2\pi^2 t} \sin n\pi x.$$

The coefficients c_n are determined as always by the initial conditions, and the rates of decay are the eigenvalues: $\lambda_n = -n^2\pi^2$.

EXERCISE 5.4.1 Compute the 2 by 2 matrix e^{At} in the example above and show that its entries are positive for $t > 0$. Any experiment that starts with positive concentrations will stay that way: If $u_0 > 0$, then $e^{At}u_0 > 0$.†

---

† This always happens when the off-diagonal entries of A are positive.

EXERCISE 5.4.2 Suppose the time direction is reversed in the diffusion equation: $du/dt = Au$ becomes $du/d(-t) = Au$, or

$$\frac{du}{dt} = Bu = \begin{bmatrix} 2 & -1 \\ -1 & 2 \end{bmatrix} u_0, \qquad u_0 = \begin{bmatrix} 1 \\ 0 \end{bmatrix}.$$

Compute $u(t)$, and show that now it blows up instead of decaying as $t \to +\infty$. (In the continuous case this blowup occurs instantaneously as we leave $t = 0$; *the heat equation is irreversible in time*, and you cannot undo the diffusion of molecules.)

EXERCISE 5.4.3 To produce a continuous Markov process, we block out the infinite segments S_0 and S_3 and switch to the equations

$$\frac{dv}{dt} = w - v$$

$$\text{or} \quad \frac{du}{dt} = \begin{bmatrix} -1 & 1 \\ 1 & -1 \end{bmatrix} u = Au.$$

$$\frac{dw}{dt} = v - w$$

Find the general solution, the solution that matches initial concentrations of $v = 3$ and $w = 1$, and the steady state as $t \to \infty$. Note that $\lambda = 0$ in a differential equation corresponds to $\lambda = 1$ in a difference equation ($e^{0t} = 1$) and the eigenvector governs the steady state.

EXERCISE 5.4.4 Derive the formula $u(t) = Se^{\Lambda t}S^{-1}u_0$ in a different way, by a change of variables $v = S^{-1}u$ in the differential equation $du/dt = Au$. Find the new v_0, solve the equation for v, and convert back to u.

EXERCISE 5.4.5 Assuming you can diagonalize A, show from (25) that $e^{tA}e^{sA} = e^{(t+s)A}$. But show for a 2 by 2 counterexample that in general $e^B e^C \neq e^{B+C}$; the scalar law is broken in the matrix case.

Stability of Differential Equations

Just as for difference equations, it is the eigenvalues that decide how $u(t)$ behaves as $t \to \infty$. As long as A can be diagonalized, there will be n pure exponential solutions to the differential equation, and any specific solution $u(t)$ is some combination

$$u(t) = Se^{\Lambda t}S^{-1}u_0 = c_1 e^{\lambda_1 t}x_1 + \cdots + c_n e^{\lambda_n t}x_n.$$

Stability is governed by the factors $e^{\lambda_i t}$. If they all approach zero, then $u(t)$ approaches zero; if they all stay bounded, then $u(t)$ stays bounded; and if one of them blows up, then except for very special starting conditions the solution $u(t)$ will blow up. Furthermore, since the size (or modulus) of $e^{\lambda t}$ depends only on the real part of λ, *it is only the real parts that govern stability*: If $\lambda = \alpha + i\beta$, then

$$e^{\lambda t} = e^{\alpha t}e^{i\beta t} = e^{\alpha t}(\cos \beta t + i \sin \beta t), \qquad \text{and} \qquad |e^{\lambda t}| = e^{\alpha t}.$$

This decays for $\alpha < 0$, it is constant for $\alpha = 0$, and it explodes for $\alpha > 0$, while the imaginary parts β are producing pure oscillations. For any diagonalizable A, this proves:

5J   The differential equation $du/dt = Au$ is **stable** and $e^{At} \to 0$ whenever all Re $\lambda_i < 0$; it is **neutrally stable** and e^{At} is bounded whenever all Re $\lambda_i \leq 0$; and it is **unstable** and e^{At} is unbounded when at least one eigenvalue of A has Re $\lambda_i > 0$.

EXAMPLE 1

$$\frac{du}{dt} = Au = \begin{bmatrix} 0 & -1 \\ 1 & 0 \end{bmatrix} u, \qquad u_0 = \begin{bmatrix} 1 \\ 0 \end{bmatrix}.$$

The eigenvalues of A satisfy

$$|A - \lambda I| = \begin{vmatrix} -\lambda & -1 \\ 1 & -\lambda \end{vmatrix} = \lambda^2 + 1 = 0, \qquad \text{or} \qquad \lambda = \pm i.$$

They are purely imaginary, so the solution should be neutrally stable. In fact e^{At} is a rotation matrix, and each component of the solution is in simple harmonic motion:

$$e^{At} = \begin{bmatrix} \cos t & -\sin t \\ \sin t & \cos t \end{bmatrix}, \qquad \text{and} \qquad u(t) = e^{At} u_0 = \begin{bmatrix} \cos t \\ \sin t \end{bmatrix}. \qquad (27)$$

The equation just describes a point moving around in a circle.†

EXAMPLE 2   The diffusion equation is stable, with $\lambda = -1$ and $\lambda = -3$.

EXAMPLE 3   The continuous Markov process in the previous exercises is only neutrally stable, with $\lambda = 0$ and $\lambda = -2$. This is a diffusion equation in which the ends are insulated: Nothing leaves the system and goes into the infinite segments outside.

EXAMPLE 4   In nuclear engineering a reactor is called *critical* when it is neutrally stable; the fission balances the decay. Slower fission makes it stable, or *subcritical*, and eventually it runs down; with unstable fission it becomes a bomb.

Just as for difference equations, a problem can be stable and still permit $\| u(t) \|$ to increase for a little while before it decays. I believe that this temporary growth is unusual in a genuine application; the decay ordinarily begins

† All exponentials of skew-symmetric matrices are orthogonal matrices.

right at $t = 0$. Given the derivative of an inner product

$$\frac{d}{dt} x^T y = \left[\frac{dx}{dt}\right]^T y + x^T \left[\frac{dy}{dt}\right], \tag{28}$$

we can take $x = y = u$, and find the quantity whose sign decides between growth and decay:

$$\frac{d}{dt} \| u(t) \|^2 = \left[\frac{du}{dt}\right]^T u + u^T \left[\frac{du}{dt}\right] = u^T(A^T + A)u. \tag{29}$$

5K If $u^T(A^T + A)u < 0$ for all nonzero vectors u, which means that every eigenvalue of $A^T + A$ is negative, then the solution decays at every instant. If $A^T + A = 0$, then the length $\| u \|^2$ is a constant in time, there is no dissipation of energy, and the system is conservative.

Example 1 was conservative; if a point goes around a circle, then certainly its length $\| u \|$ is a constant.

To find a condition that is both necessary and sufficient for stability—in other words, a condition that is equivalent to Re $\lambda_i < 0$—there are two possibilities. One is to go back to Routh and Hurwitz who, in the nineteenth century, found a series of inequalities on the entries a_{ij}. I do not think this approach is much good for a large matrix; the computer can probably find the eigenvalues with more certainty than it can test these inequalities. The other possibility was discovered by Lyapunov and published in 1897, two years after the work of Hurwitz. *It is to find a weighting matrix W so that the weighted length $\| Wu(t) \|$ is always decreasing.* If there exists such a W, then $u' = Au$ must be stable; $\| Wu \|$ will decrease steadily to zero, and after a few ups and downs u must get there too. The real value of Lyapunov's method is in the case of a nonlinearity, which may make the equation impossible to solve but still leave a decreasing $\| Wu(t) \|$—so that stability can be proved without knowing a formula for $u(t)$.

EXERCISE 5.4.6 Verify that (28) is the derivative of the inner product

$$x^T y = x_1(t)y_1(t) + \cdots + x_n(t)y_n(t).$$

EXERCISE 5.4.7 From the eigenvalues of $A = \begin{bmatrix} 0 & 4 \\ -1 & 0 \end{bmatrix}$, decide the stability of $du/dt = Au$.

EXERCISE 5.4.8 From the eigenvalues of

$$A = \begin{bmatrix} -1 & 3 \\ 0 & -1 \end{bmatrix} \quad \text{and} \quad A + A^T = \begin{bmatrix} -2 & 3 \\ 3 & -2 \end{bmatrix},$$

decide whether the solutions of $du/dt = Au$ must decrease immediately at $t = 0$, and whether they must decrease as $t \to \infty$.

EXERCISE 5.4.9 Decide on the stability or instability of $dv/dt = w$, $dw/dt = v$.

Second-Order Equations

The laws of diffusion led to a first-order system $du/dt = Au$. So do a lot of other applications, in chemistry, in biology, and elsewhere, but the most important law of physics does not. It is Newton's law $F = ma$, and the acceleration a is a second derivative. Inertial terms produce second-order equations (we have to solve $d^2u/dt^2 = Au$ instead of $du/dt = Au$) and the goal is to understand how this change to second derivatives alters the behavior of the solution.†

The comparison will be perfect if we keep the same A:

$$\frac{d^2u}{dt^2} = Au = \begin{bmatrix} -2 & 1 \\ 1 & -2 \end{bmatrix} u. \tag{30}$$

Two initial conditions have to be specified at $t = 0$ in order to get the system started—the "displacement" $u = u_0$ and the "velocity" $du/dt = u_0'$. To match these conditions, there will be not n but $2n$ pure exponential solutions to a system of n equations.

Suppose we use ω rather than λ, and write these special solutions as $u = e^{i\omega t}x$. Substituting this exponential into the differential equation, it must satisfy

$$\frac{d^2}{dt^2}(e^{i\omega t}x) = A(e^{i\omega t}x), \quad \text{or} \quad -\omega^2 x = Ax. \tag{31}$$

The vector x must be an eigenvector of A, exactly as before. The corresponding eigenvalue is now $-\omega^2$, so the frequency ω is connected to the decay rate λ by the law $-\omega^2 = \lambda$. Every special solution $e^{\lambda t}x$ of the first-order equation leads to two special solutions $e^{i\omega t}x$ of the second-order equation, and the two exponents are $\omega = \pm\sqrt{-\lambda}$. This breaks down only when $\lambda = 0$, which has just one square root; if the eigenvector is x, the two special solutions are x and tx.

For a genuine diffusion matrix, the eigenvalues λ are all negative and therefore the frequencies ω are all real: Pure diffusion is converted into pure oscillation. The factors $e^{i\omega t}$ produce neutral stability, the solution neither grows or decays, and in fact the total energy stays precisely constant; it just keeps passing around the system. The general solution to $d^2u/dt^2 = Au$, if A has negative eigenvalues $\lambda_1, \ldots, \lambda_n$ and if $\omega_j = \sqrt{-\lambda_j}$, is

$$u(t) = (c_1 e^{i\omega_1 t} + d_1 e^{-i\omega_1 t})x_1 + \cdots + (c_n e^{i\omega_n t} + d_n e^{-i\omega_n t})x_n. \tag{32}$$

As always, the constants are found from the initial conditions. This is easier to do (at the expense of one extra formula) by switching from oscillating exponentials to the more familiar sine and cosine:

$$u(t) = (a_1 \cos \omega_1 t + b_1 \sin \omega_1 t)x_1 + \cdots + (a_n \cos \omega_n t + b_n \sin \omega_n t)x_n. \tag{33}$$

---

† Fourth derivatives are also possible, in the bending of beams, but nature seems to resist going higher than four.

Now the initial displacement is easily distinguished from the initial velocity: $t = 0$ means that $\sin \omega t = 0$ and $\cos \omega t = 1$, leaving only

$$u_0 = a_1 x_1 + \cdots + a_n x_n , \qquad \text{or} \qquad u_0 = Sa, \qquad \text{or} \qquad a = S^{-1} u_0 .$$

The displacement is matched by the a's, and the velocity by the b's: Differentiating $u(t)$ and setting $t = 0$, the b's are determined by

$$u_0' = b_1 \omega_1 x_1 + \cdots + b_n \omega_n x_n .$$

Substituting back into (33), the equation is solved.

We want to apply these formulas to the example. Its eigenvalues were $\lambda_1 = -1$ and $\lambda_2 = -3$, so the frequencies are $\omega_1 = 1$ and $\omega_2 = \sqrt{3}$. If the system starts from rest (the initial velocity u_0' is zero), then the terms in $b \sin \omega t$ will disappear. And if the first oscillator is given a unit displacement, the requirement $u_0 = a_1 x_1 + a_2 x_2$ leads to

$$\begin{bmatrix} 1 \\ 0 \end{bmatrix} = a_1 \begin{bmatrix} 1 \\ 1 \end{bmatrix} + a_2 \begin{bmatrix} 1 \\ -1 \end{bmatrix}, \qquad \text{or} \qquad a_1 = a_2 = \tfrac{1}{2}.$$

Therefore the solution is

$$u(t) = \tfrac{1}{2} \cos t \begin{bmatrix} 1 \\ 1 \end{bmatrix} + \tfrac{1}{2} \cos \sqrt{3} t \begin{bmatrix} 1 \\ -1 \end{bmatrix}.$$

Suppose we interpret this solution physically. The system represents two unit masses, connected to each other and to stationary walls by three identical springs (Fig. 5.2). The first mass is pushed to $v_0 = 1$, the second mass is held in place, and at $t = 0$ we let go. Their motion $u(t)$ becomes an average of two pure oscillations, corresponding to the two eigenvectors. In the first mode, the masses move exactly in unison and the spring in the middle is never stretched (Fig. 5.2a). The frequency $\omega_1 = 1$ is the same as for a single spring and a single mass. In the faster mode $x_2 = (1, -1)$, with components of opposite sign and with frequency $\sqrt{3}$, the masses move in opposite directions but with equal speeds (Fig. 5.2b). The general solution is a combination of these two normal modes, and our particular solution is half of each.

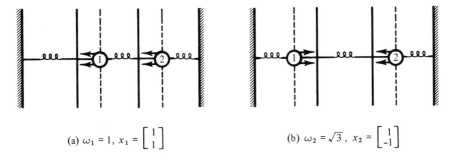

(a) $\omega_1 = 1$, $x_1 = \begin{bmatrix} 1 \\ 1 \end{bmatrix}$ (b) $\omega_2 = \sqrt{3}$, $x_2 = \begin{bmatrix} 1 \\ -1 \end{bmatrix}$

Fig. 5.2. The two normal modes of oscillation.

As time goes on, the motion is what we call "almost periodic." If the ratio ω_1/ω_2 had been a fraction, the two masses would eventually return to $v = 1$ and $w = 0$, and then begin the whole pattern again. A combination of $\sin 2t$ and $\sin 3t$ would have a period of 2π. But since $\sqrt{3}$ is irrational, the best we can say is that the masses will come *arbitrarily close* to reproducing the initial situation. They also come close, if we can wait long enough, to the opposite situation $v = 0$ and $w = 1$. Like a billiard ball bouncing forever on a perfectly smooth table, the total energy of the masses is fixed, and sooner or later they come arbitrarily near to any state with this energy.

Again we cannot leave the problem without drawing a parallel to the continuous case: Instead of two masses, or N masses, there is a continuum. As the discrete masses and springs merge into a solid rod, the "second differences" given by the matrix coefficients $1, -2, 1$ turn into second derivatives. This limit is described by the celebrated *wave equation* $\partial^2 u/\partial t^2 = \partial^2 u/\partial x^2$.

EXERCISE 5.4.10   Use the diagonalization on p. 178 to solve

$$\frac{d^2u}{dt^2} = \begin{bmatrix} -1 & 1 & 0 \\ 1 & -2 & 1 \\ 0 & 1 & -1 \end{bmatrix} u, \quad \text{with} \quad u_0 = \begin{bmatrix} 2 \\ -2 \\ 0 \end{bmatrix} \quad \text{and} \quad u_0' = \begin{bmatrix} 1 \\ 1 \\ 1 \end{bmatrix}.$$

EXERCISE 5.4.11   Solve

$$\frac{d^2u}{dt^2} = \begin{bmatrix} -5 & -1 \\ -1 & -5 \end{bmatrix} u, \quad \text{with} \quad u_0 = \begin{bmatrix} 1 \\ 0 \end{bmatrix} \quad \text{and} \quad u_0' = \begin{bmatrix} 0 \\ 0 \end{bmatrix}.$$

EXERCISE 5.4.12   With a friction matrix F, the two unit masses will oscillate according to $d^2u/dt^2 = -F\, du/dt + Au$. Substitute a pure exponential $e^{\rho t}x$ into the equation and establish a *quadratic eigenvalue problem* for ρ.

EXERCISE 5.4.13   The solution to $u'' = Au$ is still a mixture of the eigenvectors of A, which go into the columns of S. Instead of $u = Se^{\Lambda t}S^{-1}u_0$, which appeared in the case $u' = Au$, show that the initial conditions and differential equation are satisfied by

$$u(t) = S \begin{bmatrix} \cos \omega_1 t & & \\ & \ddots & \\ & & \cos \omega_n t \end{bmatrix} S^{-1}u_0 + S \begin{bmatrix} (\sin \omega_1 t)/\omega_1 & & \\ & \ddots & \\ & & (\sin \omega_n t)/\omega_n \end{bmatrix} S^{-1}u_0'.$$

EXERCISE 5.4.14   For the example in the text, find the motion of the second mass if the first one is hit at $t = 0$: $u_0 = \begin{bmatrix} 0 \\ 0 \end{bmatrix}$ and $u_0' = \begin{bmatrix} 1 \\ 0 \end{bmatrix}$.

5.5 ■ THE COMPLEX CASE: HERMITIAN AND UNITARY MATRICES

It is no longer possible to work only with real vectors and real matrices. In the first half of this book, when the basic problem was $Ax = b$, it was certain that

x would be real whenever A and b were. Therefore there was no need for complex numbers; they could have been permitted, but would have contributed nothing new. Now we cannot avoid them. A real matrix has real coefficients in its characteristic polynomial, but the eigenvalues (as in the case when $\lambda^2 + 1 = 0$) may fail to be real.

We shall therefore introduce the space $\mathbf{C}^n$ of vectors with n complex components. Addition and matrix multiplication follow the same rules as before. But *the length of a vector has to be changed from* $\| x \|^2 = x_1^2 + \cdots + x_n^2$, or the vector with components $(1, i)$ will have zero length: $1^2 + i^2 = 0$. This change in the computation of length forces a whole series of other changes. The inner product of two vectors, the transpose of a matrix, the definitions of symmetric, skew-symmetric, and orthogonal matrices, all need to be modified in the presence of complex numbers. In every case, the new definition coincides with the old when the vectors and matrices are real.

We have listed all these changes on p. 220, and that list virtually amounts to a dictionary for translating between the real and the complex case. We hope it will be useful to the reader. It also includes, for each class of matrices, the best information known about the location of their eigenvalues. We particularly want to find out about symmetric matrices: Where are their eigenvalues, and what is special about their eigenvectors? For practical purposes, those are the most important questions in the theory of eigenvalues, and therefore this is the most important section.

Complex Numbers and Their Conjugates

Probably the reader is already acquainted with complex numbers; but since only the most basic facts are needed, a brief review is easy to give.† Everyone knows that whatever i is, it satisfies the equation $i^2 = -1$. It is a pure imaginary number, and so are its multiples ib; b is real. The sum of a real and an imaginary number is a complex number $a + ib$; and it is plotted in a natural way on the complex plane (Fig. 5.3).

The real numbers (for which $b = 0$) and the imaginary numbers ($a = 0$) are included as special cases of complex numbers; they lie on one coordinate axis or the other. Two complex numbers are added by

$$(a + ib) + (c + id) = (a + c) + i(b + d),$$

and multiplied using the rule that $i^2 = -1$:

$$(a + ib)(c + id) = ac + ibc + iad + i^2bd = (ac - bd) + i(bc + ad).$$

The *complex conjugate* of $a + ib$ is the number $a - ib$, that is, with the sign of the imaginary part reversed. Geometrically, it is the mirror image on the other side of the real axis; any real number is its own conjugate. The conjugate

† The important ideas are the complex conjugate and the modulus.

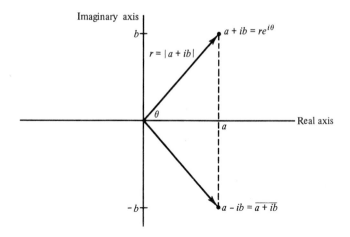

Fig. 5.3.   The complex plane.

is denoted by a bar, $\overline{a + ib} = a - ib$, and it has three important properties:

(1)   The conjugate of a product equals the product of the conjugates:

$$\overline{(a + ib)(c + id)} = (ac - bd) - i(bc + ad) = (\overline{a + ib})(\overline{c + id}). \quad (34)$$

(2)   The conjugate of a sum equals the sum of the conjugates:

$$\overline{(a + c) + i(b + d)} = (a + c) - i(b + d) = (\overline{a + ib}) + (\overline{c + id}).$$

(3)   Multiplying any $a + ib$ by its own conjugate $a - ib$ produces a real number, which is the square of the hypotenuse in Fig. 5.3:

$$(a + ib)(a - ib) = a^2 + b^2 = r^2. \quad (35)$$

This distance r is called the *modulus* of the original $a + ib$, and (like the absolute value of a real number, which also is never negative) it is denoted by vertical bars: $|a + ib| = r = \sqrt{a^2 + b^2}$.

Finally, trigonometry connects the sides to the hypotenuse by

$$a = \sqrt{a^2 + b^2}\cos\theta, \qquad b = \sqrt{a^2 + b^2}\sin\theta.$$

By combining these equations, we move into polar coordinates:

$$a + ib = \sqrt{a^2 + b^2}(\cos\theta + i\sin\theta) = re^{i\theta}. \quad (36)$$

There is an important special case, when the modulus r equals one. Then the complex number is just $e^{i\theta} = \cos\theta + i\sin\theta$, and it falls on the *unit circle* in the complex plane. As θ varies from 0 to 2π, this number $e^{i\theta}$ circles around the origin at the constant radial distance $|e^{i\theta}| = \sqrt{\cos^2\theta + \sin^2\theta} = 1$.

EXERCISE 5.5.1   For the complex numbers $3 + 4i$ and $1 - i$

(a)   find their positions in the complex plane;
(b)   find their sum and product;
(c)   find their conjugates and their moduli.

Do they lie inside or outside the unit circle?

EXERCISE 5.5.2   What can you say about

(i)   the sum of a complex number and its conjugate?
(ii)   the conjugate of a number on the unit circle?
(iii)   the product of two numbers on the unit circle?
(iv)   the sum of two numbers on the unit circle?

EXERCISE 5.5.3   Verify that multiplying two numbers just multiplies their moduli:

$$| (a + ib)(c + id) | = | a + ib | | c + id |, \quad \text{or} \quad | (re^{i\theta})(Re^{i\phi}) | = rR.$$

Lengths and Transposes in the Complex Case

We return to linear algebra, and make the conversion from real to complex. The first step is to admit complex vectors, and that is no problem: By definition, *the space $\mathbf{C}^n$ contains all vectors x with n complex components*:

$$x = \begin{bmatrix} x_1 \\ x_2 \\ \vdots \\ x_n \end{bmatrix}, \quad x_j = a_j + ib_j.$$

Vectors x and y are still added component by component, but scalar multiplication is now done with complex numbers. As before, the vectors $v_1, \ldots, v_k$ are linearly dependent if some nontrivial combination $c_1 v_1 + \cdots + c_k v_k$ produces the zero vector; the c_j may now be complex. The unit coordinate vectors are still in $\mathbf{C}^n$; they are still independent; and they still form a basis. Therefore $\mathbf{C}^n$ is also a vector space of dimension n.

We have already emphasized that the definition of length has to be changed; the square of a complex number is not necessarily positive, and $\| x \|^2 = x_1^2 + \cdots + x_n^2$ is of no use. The new definition is completely natural: x_j^2 is replaced by its modulus $| x_j |^2$, and the length satisfies

$$\| x \|^2 = | x_1 |^2 + \cdots + | x_n |^2. \tag{37}$$

In two dimensions,

$$x = \begin{bmatrix} 1 \\ i \end{bmatrix} \quad \text{and} \quad \| x \|^2 = 2;$$

$$y = \begin{bmatrix} 2 + i \\ 2 - 4i \end{bmatrix} \quad \text{and} \quad \| y \|^2 = 25.$$

For real vectors there was a close connection between the length and the inner product: $\| x \|^2 = x^{\mathrm{T}}x$. This connection we want to preserve. Therefore the inner product must be modified to match the new definition of length, and the standard modification is *to conjugate the first vector in the inner product*. This means that x is replaced by $\bar{x}$, and the inner product of x and y becomes

$$\bar{x}^{\mathrm{T}}y = \bar{x}_1 y_1 + \cdots + \bar{x}_n y_n. \tag{38}$$

A typical example in $\mathbf{C}^2$ is

$$x = \begin{bmatrix} 1 + i \\ 3i \end{bmatrix}, \qquad y = \begin{bmatrix} 4 \\ 2 - i \end{bmatrix},$$

$$\bar{x}^{\mathrm{T}}y = (1 - i)4 + (-3i)(2 - i) = 1 - 10i.$$

And if we take the inner product of x with itself, we are back to the square of the length:

$$\bar{x}^{\mathrm{T}}x = (\overline{1 + i})(1 + i) + (\overline{3i})(3i) = 2 + 9 = \| x \|^2.$$

Note that $\bar{y}^{\mathrm{T}}x$ is different from $\bar{x}^{\mathrm{T}}y$; from now on we have to watch the order of the vectors in inner products. And there is a further novelty: If x is changed to cx, then the inner product of x and y is multiplied not by c, but by $\bar{c}$.

This leaves only one more change to make. It is a change in notation more than anything else, and it condenses two symbols into one: Instead of a bar for the conjugate and a T for the transpose, these operations are combined into the **conjugate transpose**, and denoted by a superscript H. Thus $\bar{x}^{\mathrm{T}} = x^{\mathrm{H}}$, and the same notation applies to matrices: The conjugate transpose of A is

$$\bar{A}^{\mathrm{T}} = A^{\mathrm{H}}, \qquad \text{with entries} \qquad (A^{\mathrm{H}})_{ij} = \overline{A_{ji}}. \tag{39}$$

If A is an m by n matrix, then A^{H} is n by m. For example,

$$\begin{bmatrix} 2 + i & 3i \\ 4 - i & 5 \\ 0 & 0 \end{bmatrix}^{\mathrm{H}} = \begin{bmatrix} 2 - i & 4 + i & 0 \\ -3i & 5 & 0 \end{bmatrix}.$$

This symbol A^{H} gives official recognition to the fact that, with complex entries, it is very seldom that we want only the transpose of A. It is the conjugate transpose, or *Hermitian transpose*, which becomes appropriate in virtually every case.† The modifications required by complex numbers are easily summarized:

5L   (i)   The inner product of x and y is $x^{\mathrm{H}}y$, and these vectors are orthogonal if $x^{\mathrm{H}}y = 0$.
   (ii)   The length of x is $\| x \| = (x^{\mathrm{H}}x)^{1/2}$.
   (iii)   The rule $(AB)^{\mathrm{T}} = B^{\mathrm{T}}A^{\mathrm{T}}$, after conjugating every entry, turns into $(AB)^{\mathrm{H}} = B^{\mathrm{H}}A^{\mathrm{H}}$.

EXERCISE 5.5.4   Find the lengths and the inner product of

$$x = \begin{bmatrix} 2 - 4i \\ 4i \end{bmatrix} \qquad \text{and} \qquad y = \begin{bmatrix} 2 + 4i \\ 4 \end{bmatrix}.$$

† The matrix A^{H} is often referred to as "A Hermitian." Unfortunately, you have to listen closely to distinguish that name from the phrase "A is Hermitian," which means that A equals A^{H}.

EXERCISE 5.5.5   Write out the matrix A^H and compute $C = A^H A$ if

$$A = \begin{bmatrix} 1 & i & 0 \\ i & 0 & 1 \end{bmatrix}.$$

What is the relation between C and C^H? Does it hold whenever C is constructed from some $A^H A$?

EXERCISE 5.5.6   (i)   With the preceding A, use elimination to solve $Ax = 0$.
(ii)   Show that the nullspace that you just computed is orthogonal to $\mathfrak{R}(A^H)$ *and not to the usual row space* $\mathfrak{R}(A^T)$. The four fundamental spaces in the complex case are $\mathfrak{N}(A)$ and $\mathfrak{R}(A)$, as before, and then $\mathfrak{N}(A^H)$ and $\mathfrak{R}(A^H)$.

Hermitian Matrices

We spoke in earlier chapters about symmetric matrices: $A = A^T$. Now, in the presence of matrices with complex entries, this idea of symmetry has to be extended. The right generalization is not to matrices that equal their transpose, but to *matrices that equal their conjugate transpose*. These are the Hermitian matrices, and a typical example is

$$A = \begin{bmatrix} 2 & 3 - 3i \\ 3 + 3i & 5 \end{bmatrix} = A^H. \qquad (40)$$

Notice that the diagonal entries must be real; they have to be unchanged by the process of conjugation. Each off-diagonal entry is matched with its mirror image across the main diagonal, and the two are complex conjugates of one another. In every case $a_{ij} = \overline{a_{ji}}$, and this example will illustrate very clearly the four basic properties of Hermitian matrices.

Our main goal is to establish those four properties, and it needs to be emphasized again that they apply equally well to real symmetric matrices. The latter represent a special case of Hermitian matrices, and the most important case.

Property 1   If $A = A^H$, then for all complex vectors x, $x^H A x$ is real.

There is a contribution to $x^H A x$ from every entry of A,

$$x^H A x = \begin{bmatrix} \bar{u} & \bar{v} \end{bmatrix} \begin{bmatrix} 2 & 3 - 3i \\ 3 + 3i & 5 \end{bmatrix} \begin{bmatrix} u \\ v \end{bmatrix}$$

$$= 2\bar{u}u + 5\bar{v}v + (3 - 3i)\bar{u}v + (3 + 3i)u\bar{v}.$$

Each of the "diagonal terms" is real, because $2\bar{u}u = 2 \, | \, u \, |^2$ and $5\bar{v}v = 5 \, | \, v \, |^2$. The off-diagonal terms are complex conjugates of one another, so they combine to give twice the real part of $(3 - 3i)\bar{u}v$. Therefore the whole expression $x^H A x$ is real.

For a proof in general, we can compute $(x^H A x)^H$. We should get the conjugate of the 1 by 1 matrix $x^H A x$, but we actually get the same number back again: $(x^H A x)^H = x^H A^H x^{HH} = x^H A x$. So that number must be real.

Property 2   Every eigenvalue of a Hermitian matrix is real.

Proof   Suppose λ is an eigenvalue, and x is a corresponding nonzero eigenvector: $Ax = \lambda x$. Then the trick is to multiply by x^H: $x^H A x = \lambda x^H x$. The left side is real by Property 1, and the right side $x^H x = \| x \|^2$ is real and positive, because $x \neq 0$. Therefore λ must be real. In our example,

$$| A - \lambda I | = \begin{vmatrix} 2 - \lambda & 3 - 3i \\ 3 + 3i & 5 - \lambda \end{vmatrix} = \lambda^2 - 7\lambda + 10 - | 3 - 3i |^2$$

$$= \lambda^2 - 7\lambda - 8 = (\lambda - 8)(\lambda + 1). \tag{41}$$

Property 3   The eigenvectors of a Hermitian matrix, if they correspond to different eigenvalues, are orthogonal to one another.

Again the proof starts with the information given, $Ax = \lambda x$ and $Ay = \mu y$ with $\lambda \neq \mu$, and requires a small trick. The Hermitian transpose of $Ax = \lambda x$ is normally $x^H A^H = \bar{\lambda} x^H$; but because $A = A^H$, and λ is real by Property 2, this is really $x^H A = \lambda x^H$. Multiplying this equation on the right by y, and the other equation on the left by x^H, we find

$$x^H A y = \lambda x^H y \quad \text{and} \quad x^H A y = \mu x^H y.$$

Therefore $\lambda x^H y = \mu x^H y$, and since $\lambda \neq \mu$, we conclude that $x^H y = 0$: x is *orthogonal to y.* In our example, with $Ax = 8x$ and $Ay = -y$, the eigenvectors are computed in the usual way from

$$(A - 8I)x = \begin{bmatrix} -6 & 3 - 3i \\ 3 + 3i & -3 \end{bmatrix} \begin{bmatrix} x_1 \\ x_2 \end{bmatrix} = \begin{bmatrix} 0 \\ 0 \end{bmatrix}, \quad x = \begin{bmatrix} 1 \\ 1 + i \end{bmatrix}$$

$$(A + I)y = \begin{bmatrix} 3 & 3 - 3i \\ 3 + 3i & 6 \end{bmatrix} \begin{bmatrix} y_1 \\ y_2 \end{bmatrix} = \begin{bmatrix} 0 \\ 0 \end{bmatrix}, \quad y = \begin{bmatrix} 1 - i \\ -1 \end{bmatrix}.$$

These two eigenvectors are orthogonal:

$$x^H y = \begin{bmatrix} 1 & 1 - i \end{bmatrix} \begin{bmatrix} 1 - i \\ -1 \end{bmatrix} = 0.$$

Of course any multiples x/α and y/β would be equally good as eigenvectors. Suppose we pick $\alpha = \| x \|$ and $\beta = \| y \|$, so that x/α and y/β are unit vectors; the eigenvectors have been normalized to have length one. Since they were already orthogonal, they are now *orthonormal.* If they are chosen to be the columns of S, then (as always, when the eigenvectors are the columns) we have $S^{-1}AS = \Lambda$. *The diagonalizing matrix has orthonormal columns.* If the original A is real and symmetric, then its eigenvalues and eigenvectors are real; therefore normalizing the eigenvectors turns S into an orthogonal matrix. A real symmetric matrix can be diagonalized by an orthogonal matrix Q.

In the complex case, we need a new name and a new symbol for a matrix with orthonormal columns: It is called a **unitary matrix**, and denoted by U. In exact analogy with real orthogonal matrices, which satisfy $Q^T Q = I$ and therefore $Q^T = Q^{-1}$, the property of orthonormal columns translates into

$$U^H U = \begin{bmatrix} \text{---} \bar{x}_1 \text{---} \\ \text{---} \bar{x}_2 \text{---} \\ \\ \text{---} \bar{x}_n \text{---} \end{bmatrix} \begin{bmatrix} & & & \\ x_1 & x_2 & \cdots & x_n \\ & & & \end{bmatrix} = I, \quad \text{or} \quad U^H = U^{-1}. \quad (42)$$

This leads to the last special property of Hermitian matrices:

Property 4 If $A = A^H$, then there is a diagonalizing matrix which is also unitary, $S = U$: Its columns are orthonormal, and

$$U^{-1}AU = U^H A U = \Lambda. \quad (43)$$

Therefore any Hermitian matrix can be decomposed into

$$A = U\Lambda U^H = \lambda_1 x_1 x_1^H + \lambda_2 x_2 x_2^H + \cdots + \lambda_n x_n x_n^H. \quad (44)$$

This decomposition is known as the **spectral theorem**. It expresses A as a combination of the one-dimensional projections $x_i x_i^H$, which are just like the projections aa^T of Chapter 3. They split any vector b into its components $p = x_i(x_i^H b)$ in the directions of the unit eigenvectors, which are a set of mutually perpendicular axes. Then these individual projections p are weighted by the λ_i and reassembled to form

$$Ab = \lambda_1 x_1(x_1^H b) + \cdots + \lambda_n x_n(x_n^H b). \quad (45)$$

If every $\lambda_i = 1$, we have reassembled b itself; $A = UIU^H$ is the identity. In any case we can verify (44) and (45) directly from matrix multiplication:

$$Ab = U\Lambda U^H b = \begin{bmatrix} & & \\ x_1 & \cdots & x_n \\ & & \end{bmatrix} \begin{bmatrix} \lambda_1 & & \\ & \ddots & \\ & & \lambda_n \end{bmatrix} \begin{bmatrix} \text{---} \bar{x}_1 \text{---} \\ \\ \text{---} \bar{x}_n \text{---} \end{bmatrix} \begin{bmatrix} \\ b \\ \end{bmatrix}$$

$$= \begin{bmatrix} & & \\ x_1 & \cdots & x_n \\ & & \end{bmatrix} \begin{bmatrix} \lambda_1 x_1^H b \\ \vdots \\ \lambda_n x_n^H b \end{bmatrix} = \lambda_1 x_1 x_1^H b + \cdots + \lambda_n x_n x_n^H b.$$

In our example, both eigenvectors $x = \begin{bmatrix} 1 \\ 1+i \end{bmatrix}$ and $y = \begin{bmatrix} 1-i \\ -1 \end{bmatrix}$ have length $\sqrt{3}$, and a normalization produces the unitary matrix that diagonalizes A:

$$U = \frac{1}{\sqrt{3}} \begin{bmatrix} 1 & 1-i \\ 1+i & -1 \end{bmatrix}, \qquad U^{-1}AU = \begin{bmatrix} 8 & 0 \\ 0 & -1 \end{bmatrix}.$$

Then the spectral decomposition $U\Lambda U^{\mathrm{H}} = \lambda_1 x_1 x_1^{\mathrm{H}} + \lambda_2 x_2 x_2^{\mathrm{H}}$ becomes

$$\frac{8}{3} \begin{bmatrix} 1 \\ 1+i \end{bmatrix} \begin{bmatrix} 1 & 1-i \end{bmatrix} - \frac{1}{3} \begin{bmatrix} 1-i \\ -1 \end{bmatrix} \begin{bmatrix} 1+i & -1 \end{bmatrix} = \begin{bmatrix} 2 & 3-3i \\ 3+3i & 5 \end{bmatrix} = A.$$

EXERCISE 5.5.7 Write down any non-Hermitian matrix, even a real one, and find an x such that $x^{\mathrm{H}}Ax$ is not real. (Only Hermitian matrices have Property 1.)

EXERCISE 5.5.8 Find the eigenvalues and unit eigenvectors of $A = \begin{bmatrix} 0 & i \\ -i & 0 \end{bmatrix}$. Compute the matrices $\lambda_1 x_1 x_1^{\mathrm{H}}$ and $\lambda_2 x_2 x_2^{\mathrm{H}}$, and verify that their sum is A and their product is zero. (Why is $(\lambda_1 x_1 x_1^{\mathrm{H}})(\lambda_2 x_2 x_2^{\mathrm{H}}) = 0$?)

EXERCISE 5.5.9 Using the eigenvalues and eigenvectors computed on p. 178, find the unitary U that diagonalizes A and complete the spectral decomposition (into three matrices $\lambda_i x_i x_i^{\mathrm{H}}$) of

$$A = \begin{bmatrix} 1 & -1 & 0 \\ -1 & 2 & -1 \\ 0 & -1 & 1 \end{bmatrix}.$$

EXERCISE 5.5.10 Prove that the determinant of any Hermitian matrix is real.

The spectral theorem, and the rest of Property 4, have actually been proved only when the eigenvalues of A are distinct. (Then there are certainly n independent eigenvectors, and A can be safely diagonalized.) Nevertheless it is true (see Section 5.6) that *even with repeated eigenvalues, a Hermitian matrix still has a complete set of orthonormal eigenvectors.* Therefore in every case A can be diagonalized by a unitary U.

It is only Hermitian matrices that combine real eigenvalues with orthonormal eigenvectors. If $U^{-1}AU$ equals a real diagonal matrix Λ, and U is unitary, then A is necessarily Hermitian:

$$A^{\mathrm{H}} = (U\Lambda U^{-1})^{\mathrm{H}} = U\Lambda U^{\mathrm{H}} = A.$$

To any mathematician, this decomposition $A = U\Lambda U^{\mathrm{H}}$ looks absolutely cut-and-dried; it is fundamental, but no one would call it controversial. Fortunately, the psychologists do not agree with that, or with each other. They manage to generate plenty of argument about the spectral theorem, and we shall try for a brief explanation. ***However, the next two pages can be omitted entirely with no loss in continuity.***

Factor Analysis and Principal Component Analysis

The goal is to find some pattern, or some natural structure, within a set of experimental data—for example, a set of test scores in physics, mathematics, history, and English. This general field is called multivariate analysis. Among its subfields there are some which organize the experimental results for external purposes, to predict something outside the data, and others which concentrate entirely on its internal structure.

(1) The prediction problem is typical of *regression analysis*: We may find out the IQ of each student, and ask whether the test scores could have been used (or could be used in the future) to predict the IQ. In other words, we look for a linear combination of the four test scores equaling the IQ. There are four unknown coefficients in this combination, and there is one equation for every student. This is a classical problem in least squares, with more equations than unknowns, and eigenvalues are not involved.†

(2) For the internal structure, suppose that the data are boiled down into a 4 by 4 *correlation matrix R*. The diagonal entries equal 1 (each variate x_i is perfectly correlated with itself), and the off-diagonal entry r_{ij} gives the correlation between the results of tests i and j, say between physics and mathematics. Then one kind of analysis, called *principal component analysis*, looks for combinations of these four tests which are not correlated. In the language of linear algebra, we want vectors that are mutually orthogonal. For this purpose, the spectral decomposition of the real symmetric matrix R is the perfect tool: If the eigenvectors are $q_1, \ldots, q_4$, then they fit into an orthogonal matrix Q that diagonalizes R: $Q^{-1}RQ = \Lambda$, or

$$
R = Q\Lambda Q^{\mathrm{T}} = \begin{bmatrix} q_1 & \cdots & q_4 \end{bmatrix} \begin{bmatrix} \lambda_1 & & \\ & \ddots & \\ & & \lambda_4 \end{bmatrix} \begin{bmatrix} q_1^{\mathrm{T}} \\ \vdots \\ q_4^{\mathrm{T}} \end{bmatrix}
$$

$$
= \begin{bmatrix} \sqrt{\lambda_1}\, q_1 & \cdots & \sqrt{\lambda_4}\, q_4 \end{bmatrix} \begin{bmatrix} \sqrt{\lambda_1}\, q_1^{\mathrm{T}} \\ \vdots \\ \sqrt{\lambda_4}\, q_4^{\mathrm{T}} \end{bmatrix}.
$$

Q is just our unitary U, which in this application is real. Each vector $\sqrt{\lambda_i}\, q_i$ provides a combination of the four tests, and these four special combinations are uncorrelated. So far, no great controversy.

(3) We turn now to *factor analysis*, which is also devoted to accounting for the off-diagonal terms in the correlation matrix. It tries to do so with a

---

† Our example is sort of backward since it is more normal to use the IQ in predicting something else. But perhaps the reversibility of these predictions, which is almost the reversibility of past and future, has some significance.

comparatively small number of factors f_i—which are again linear combinations of the original tests. But *these factors are not expected to account fully for the diagonal entries* $r_{ii} = 1$. If one test shows a very narrow spread of grades, because the questions are all too hard or too easy, and another shows a wide variation, then factor analysis proposes to ignore the difference. It is concerned with getting the interdependence right, and an ideal situation would be one in which the correlation matrix decomposes into

$$R = \begin{bmatrix} 1 & .74 & .24 & .24 \\ .74 & 1 & .24 & .24 \\ .24 & .24 & 1 & .74 \\ .24 & .24 & .74 & 1 \end{bmatrix} = \begin{bmatrix} .7 & .5 \\ .7 & .5 \\ .7 & -.5 \\ .7 & -.5 \end{bmatrix} \begin{bmatrix} .7 & .7 & .7 & .7 \\ .5 & .5 & -.5 & -.5 \end{bmatrix} + .26I$$

In this case two factors account for all the correlations. The first is a column with equal entries .7, which would be interpreted as a general intelligence factor: Good scores in mathematics tend to go with good scores in English. The second factor distinguishes the very strong correlation between mathematics and physics, and between English and history, from the weaker correlation between the two groups. It is a kind of "literacy versus numeracy" factor. These are the right variables to use in interpreting the data, and their components .7, .5, and $-.5$ are the *loadings* of the factors on the individual tests. In all four tests, the factors account for .74 of the individual variances on the main diagonal; these are the "communalities," the diagonal entries of FF^T in the decomposition $R = FF^\mathrm{T} + D$.

The arguments come from two sources. First, the factor matrix F and the diagonal D are by no means unique. We could have factored the same R into

$$R = \begin{bmatrix} .6 & \sqrt{.38} & 0 \\ .6 & \sqrt{.38} & 0 \\ .4 & 0 & \sqrt{.58} \\ .4 & 0 & \sqrt{.58} \end{bmatrix} \begin{bmatrix} .6 & .6 & .4 & .4 \\ \sqrt{.38} & \sqrt{.38} & 0 & 0 \\ 0 & 0 & \sqrt{.58} & \sqrt{.58} \end{bmatrix} + D.$$

Here we changed the number of factors, which no one knows in advance. In fact the communalities are equally unknown, so we could have changed D and produced a million totally different decompositions. It is completely unclear how much weight to assign to general intelligence, and how to separate literacy from numeracy.

The other source of argument, once D is estimated and p factors are being looked for, is the possibility of "rotating the factors." For any orthogonal matrix Q of order p, the factor matrix $\tilde{F} = FQ$ accounts for exactly the same correlations $\tilde{F}\tilde{F}^\mathrm{T} = FQQ^\mathrm{T}F^\mathrm{T} = FF^\mathrm{T}$ that F does. Therefore F and $\tilde{F}$ are completely interchangeable. In a typical problem, the original factors may have substantial loadings on dozens of variables, and such a factor is practically impossible to interpret. It has a mathematical meaning, as a vector f_i, but no useful meaning to a social scientist. Therefore he tries to choose a rotation that produces a simple structure, with large loadings in a few components and negligible loadings elsewhere. Even oblique factors are used in order to obtain positive loadings; the orthogonality $f_i^T f_j = 0$ was lost in our second decomposition of R. Thus two experts, each estimating D, p, and Q, could very easily produce *completely different interpretations of the same data*. But the technique is so much needed that, even starting as an unwelcome black sheep of multivariate analysis, it has spread from psychology into biology and economics and the social sciences.

EXERCISE 5.5.11   Find another $FF^T + D$ decomposition of the same R. Does it have any meaning?

Unitary and Skew-Hermitian Matrices

May we propose an analogy? *A Hermitian matrix can be compared to a real number, a skew-Hermitian matrix to a pure imaginary number, and a unitary matrix to a number on the unit circle*, that is, a complex number $e^{i\theta}$ of modulus one. For the eigenvalues of these matrices, the proposed comparison is more than an analogy: The λ's are themselves real if $A^H = A$, or purely imaginary if $K^H = -K$, or on the unit circle if $U^H = U^{-1}$. For all these matrices the eigenvectors are orthogonal, and can be chosen orthonormal.†

Since the properties of a Hermitian matrix are already established, it remains only to deal with K and U. We look for a similar set of four fundamental properties, and a skew-Hermitian K is so closely linked to a Hermitian A that no work is required:

5M   If K is skew-Hermitian, so that $K^H = -K$, then the matrix $A = iK$ is Hermitian:

$$A^H = (i)^H(K)^H = (-i)(-K) = A.$$

Similarly, if A is Hermitian, then $K = iA$ is skew-Hermitian.

The Hermitian example on the previous pages would lead to

$$K = iA = \begin{bmatrix} 2i & 3 + 3i \\ -3 + 3i & 5i \end{bmatrix} = -K^H.$$

The diagonal entries are always multiples of i (allowing zero).

---

† In the next section we identify the more inclusive class of matrices corresponding to the set of all complex numbers. A matrix without orthogonal eigenvectors belongs to none of these classes, and is outside the whole analogy.

This simple relationship between K and A converts Properties 1–4 into the following:

(**1′**)   For any x, the combination $x^H K x$ is pure imaginary.

(**2′**)   Every eigenvalue of K is pure imaginary.

(**3′**)   Eigenvectors corresponding to different eigenvalues are orthogonal.

(**4′**)   There is a unitary U such that $U^{-1} K U = \Lambda$.

EXERCISE 5.5.12   If K is skew-Hermitian, how do you know that $K - I$ is nonsingular?

EXERCISE 5.5.13   Construct any real K and real x, and verify that $x^H K x = 0$. Explain how this follows from Property $1′$.

EXERCISE 5.5.14   Every matrix Z can be split into a Hermitian and a skew-Hermitian part, $Z = A + K$, just as a complex number z is split into $a + ib$. The real part of z is half of $z + \bar{z}$, and the "real part" of Z is half of $Z + Z^H$. Find a similar formula for the "imaginary part" K, and split

$$Z = \begin{bmatrix} 3 + i & 4 + 2i \\ 0 & 5 \end{bmatrix}$$

into $A + K$.

Now we turn to unitary matrices. They are already defined as *square matrices with orthonormal columns*:

$$U^H U = I, \quad \text{or} \quad U U^H = I, \quad \text{or} \quad U^H = U^{-1}. \tag{46}$$

This leads directly to the first of their properties: Multiplication by U has no effect on inner products or angles or lengths. That is a property that we already knew for orthogonal matrices (which are real unitary matrices, $U = Q$). Now we pass to the complex case, and the proof is again on one line:

Property 1″ $(Ux)^H(Uy) = x^H U^H U y = x^H y$, and $\|Ux\|^2 = \|x\|^2$.†

The next property locates the eigenvalues of U, in analogy with 2 and $2′$; for a unitary matrix, each λ is on the unit circle.

Property 2″   Every eigenvalue of U has modulus $|\lambda| = 1$.

This follows directly from $Ux = \lambda x$, by comparing the lengths of the two sides: $\|Ux\| = \|x\|$ because of $1″$, and $\|\lambda x\| = |\lambda| \|x\|$. Therefore $|\lambda| = 1$.

† Lengths are preserved whenever inner products are; just choose $y = x$.

Property 3″ Eigenvectors corresponding to different eigenvalues are orthogonal.

The proof starts from $Ux = \lambda x$, $Uy = \sigma y$, and their inner product

$$(Ux)^H Uy = (\lambda x)^H \sigma y, \quad \text{or} \quad x^H y = (\bar{\lambda}\sigma) x^H y. \quad (47)$$

If σ equaled λ, we would have $\bar{\lambda}\sigma = \bar{\lambda}\lambda = |\lambda|^2 = 1$. But since σ differs from λ, we know that $\bar{\lambda}\sigma \neq 1$—and therefore that $x^H y = 0$.

Normalizing these eigenvectors to have unit length, and placing them into the columns of S, this diagonalizing matrix becomes itself a unitary matrix—not to be confused with the U it diagonalizes!

Property 4″ If U is unitary, it has a diagonalizing matrix S that is also unitary: $S^{-1}US = S^H US = \Lambda$.

All four properties are illustrated by the rotation matrices

$$U(t) = \begin{bmatrix} \cos t & -\sin t \\ \sin t & \cos t \end{bmatrix}.$$

Their eigenvalues are e^{it} and e^{-it}, both of modulus one. The eigenvectors are $\begin{bmatrix} 1 \\ -i \end{bmatrix}$ and $\begin{bmatrix} 1 \\ i \end{bmatrix}$, which are orthogonal. After normalization they fit into a unitary S:

$$U = S\Lambda S^{-1} = \begin{bmatrix} 1/\sqrt{2} & 1/\sqrt{2} \\ -i/\sqrt{2} & i/\sqrt{2} \end{bmatrix} \begin{bmatrix} e^{it} & \\ & e^{-it} \end{bmatrix} \begin{bmatrix} 1/\sqrt{2} & i/\sqrt{2} \\ 1/\sqrt{2} & -i/\sqrt{2} \end{bmatrix}.$$

Certainly rotations preserve lengths, $\|Ux\| = \|x\|$, and the example on p. 202 identified this particular U as the exponential of a skew-symmetric matrix K:

$$\text{if} \quad K = \begin{bmatrix} 0 & -1 \\ 1 & 0 \end{bmatrix}, \quad \text{then} \quad e^{Kt} = \begin{bmatrix} \cos t & -\sin t \\ \sin t & \cos t \end{bmatrix}.$$

We promised at that time to link every K to a unitary U:

5N If K is skew-Hermitian, then e^{Kt} is unitary. Therefore the solution to $du/dt = Ku$ keeps the same length as its initial value u_0: $\|u(t)\| = \|e^{Kt}u_0\| = \|u_0\|$.

The simplest proof is to diagonalize K: $K = S\Lambda S^{-1}$ for some unitary S and pure imaginary Λ. Then $e^{\Lambda t}$ is a diagonal matrix with entries $e^{\lambda_1 t}, \ldots, e^{\lambda_n t}$, all of modulus one; such a matrix is unitary. Therefore $e^{Kt} = Se^{\Lambda t}S^{-1}$ is a product of three unitary matrices. According to the next exercise, the product must itself be unitary.

EXERCISE 5.5.15   Show that if U and V are unitary, so is UV. Use the criterion $U^H U = I$.

EXERCISE 5.5.16   Show that the determinant of a unitary matrix has modulus one, $|\det U| = 1$, but the determinant is not necessarily equal to 1. Describe all 2 by 2 matrices that are both diagonal and unitary.

EXERCISE 5.5.17   Find a third column so that

$$U = \begin{bmatrix} 1/\sqrt{3} & 1/\sqrt{2} & \\ 1/\sqrt{3} & 0 & \\ 1/\sqrt{3} & -1/\sqrt{2} & \end{bmatrix}$$

is unitary. How much freedom is there in this choice?

EXERCISE 5.5.18   Diagonalize the matrix $K = \begin{bmatrix} i & i \\ i & i \end{bmatrix}$, compute $e^{Kt} = Se^{\Lambda t}S^{-1}$, and verify that this exponential is unitary. What is its derivative at $t = 0$?

EXERCISE 5.5.19   Describe all 3 by 3 matrices that are simultaneously Hermitian, unitary, and diagonal. How many are there?

Real versus Complex

$\mathbf{R}^n$ = space of vectors with $n \leftrightarrow \mathbf{C}^n$ = space of vectors with n
real components   complex components

length: $\| x \|^2 = x_1^2 + \cdots + x_n^2 \leftrightarrow$ length: $\| x \|^2 = |x_1|^2 + \cdots + |x_n|^2$

transpose: $A^T_{ij} = A_{ji} \leftrightarrow$ Hermitian transpose: $A^H_{ij} = \overline{A_{ji}}$

$(AB)^T = B^T A^T \leftrightarrow (AB)^H = B^H A^H$

inner product: $x^T y = x_1 y_1 + \cdots + x_n y_n \leftrightarrow$ inner product: $x^H y = \bar{x}_1 y_1 + \cdots + \bar{x}_n y_n$

$(Ax)^T y = x^T (A^T y) \leftrightarrow (Ax)^H y = x^H (A^H y)$

orthogonality: $x^T y = 0 \leftrightarrow$ orthogonality: $x^H y = 0$

symmetric matrices: $A^T = A \leftrightarrow$ Hermitian matrices: $A^H = A$

$x^H A x$ is real, every eigenvalue is real, and $A = U \Lambda U^{-1} = U \Lambda U^H$

skew-symmetric matrices: $K^T = -K \leftrightarrow$ skew-Hermitian matrices: $K^H = -K$

$x^H K x$ is imaginary, every eigenvalue is imaginary, and $K = iA$

orthogonal matrices: $Q^T Q = I$, or $Q^T = Q^{-1} \leftrightarrow$ unitary matrices: $U^H U = I$, or $U^H = U^{-1}$

$(Qx)^T (Qy) = x^T y$ and $\| Qx \| = \| x \| \leftrightarrow (Ux)^H (Uy) = x^H y$ and $\| Ux \| = \| x \|$

The columns, rows, and eigenvectors are orthonormal, and every $|\lambda| = 1$

SIMILARITY TRANSFORMATIONS AND TRIANGULAR FORMS ■ 5.6

Virtually every step in this chapter has involved the combination $S^{-1}AS$, with the eigenvectors of A placed in the columns of S. When A was Hermitian or skew-Hermitian, we wrote U instead of S as a reminder that the eigenvectors were orthonormal; but it was still the matrix of eigenvectors. Now, in this last section, we look at other combinations $M^{-1}AM$—formed in the same way as $S^{-1}AS$, but using any nonsingular matrix M. The eigenvector matrix S may fail to exist, or we may not know it, or we may not want to use it.

It is worth remembering how these combinations arise. Given a differential or difference equation for the unknown u, suppose a "change of variables" $u = Mv$ introduces the new unknown v. Then

$$\frac{du}{dt} = Au \qquad \text{becomes} \qquad M\frac{dv}{dt} = AMv, \quad \text{or} \quad \frac{dv}{dt} = M^{-1}AMv$$

$$u_{n+1} = Au_n \qquad \text{becomes} \qquad Mv_{n+1} = AMv_n, \quad \text{or} \quad v_{n+1} = M^{-1}AMv_n.$$

The new matrix in the equation is $M^{-1}AM$. In the special case $M = S$ the system is uncoupled and the normal modes evolve independently. In the language of Appendix A, the eigenvectors are being chosen as a new basis for the space, and the underlying transformation is represented by the diagonal matrix $S^{-1}AS = \Lambda$. This is the maximum simplification, but other and less drastic simplifications are also useful; even for $M \neq S$ we can hope that $M^{-1}AM$ will be easier to work with than A itself.

First, suppose we look at all possible M. This produces a whole family of matrices $M^{-1}AM$, which are said to be **similar** to each other (and to A); the change from A to $M^{-1}AM$ is called a similarity transformation. Any one of these matrices can be made to appear in the differential and difference equations, by the simple change $u = Mv$, so they ought to have something in common and they do: **Similar matrices share the same eigenvalues**.

5O If $B = M^{-1}AM$, then A and B have the same eigenvalues with the same multiplicities.

The proof is very simple. The eigenvalues of B are the roots of

$$\det(B - \lambda I) = \det(M^{-1}AM - \lambda I) = \det(M^{-1}(A - \lambda I)M)$$

$$= \det M^{-1} \det(A - \lambda I) \det M = \det(A - \lambda I). \qquad (48)$$

Since A and B have the same characteristic polynomial, they have the same eigenvalues. If x is the eigenvector for a given λ, then

$$Ax = \lambda x, \qquad \text{or} \qquad MBM^{-1}x = \lambda x, \qquad \text{or} \qquad B(M^{-1}x) = \lambda(M^{-1}x).$$

This proves again that λ is also an eigenvalue of B—and the eigenvector is $M^{-1}x$.

The combinations $M^{-1}AM$ did not arise in solving $Ax = b$; there the basic operation was to multiply A (on the left side only!) by a matrix that subtracts a multiple of one row from another. Such a transformation preserved the nullspace and row space of A; it had nothing to do with the eigenvalues. In contrast, similarity transformations leave the eigenvalues unchanged, and in fact those eigenvalues are actually calculated by a sequence of simple similarities. The matrix goes gradually toward a triangular form, and the eigenvalues gradually appear on the main diagonal. (Such a sequence is described in Chapter 7, and one step is illustrated in the second exercise below.) This is very much better than trying to compute the polynomial $\det(A - \lambda I)$, whose roots should be the eigenvalues; in practice it is impossible to concentrate all that information into the polynomial and then get it out again.

EXERCISE 5.6.1 Show that if B is similar to A by way of M, and C is similar to B by way of M', then C is similar to A. Which matrices are similar to the identity?

EXERCISE 5.6.2 Consider any A and a special M:

$$A = \begin{bmatrix} a & b & c \\ d & e & f \\ g & h & i \end{bmatrix}, \qquad M = \begin{bmatrix} \cos\theta & -\sin\theta & 0 \\ \sin\theta & \cos\theta & 0 \\ 0 & 0 & 1 \end{bmatrix}.$$

Choose the rotation angle θ so as to annihilate the $(3, 1)$ entry of $M^{-1}AM$.

Note This "annihilation" is not so easy to continue, because the rotations that produce zeros in place of d and h will spoil the new zero in the corner. For matrices of any size, we have to leave one diagonal below the main one, and finish the eigenvalue calculation in a different way. Otherwise the zeros of a polynomial could be found by using only the square roots that determine θ, and this is impossible.

EXERCISE 5.6.3 (a) If A is invertible, find an M which shows that AB and BA are similar, and therefore have the same eigenvalues.

(b) Deduce that AB and BA have the same trace, and then prove this directly by adding up their main diagonals when

$$A = \begin{bmatrix} a & b \\ c & d \end{bmatrix}, \qquad B = \begin{bmatrix} q & r \\ s & t \end{bmatrix}.$$

(c) Show from the traces that $AB - BA = I$ cannot happen for any A and B. In Hilbert space this is a fundamental equation of quantum mechanics, and it does have solutions:

$$Au = \frac{du}{dx}, \qquad Bu = xu, \qquad (AB - BA)u = \frac{d}{dx}(xu) - x\frac{du}{dx} = u.†$$

† Linear algebra in infinite dimensions was turned by von Neumann into the mathematical foundation of quantum mechanics; the momentum A and position B are one of the pairs in Heisenberg's uncertainty principle.

Triangular Forms with a Unitary M

Our first move beyond the usual case $M = S$ is a little bit crazy: Instead of allowing a more general M, we go the other way and *restrict it to be unitary*. The problem is to find some reasonably simple form that $M^{-1}AM$ can achieve under this restriction. Unless the eigenvectors are orthogonal, a diagonal Λ is impossible; but the following "Schur's lemma" produces a form which is very useful—at least to the theory.†

5P For any square matrix A, there is a unitary matrix $M = U$ such that $U^{-1}AU = T$ is upper triangular. The eigenvalues of A must be shared by the similar matrix T, and appear along its main diagonal.

Proof Any matrix, say any 4 by 4 matrix, has at least one eigenvalue λ_1 ; in the worst case, it could be repeated four times. Therefore A has at least one eigenvector x. We normalize x to be a unit vector x_1 , and place it in the first column of U. At this stage the other three columns are impossible to determine, so we complete the matrix in any way which leaves it unitary, and call it U_1. (The Gram–Schmidt process guarantees that this can be done.) The product $U_1^{-1}AU_1$ has at least its first column in the right form: $Ax_1 = \lambda_1 x_1$ means that

$$AU_1 = U_1 \begin{bmatrix} \lambda_1 & * & * & * \\ 0 & * & * & * \\ 0 & * & * & * \\ 0 & * & * & * \end{bmatrix}, \quad \text{or} \quad U_1^{-1}AU_1 = \begin{bmatrix} \lambda_1 & * & * & * \\ 0 & * & * & * \\ 0 & * & * & * \\ 0 & * & * & * \end{bmatrix}.$$

At the second step, we work with the 3 by 3 matrix now in the lower right corner. This matrix has an eigenvalue λ_2 and a unit eigenvector x_2 , which can be made into the first column of a 3 by 3 unitary matrix M_2. Then

$$U_2 = \begin{bmatrix} 1 & 0 & 0 & 0 \\ 0 & & & \\ 0 & & M_2 & \\ 0 & & & \end{bmatrix}, \quad \text{and} \quad U_2^{-1}(U_1^{-1}AU_1)U_2 = \begin{bmatrix} \lambda_1 & * & * & * \\ 0 & \lambda_2 & * & * \\ 0 & 0 & * & * \\ 0 & 0 & * & * \end{bmatrix}.$$

Finally, at the last step, an eigenvector of the 2 by 2 matrix in the lower right corner goes into a unitary M_3, which is put into the corner of U_3, and

$$U_3^{-1}(U_2^{-1}U_1^{-1}AU_1U_2)U_3 = \begin{bmatrix} \lambda_1 & * & * & * \\ 0 & \lambda_2 & * & * \\ 0 & 0 & \lambda_3 & * \\ 0 & 0 & 0 & * \end{bmatrix} = T.$$

† The rest of this chapter is devoted more to theory than to applications.

The product $U = U_1 U_2 U_3$ is again unitary (Exercise 5.5.15), so we have the required $U^{-1}AU = T$.†

Because this lemma applies to all matrices, it often allows us to escape the hypothesis that A is diagonalizable. For example, we could use it to prove that *the powers A^k approach zero when all $| \lambda_i | < 1$*, and *the exponentials e^{At} approach zero when all* $\mathrm{Re}\, \lambda_i < 0$—even without the full set of eigenvectors which the theory of stability assumed in Sections 5.3 and 5.4.

EXERCISE 5.6.4   Repeat the steps of the proof to find $U^{-1}AU = T$ if $A = \begin{bmatrix} 4 & -5 \\ 2 & -3 \end{bmatrix}$.

EXERCISE 5.6.5   Find a triangular form $U^{-1}AU = T$ for

$$A = \begin{bmatrix} 1 & -1 \\ 1 & -1 \end{bmatrix}.$$

EXERCISE 5.6.6   Prove the Cayley–Hamilton theorem, that any matrix satisfies its own characteristic equation: If $p(\lambda) = \det(A - \lambda I)$, then $p(A) = 0$. Hint: Prove that $p(T) = 0$, say in the 3 by 3 case, by carefully multiplying out the triangular product $p(T) = (T - \lambda_1)(T - \lambda_2)(T - \lambda_3)$. Then substitute $A = UTU^{-1}$ into $p(A) = (A - \lambda_1)(A - \lambda_2)(A - \lambda_3)$, and find $p(A) = 0$.

Diagonalizing Hermitian Matrices; the Spectral Theorem

As an application of this triangular form, we want to show that any Hermitian matrix, whether its eigenvalues are distinct or not, has a complete set of orthonormal eigenvectors. The argument has two steps:

(i)   If A is Hermitian and U is unitary, then $U^{-1}AU$ is automatically Hermitian:

$$(U^{-1}AU)^{\mathrm{H}} = U^{\mathrm{H}}A^{\mathrm{H}}(U^{-1})^{\mathrm{H}} = U^{-1}AU.$$

(ii)   The matrix $T = U^{-1}AU$ produced by Theorem 5P is therefore both upper triangular and Hermitian, so *it has to be diagonal*. T must be the same as Λ, when A is Hermitian.

5Q   Any Hermitian matrix A can be diagonalized by a suitable unitary U.

Remark 1   The same is true for any skew-Hermitian K, since $K = iA$.

Remark 2   If A is real and symmetric, then so are its eigenvectors (or at least they can be *chosen* real). Therefore U is real as well as unitary—in other words, it is an orthogonal matrix Q.

Remark 3   It is certainly reasonable, for Hermitian matrices, that even with repeated eigenvalues there is a full set of orthogonal eigenvectors. We can

---

† For a larger matrix the proof would go by mathematical induction.

think of A as the limit of Hermitian matrices with *distinct* eigenvalues, and as the limit approaches the eigenvectors stay perpendicular. In contrast, the non-Hermitian matrices

$$A(\theta) = \begin{bmatrix} 0 & \cos\theta \\ 0 & \sin\theta \end{bmatrix}$$

have eigenvectors $\begin{bmatrix} 1 \\ 0 \end{bmatrix}$ and $\begin{bmatrix} \cos\theta \\ \sin\theta \end{bmatrix}$. As $\theta \to 0$, the second eigenvector approaches the first—which is the *only* eigenvector of the nondiagonalizable matrix $\begin{bmatrix} 0 & 1 \\ 0 & 0 \end{bmatrix}$.

Remark 4   The spectral theorem is now proved even for a matrix

$$A = \begin{bmatrix} 0 & 1 & 0 \\ 1 & 0 & 0 \\ 0 & 0 & 1 \end{bmatrix} = A^H$$

which has repeated eigenvalues: $\lambda_1 = \lambda_2 = 1$, $\lambda_3 = -1$. One choice of ortho-normal eigenvectors is

$$x_1 = \frac{1}{\sqrt{2}} \begin{bmatrix} 1 \\ 1 \\ 0 \end{bmatrix}, \qquad x_2 = \begin{bmatrix} 0 \\ 0 \\ 1 \end{bmatrix}, \qquad x_3 = \frac{1}{\sqrt{2}} \begin{bmatrix} 1 \\ -1 \\ 0 \end{bmatrix}.$$

These are the columns of the unitary U, and $A = U\Lambda U^H$ becomes

$$A = \sum \lambda_i x_i x_i^H = \lambda_1 \begin{bmatrix} \frac{1}{2} & \frac{1}{2} & 0 \\ \frac{1}{2} & \frac{1}{2} & 0 \\ 0 & 0 & 0 \end{bmatrix} + \lambda_2 \begin{bmatrix} 0 & 0 & 0 \\ 0 & 0 & 0 \\ 0 & 0 & 1 \end{bmatrix} + \lambda_3 \begin{bmatrix} \frac{1}{2} & -\frac{1}{2} & 0 \\ -\frac{1}{2} & \frac{1}{2} & 0 \\ 0 & 0 & 0 \end{bmatrix}.$$

But since $\lambda_1 = \lambda_2$, the first two projections (each of rank one) combine to give a projection P_1 of rank two, and A is

$$\begin{bmatrix} 0 & 1 & 0 \\ 1 & 0 & 0 \\ 0 & 0 & 1 \end{bmatrix} = \lambda_1 P_1 + \lambda_3 P_3 = (+1) \begin{bmatrix} \frac{1}{2} & \frac{1}{2} & 0 \\ \frac{1}{2} & \frac{1}{2} & 0 \\ 0 & 0 & 1 \end{bmatrix} + (-1) \begin{bmatrix} \frac{1}{2} & -\frac{1}{2} & 0 \\ -\frac{1}{2} & \frac{1}{2} & 0 \\ 0 & 0 & 0 \end{bmatrix}.$$

There is a whole plane of eigenvectors corresponding to $\lambda = 1$; our x_1 and x_2 were a more or less arbitrary choice. Therefore the separate $x_1 x_1^H$ and $x_2 x_2^H$ were equally arbitrary, and it is only their sum—the projection P_1 onto the whole plane—which is uniquely determined. *Every Hermitian matrix with k different eigenvalues has its own "spectral decomposition" into $A = \lambda_1 P_1 + \cdots + \lambda_k P_k$, where P_i is the projection onto the eigenspace for λ_i.* Since there is a full set of eigenvectors, the projections add up to the identity. And since the eigenspaces are orthogonal, one projection followed by another must produce zero: $P_j P_i = 0$.

EXERCISE 5.6.7   (a)   Find a unitary U so that $U^{-1}AU = \Lambda$ if

$$A = \begin{bmatrix} 1 & 1 & 1 \\ 1 & 1 & 1 \\ 1 & 1 & 1 \end{bmatrix} \quad \text{and} \quad \Lambda = \begin{bmatrix} 0 & 0 & 0 \\ 0 & 0 & 0 \\ 0 & 0 & 3 \end{bmatrix}.$$

Then find a second pair of orthonormal eigenvectors x_1, x_2 for $\lambda = 0$.
  (b)   Verify that $P = x_1 x_1^{\mathrm{H}} + x_2 x_2^{\mathrm{H}}$ is the same for both pairs.

EXERCISE 5.6.8   Prove that every *unitary* matrix A is diagonalizable, in two steps:

  (i)   If A is unitary, and U is too, then so is $T = U^{-1}AU$.
  (ii)   An upper triangular T that is unitary must be diagonal.

It follows that the triangular T in the theorem is Λ, and any unitary matrix (distinct eigenvalues or not) has a complete set of orthonormal eigenvectors: $U^{-1}AU = \Lambda$. All eigenvalues satisfy $|\lambda| = 1$.

We are very close to answering a natural and important question, and might as well push on the rest of the way: For which matrices is the triangular T the same as the diagonal Λ? Hermitian, skew-Hermitian, and unitary matrices are already located in this class; they correspond to numbers on the real axis, the pure imaginary axis, and the unit circle. Now we want the whole class, corresponding to all complex numbers.

5R   The matrix N is ***normal*** if it commutes with N^{H}: $NN^{\mathrm{H}} = N^{\mathrm{H}}N$. For such matrices, and no others, the triangular $T = U^{-1}NU$ is the diagonal matrix Λ. Normal matrices are exactly those that possess a complete set of orthonormal eigenvectors.

Note that Hermitian (or skew-Hermitian) matrices are certainly normal: If $A = A^{\mathrm{H}}$, then AA^{H} and $A^{\mathrm{H}}A$ both equal A^2. Unitary matrices are also normal: UU^{H} and $U^{\mathrm{H}}U$ both equal the identity. In these special cases we proved that $T = \Lambda$ in two steps, and the same two steps will work for any normal matrix:

  (i)   If N is normal, then so is $T = U^{-1}NU$:

$$TT^{\mathrm{H}} = U^{-1}NUU^{\mathrm{H}}N^{\mathrm{H}}U = U^{-1}NN^{\mathrm{H}}U = U^{-1}N^{\mathrm{H}}NU$$
$$= U^{\mathrm{H}}N^{\mathrm{H}}UU^{-1}NU = T^{\mathrm{H}}T.$$

  (ii)   An upper triangular T that is normal must automatically be diagonal (Exercise 5.6.10).

Thus if N is normal, the triangular $U^{-1}NU$ must be diagonal, and since it has the same eigenvalues as N it must be Λ. The eigenvectors of N are the columns of U, and they are orthonormal. In fact N allows the same spectral theorem as a Hermitian matrix, $N = U\Lambda U^{-1} = \Sigma\lambda_i x_i x_i^{\mathrm{H}}$; the only difference is that the eigenvalues λ need not be real.

EXERCISE 5.6.9   Find a normal matrix that is not Hermitian, skew-Hermitian, unitary, or diagonal.

EXERCISE 5.6.10   Suppose T is a 3 by 3 upper triangular matrix, with entries t_{ij}. Compare the entries of TT^H and T^HT, and show that if they are equal then T must be diagonal.

EXERCISE 5.6.11   Prove that a matrix with orthonormal eigenvectors has to be normal, as claimed in 5R: If $U^{-1}NU = \Lambda$, then $NN^H = N^HN$.

The Jordan Form

So far in this section, we have done the best we could with unitary similarities; requiring M to be a unitary matrix U, we got $M^{-1}AM$ into a triangular form T. Now we lift this restriction on M. Any matrix is allowed, and the goal is to make $M^{-1}AM$ as *nearly diagonal as possible*.

The result of this supreme effort at diagonalization is the Jordan form J. In case the matrix A has a full set of independent eigenvectors, we take $M = S$ and arrive at $J = S^{-1}AS = \Lambda$; the Jordan form coincides with the diagonal matrix Λ. This is impossible for a defective matrix, and for every missing eigenvector the Jordan form will have a 1 just above its main diagonal. The eigenvalues must appear on the diagonal itself, because J is triangular. And distinct eigenvalues can be decoupled just as in the diagonal case; it is only a repeated λ that may (or may not!) require an off-diagonal entry in J.

5S   If A has s independent eigenvectors, it is similar to a matrix of the form

$$J = M^{-1}AM = \begin{bmatrix} J_1 & & & \\ & \cdot & & \\ & & \cdot & \\ & & & J_s \end{bmatrix}.$$

Each Jordan block J_i is a triangular matrix with only a single eigenvalue, and only one eigenvector:

$$J_i = \begin{bmatrix} \lambda_i & 1 & & \\ & \cdot & \cdot & \\ & & \cdot & 1 \\ & & & \lambda_i \end{bmatrix}.$$

The same eigenvalue λ_i may appear in several blocks, if it corresponds to several independent eigenvectors. Two matrices are similar if and only if they share the same J.

Many authors have made this theorem the climax of their linear algebra course. Frankly, I think that is a mistake. It is certainly true that not all matrices are diagonalizable, and that the Jordan form is the most general case; but for that very reason its construction is both technical and extremely unstable. (A slight change in A can put back all the missing eigenvectors, and remove the off-diagonal 1's.) Therefore the right place for the details is in the appendix,† and the best way to start on the Jordan form is to look at some specific and manageable examples.

We introduce two of them:

$$A = \begin{bmatrix} 0 & 1 & 2 \\ 0 & 0 & 1 \\ 0 & 0 & 0 \end{bmatrix} \quad \text{and} \quad B = \begin{bmatrix} 0 & 0 & 1 \\ 0 & 0 & 0 \\ 0 & 0 & 0 \end{bmatrix}.$$

Since zero is a triple eigenvalue for both matrices, it will appear in every Jordan block; either there is a single 3 by 3 block, or a 2 by 2 and a 1 by 1 block, or three 1 by 1 blocks. Therefore the possible Jordan forms are

$$J_1 = \begin{bmatrix} 0 & 1 & 0 \\ 0 & 0 & 1 \\ 0 & 0 & 0 \end{bmatrix}, \quad J_2 = \begin{bmatrix} 0 & 1 & 0 \\ 0 & 0 & 0 \\ 0 & 0 & 0 \end{bmatrix}, \quad \text{and} \quad J_3 = \begin{bmatrix} 0 & 0 & 0 \\ 0 & 0 & 0 \\ 0 & 0 & 0 \end{bmatrix}.$$

In the case of A, the only eigenvector is $(1, 0, 0)^\mathsf{T}$. Therefore its Jordan form has only one block, and according to the main theorem 5S, A must be similar to J_1. The matrix B has the additional eigenvector $(0, 1, 0)^\mathsf{T}$, and therefore its Jordan form is J_2; there must be two blocks along the diagonal. As for J_3, it is in a class by itself; the only matrix similar to the zero matrix is $M^{-1}0M = 0$.

In these examples, a simple count of the eigenvectors was enough to determine J—and that is always possible when there is nothing more complicated than a triple eigenvalue. But as a general theory, this counting technique is exploded by the last exercise.

EXERCISE 5.6.12 Find the matrix M that achieves $M^{-1}BM = J$.

EXERCISE 5.6.13 Show, by trying for an M and failing, that no two of these Jordan forms are similar: $J_1 \neq M^{-1}J_2M$, and $J_1 \neq M^{-1}J_3M$, and $J_2 \neq M^{-1}J_3M$.

EXERCISE 5.6.14 Write out all possible Jordan forms for a 4 by 4 matrix that has zero as a quadruple eigenvalue. (By convention, the blocks get smaller as we move down the matrix J.) If there are two independent eigenvectors, show that there are two different possibilities for J.

---

† Every author tries to make these details easy to follow, and I believe Filippov's new proof is the best. It is almost simple enough to reverse our decision and bring the construction of J back from the appendix.

Table of Similarity Transformations

1. A is **diagonalizable**: The columns of S are the eigenvectors of A, and $S^{-1}AS = \Lambda$ is *diagonal*.
2. A is **arbitrary**: The columns of M are the eigenvectors and generalized eigenvectors of A, and the Jordan form $M^{-1}AM = J$ is *block diagonal*.
3. A is **arbitrary**, and U is unitary: U can be chosen so that $U^{-1}AU = T$ is upper triangular.
4. A is **normal**, $AA^H = A^H A$: U can be chosen so that $U^{-1}AU = \Lambda$.
   Special cases, all with orthonormal eigenvectors:

   a.   If A is Hermitian, then Λ is real.
   a'.   If A is real symmetric, then Λ is real and $U = Q$ is orthogonal.
   b.   If A is skew-Hermitian, then Λ is imaginary.
   c.   If A is unitary, then all $|\lambda_i| = 1$.

REVIEW EXERCISES

5.1   Find the eigenvalues and eigenvectors of

$$A = \begin{bmatrix} 1 & 0 \\ 2 & 3 \end{bmatrix},$$

and choose an S so that $S^{-1}AS$ is diagonal.

5.2   Find the characteristic polynomial $\det(A - \lambda I)$, and the eigenvalues, of

$$A = \begin{bmatrix} 0 & 1 & 0 & 0 \\ 0 & 0 & 1 & 0 \\ 0 & 0 & 0 & 1 \\ 1 & 0 & 0 & 0 \end{bmatrix}.$$

5.3   If A has eigenvalues 0 and 1, corresponding to the eigenvectors

$$\begin{bmatrix} 1 \\ 2 \end{bmatrix} \quad \text{and} \quad \begin{bmatrix} 2 \\ -1 \end{bmatrix},$$

how can you tell in advance that A is symmetric? What are its trace and determinant? What is A?

5.4   In the previous exercise, what will be the eigenvalues and eigenvectors of A^2? What is the relation of A^2 to A?

5.5   Does there exist a matrix A such that the entire family $A + cI$ is invertible for all numbers c?

5.6   If

$$A = \begin{bmatrix} 1 & 0 \\ 2 & 1 \end{bmatrix},$$

find A^9.

5.7   Would you prefer to have interest compounded daily at 5% for the first three years and at 6% for the next three, or vice versa?

5.8   Solve

$$du/dt = \begin{bmatrix} 3 & 1 \\ 1 & 3 \end{bmatrix} u \quad \text{if } u_0 = \begin{bmatrix} 2 \\ 0 \end{bmatrix}.$$

5.9   If

$$A = \begin{bmatrix} 3 & 1 \\ 1 & 3 \end{bmatrix},$$

find e^{At}.

5.10   Find all square roots of the preceding matrix A.

5.11   True or false (with counterexample if false):

(i)   If B is formed from A by exchanging two rows, then B is similar to A.
(ii)   If an upper triangular matrix is similar to a diagonal matrix, then it is already diagonal.
(iii) Any two of the following statements imply the third: A is Hermitian, A is unitary, $A^2 = I$.
(iv)   If A and B are diagonalizable, so is $A + B$.

5.12   Find an orthogonal Q and a diagonal Λ such that $Q^T A Q = \Lambda$ if

$$A = \begin{bmatrix} 1 & 0 & 0 \\ 0 & 1 & 1 \\ 0 & 1 & 1 \end{bmatrix}.$$

5.13   Split

$$\begin{bmatrix} 1 & 1 \\ 0 & 0 \end{bmatrix}$$

into $A + K$, with $A = A^H$ and $K = -K^H$.

5.14   Write down a 3 by 3 matrix whose rows add up to one, and show that $\lambda = 1$ is an eigenvalue. What is the eigenvector?

5.15   What happens to the Fibonacci sequence if we go backward in time, and how is F_{-k} related to F_k? The law $F_{k+2} = F_{k+1} + F_k$ is still in force, so $F_{-1} = 1$.

6

POSITIVE DEFINITE
MATRICES

MINIMA, MAXIMA, AND SADDLE POINTS ■ 6.1

Up to now, we have had almost no reason to worry about the *signs of the eigen-values*. In fact, it would have been premature to ask whether λ is positive or negative before it was known to be real. But Chapter 5 established that the most important matrices—symmetric matrices in the real case, and Hermitian matrices in the complex case—do have real eigenvalues. Therefore it makes sense to ask whether they are positive, and one of our goals is the following: to find a test that can be applied directly to the symmetric matrix A, without computing its eigenvalues, which will guarantee in advance that all those eigenvalues are positive.

Before looking for such a test, we want to describe a new situation in which the signs of the eigenvalues are significant. It is completely unlike the question of stability in differential equations, where we needed negative rather than positive eigenvalues. (We should not hurry past that point, but we will: If $-A$ passes the test we are looking for, then $du/dt = Au$ has decaying solutions $e^{\lambda t}x$, with every eigenvalue $\lambda < 0$. And $d^2u/dt^2 = Au$ has pure oscillatory solutions $e^{i\omega t}x$, with $\omega = \sqrt{-\lambda}$.) The new situation is one arising in so many applications to science, to engineering, and to every problem of optimization, that we hope the reader is willing to take the background for granted and start directly with the mathematical problem.

It is the problem of identifying a minimum, and we introduce it with two examples:

$$F(x, y) = 7 + 2(x + y)^2 - y \sin y - x^3 + y^4$$

and

$$f(x, y) = 2x^2 + 4xy + y^2.$$

Does either one of these functions have a minimum at the point $x = y = 0$?

Remark 1   The zero-order terms $F(0, 0) = 7$ and $f(0, 0) = 0$ have no effect on the answer. They simply raise or lower the graphs of F and f.

Remark 2   The linear terms give a necessary condition: To have any chance of a minimum, we must have a stationary point. The first derivatives must vanish at $x = y = 0$, and they do:

$$\frac{\partial F}{\partial x} = 4(x + y) - 3x^2 = 0$$

$$\frac{\partial F}{\partial y} = 4(x + y) - y \cos y - \sin y + 4y^3 = 0,$$

$$\frac{\partial f}{\partial x} = 4x + 4y = 0 \quad \text{and} \quad \frac{\partial f}{\partial y} = 4x + 2y = 0.$$

Thus the origin is a stationary point for both F and f. Geometrically, the surface $z = F(x, y)$ is tangent to the horizontal plane $z = 7$, and the surface $z = f(x, y)$ is tangent to the plane $z = 0$. The question is whether F and f lie above those planes, as we move away from the tangency point $x = y = 0$.

Remark 3   *The quadratic terms, or in other words the second derivatives, are decisive:*

$$\frac{\partial^2 F}{\partial x^2} = 4 - 6x = 4 \qquad\qquad \frac{\partial^2 f}{\partial x^2} = 4$$

$$\frac{\partial^2 F}{\partial x \, \partial y} = \frac{\partial^2 F}{\partial y \, \partial x} = 4 \qquad\qquad \frac{\partial^2 f}{\partial x \, \partial y} = \frac{\partial^2 f}{\partial y \, \partial x} = 4$$

$$\frac{\partial^2 F}{\partial y^2} = 4 + y \sin y - 2 \cos y + 12y^2 = 2, \qquad \frac{\partial^2 f}{\partial y^2} = 2.$$

These derivatives contain the answer, and since they are the same for F and f they must contain the same answer. The two functions behave in exactly the same way near the origin, and F **has a minimum if and only if f has a minimum**.

Remark 4   The higher-degree terms in F have no effect on the question of a local minimum, but they can prevent it from being a global minimum. In our example this is certainly the case; the term $-x^3$ must sooner or later pull F

toward $-\infty$, regardless of what happens near $x = y = 0$. Such an eventuality is impossible for f, or for any other *quadratic form*, which has no higher-order terms. Every quadratic form $f = ax^2 + 2bxy + cy^2$ has a stationary point at the origin, where $\partial f/\partial x = \partial f/\partial y = 0$; and if it has a local minimum at $x = y = 0$, then that point is also a global minimum. The surface $z = f(x, y)$ will be shaped like a bowl, resting on the one point at the origin.

To summarize: The question of a local minimum for F is equivalent to the same question for f. If the stationary point of F were at $x = \alpha$, $y = \beta$ instead of $x = y = 0$, the only change would be to use the second derivatives at α, β:

$$f(x, y) = \frac{x^2}{2} \frac{\partial^2 F}{\partial x^2} (\alpha, \beta) + xy \frac{\partial^2 F}{\partial x \, \partial y} (\alpha, \beta) + \frac{y^2}{2} \frac{\partial^2 F}{\partial y^2} (\alpha, \beta). \tag{1}$$

Then f behaves near $(0, 0)$ in the same way that F behaves near (α, β).

There is one case to be excluded. It corresponds to the possibility $F'' = 0$, which is a tremendous headache even for a function of one variable. The third derivatives are drawn into the problem because the second derivatives fail to give a definite decision. To avoid that difficulty it is usual to require that the quadratic part be nonsingular: For a true minimum, f is allowed to vanish *only at* $x = y = 0$. A quadratic form that is strictly positive at all other points is called **positive definite**.

The problem now comes down to this: For a function of two variables x and y, what is the correct replacement for the condition $F'' > 0$? With only one variable, the sign of the second derivative is enough to decide between a minimum or maximum; but now we have three second derivatives, F_{xx}, $F_{xy} = F_{yx}$, and F_{yy}. These three numbers specify f, and they must determine whether or not F (as well as f) has a minimum. *What are the conditions on a, b, and c which ensure that $f = ax^2 + 2bxy + cy^2$ is positive definite?*

It is easy to find one necessary condition:

(i)   *If f is positive definite, then necessarily $a > 0$.*

We simply consider the point $x = 1$, $y = 0$, where $ax^2 + 2bxy + cy^2$ is equal to a; this must be positive if f is to be positive definite. Translating back to F, this just means that $\partial^2 F/\partial x^2 > 0$; we fix $y = 0$, let only x vary, and must have $F'' > 0$ for a minimum. Similarly, if we fix $x = 0$ and vary y, there is a condition on the coefficient c:

(ii)   *If f is positive definite, then necessarily $c > 0$.*

It is not so easy to decide whether conditions (i) and (ii), taken together, are sufficient to make f positive definite. Is it necessary to look at the cross-derivative coefficient b?

EXAMPLE $f = x^2 - 10xy + y^2$. In this case $a = 1$ and $c = 1$ are both positive. Suppose, however, that we choose the point $x = y = 1$; since $f(1, 1) = -8$, this form f is not positive definite. The conditions $a > 0$ and $c > 0$ ensure that f is increasing in the x and y directions, but it may still decrease along another line, in this case the line $x = y$. The coefficient b can overwhelm a and c. In fact, it is impossible to test for positive definiteness by looking along any finite number of fixed lines.

Evidently b enters the problem, and in our original f the coefficient b was positive. Does this fact suffice to ensure a minimum? Again the answer is no; the sign of b is of no importance. *Even though all its coefficients were positive, our original example $2x^2 + 4xy + y^2$ was not positive definite, and neither F nor f had a minimum.* Along the line $x = -y$, $f(1, -1) = 2 - 4 + 1 = -1$.

It is the size of b, compared to a and c, *that must be controlled if f is to be positive definite.* We now want to find a precise test, giving a necessary and sufficient condition for positive definiteness. The simplest technique is to "complete the square":

$$f = ax^2 + 2bxy + cy^2 = a\left(x + \frac{b}{a}y\right)^2 + \left(c - \frac{b^2}{a}\right)y^2. \tag{2}$$

Obviously the first term on the right is positive, since it is a perfect square multiplied by the positive coefficient a—necessary condition (i) is still in force. This perfect square is zero, however, when $x = b$ and $y = -a$; at that point we find $f(b, -a) = a(ac - b^2)$. Therefore a new necessary condition is required:

(iii)   *If f is positive, then necessarily $ac > b^2$.*

Notice that conditions (i) and (iii), taken together, automatically imply condition (ii): If $a > 0$, and $ac > b^2 \geq 0$, then certainly $c > 0$. The right side of (2) is guaranteed to be positive, and we have answered the question at last:

6A   The quadratic form $f = ax^2 + 2bxy + cy^2$ is positive definite if and only if $a > 0$ and $ac - b^2 > 0$. Correspondingly, F has a (nonsingular) minimum at $x = y = 0$ if and only if

$$\frac{\partial^2 F}{\partial x^2}(0, 0) > 0, \qquad \left[\frac{\partial^2 F}{\partial x^2}(0, 0)\right]\left[\frac{\partial^2 F}{\partial y^2}(0, 0)\right] > \left[\frac{\partial^2 F}{\partial x\, \partial y}(0, 0)\right]^2.$$

The conditions for a maximum are easy, since f has a maximum whenever $-f$ has a minimum. This means reversing the signs of a, b, and c, and it actually leaves the second condition $ac - b^2 > 0$ unchanged: The quadratic form is **negative definite** if and only if $a < 0$ and $ac - b^2 > 0$. The same change applies to F.

The quadratic form f is singular when $ac - b^2 = 0$; this is the case we have so far ruled out. The second term in (2) would disappear, leaving only one square $f = a(x + (b/a)y)^2$—which is either **positive semidefinite**, when $a > 0$, or **negative semidefinite**, when $a < 0$. The prefix *semi* allows the possibility that f can equal zero, as it will at the point $x = b$, $y = -a$. Geometrically the surface $z = f(x, y)$ degenerates from a genuine bowl into an infinitely long trough. (Think of the surface $z = x^2$ in three-dimensional space;

the trough runs up and down the y axis, and each cross section is the same parabola $z = x^2$.) And a still more singular quadratic form is to have zero everywhere, $a = b = c = 0$, which is both positive semidefinite and negative semidefinite. The bowl has become completely flat.

In one dimension, for a function $F(x)$, the possibilities would now be exhausted: Either there is a minimum, or a maximum, or $F'' = 0$. In two dimensions, however, a very important possibility still remains: *The combination $ac - b^2$ may be negative*. This occurred in our example, when the size of b dominated a and c; f was positive in some directions and negative in others. It is also guaranteed to occur, regardless of b, if a and c are of opposite sign. The x and y directions give opposite results—on one axis f increases, on the other it decreases. It is useful to consider the two special cases

$$f_1 = 2xy \quad \text{and} \quad f_2 = x^2 - y^2.$$

In the first, b is dominating, with $a = c = 0$; in the second, a and c are of opposite sign. Both have $ac - b^2 = -1$.

These quadratic forms are **indefinite**, because they can take either sign; both $f > 0$ and $f < 0$ are possible, depending on x and y. So we have a stationary point that is neither a maximum or a minimum. It is called a **saddle point**. (Presumably because the surface $z = f(x, y)$, say $z = x^2 - y^2$, is shaped like a saddle; it goes down in the direction of the y axis, where the legs fit, and goes up in the direction of the x axis.) You may prefer to think of a road going over a mountain pass; the top of the pass is a minimum as you look along the range of mountains, but it is a maximum as you go along the road (Fig. 6.1).

The saddles $2xy$ and $x^2 - y^2$ are practically the same; if we turn one through $45°$ we get the other. They are also almost impossible to draw.

EXERCISE 6.1.1 Show that the original quadratic $f = 2x^2 + 4xy + y^2$ has a saddle point at the origin, despite the fact that its coefficients are positive. Show how f can be rewritten as a *difference of two perfect squares*.

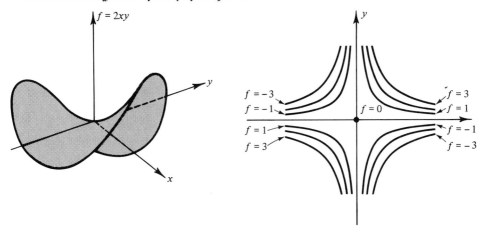

Fig. 6.1. The saddle $f = 2xy$ and its level curves.

Calculus would have been enough to find our conditions for a minimum: $F_{xx} > 0$ and $F_{xx}F_{yy} > F_{xy}^2$. But linear algebra is ready to do more, as soon as we recognize how the coefficients of f fit into a symmetric matrix A. The terms ax^2 and cy^2 appear on the diagonal, the cross derivative $2bxy$ is split between the entry above and the entry below, and the form f is identically equal to the matrix product

$$ax^2 + 2bxy + cy^2 = \begin{bmatrix} x & y \end{bmatrix} \begin{bmatrix} a & b \\ b & c \end{bmatrix} \begin{bmatrix} x \\ y \end{bmatrix}. \tag{3}$$

This identity is the key to the whole chapter. It can be rewritten as $f = x^{\mathrm{T}}Ax$; it generalizes immediately to n dimensions; and it provides a perfect shorthand for studying the problem of maxima and minima. When there are n independent variables $x_1, \ldots, x_n$, instead of simply x and y, they go into a column vector x. Then **for any symmetric matrix A, the product $f = x^{\mathrm{T}}Ax$ is a pure quadratic form.** It has a stationary point at the origin, and no higher-order terms:

$$x^{\mathrm{T}}Ax = \begin{bmatrix} x_1 & x_2 & \cdot & x_n \end{bmatrix} \begin{bmatrix} a_{11} & a_{12} & \cdot & a_{1n} \\ a_{21} & a_{22} & \cdot & a_{2n} \\ \cdot & \cdot & \cdot & \cdot \\ a_{n1} & a_{n2} & \cdot & a_{nn} \end{bmatrix} \begin{bmatrix} x_1 \\ x_2 \\ \cdot \\ x_n \end{bmatrix}$$

$$= a_{11}x_1^2 + a_{12}x_1x_2 + a_{21}x_2x_1 + \cdots + a_{nn}x_n^2$$

$$= \sum_{i=1}^{n} \sum_{j=1}^{n} a_{ij}x_ix_j. \tag{4}$$

Given a function $F(x_1, \ldots, x_n)$ that has a stationary point at the origin—all the first derivatives are zero—we test for a minimum or a maximum or a saddle point by constructing the "Hessian matrix" A from the second derivatives: $a_{ij} = \partial^2 F / \partial x_i \, \partial x_j$. A is automatically symmetric, so the terms $a_{12}x_1x_2$ and $a_{21}x_2x_1$ are the same; this matches the splitting of $2bxy$ in the two-dimensional case. *The function has a minimum when f is positive definite, or in other words when the matrix A is positive definite; this means that*

$$x^{\mathrm{T}}Ax > 0 \qquad \text{for all } x \text{ other than } x_1 = x_2 = \cdots = x_n = 0. \tag{5}$$

EXERCISE 6.1.2   Decide for or against the positive definiteness of the following matrices, and write out the corresponding f:

(a) $A = \begin{bmatrix} 1 & 3 \\ 3 & 5 \end{bmatrix}$ (b) $A = \begin{bmatrix} 1 & -1 \\ -1 & 1 \end{bmatrix}$

(c) $A = \begin{bmatrix} 1 & 0 & 0 \\ 0 & 1 & 0 \\ 0 & 0 & -1 \end{bmatrix}$ (d) $A = \begin{bmatrix} 2 & -1 & 0 \\ -1 & 2 & -1 \\ 0 & -1 & 2 \end{bmatrix}$.

In case (d), write $x^{\mathrm{T}}Ax$ as a sum of three squares. In (b), note that a singular matrix corresponds precisely to a singular quadratic form: Both are identified by the vanishing of $ac - b^2$, which is the determinant of A. Along which line does f remain identically zero?

EXERCISE 6.1.3 For a positive definite f, the curve $f(x, y) = 1$ is an ellipse. (It is also the cross section of the bowl $z = f(x, y)$ cut out by the plane $z = 1$.) If $a = c = 2$ and $b = -1$, sketch the ellipse.

EXERCISE 6.1.4 Decide between a minimum, maximum, or saddle point for the functions

(a) $F = -1 + 4(e^x - x) - 5x \sin y + 6y^2$ at the point $x = y = 0$;
(b) $F = (x^2 - 2x) \cos y$, with stationary point at $x = 1$, $y = \pi$.

TESTS FOR POSITIVE DEFINITENESS ■ 6.2

Which symmetric matrices have the property that $x^{\mathrm{T}}Ax > 0$ for all nonzero vectors x? There are four or five different ways to answer this question, and we hope to find all of them. The previous section began with some hints about the signs of eigenvalues, but that discussion was left hanging in midair. Instead, the question of eigenvalues gave place to a pair of conditions on a matrix:

$$\text{if} \quad A = \begin{bmatrix} a & b \\ b & c \end{bmatrix}, \quad \text{we need} \quad a > 0, \quad ac - b^2 > 0.$$

Our goal is to generalize those conditions to a matrix of order n, and to find the connection with the signs of the eigenvalues. In the 2 by 2 case, at least, the conditions mean that both eigenvalues are positive: Their product is the determinant $ac - b^2 > 0$, so the eigenvalues are either both positive or both negative, and it must be the former because their sum is the trace $a + c > 0$.

It is remarkable how closely these two approaches—one straightforward and computational, the other more concerned with the intrinsic properties of the matrix (its eigenvalues)—reflect the two parts of this book. In fact, looking more closely at the computational test, it is even possible to spot the appearance of the pivots. They turned up when we decomposed f into a sum of squares:

$$ax^2 + 2bxy + cy^2 = a\left(x + \frac{b}{a}y\right)^2 + \frac{ac - b^2}{a}y^2.$$

The coefficients a and $(ac - b^2)/a$ are exactly the pivots for a 2 by 2 matrix. If this relationship continues to hold for larger matrices, it will allow a very simple test for positive definiteness: We check the pivots. And at the same time it will have a very natural interpretation: $x^{\mathrm{T}}Ax$ is positive definite if and only if it can be written as a sum of n independent squares.

One more preliminary remark. The two parts of this book were linked by the theory of determinants, and therefore we ask what role determinants play in positive definiteness. Certainly it is not enough to require that $\det A > 0$;

that requirement is satisfied when $a = c = -1$ and $b = 0$, giving $A = -I$, a form that is actually negative definite. The important point is that the determinant test is applied not only to A itself, giving $ac - b^2 > 0$, but also to the 1 by 1 submatrix a in the upper left corner. The natural generalization will involve all n of the upper left submatrices

$$A_1 = [a_{11}], \quad A_2 = \begin{bmatrix} a_{11} & a_{12} \\ a_{21} & a_{22} \end{bmatrix}, \quad A_3 = \begin{bmatrix} a_{11} & a_{12} & a_{13} \\ a_{21} & a_{22} & a_{23} \\ a_{31} & a_{32} & a_{33} \end{bmatrix}, \quad \ldots, \quad A_n = A.$$

Here is the main theorem, and a detailed proof:

6B   Each of the following tests is a necessary and sufficient condition for the real symmetric matrix A to be ***positive definite***:

(I) $x^T A x > 0$ for all nonzero vectors x.
(II)   All the eigenvalues of A satisfy $\lambda_i > 0$.
(III)   All the submatrices A_k have positive determinants.
(IV)   All the pivots (without row exchanges) satisfy $d_i > 0$.

Proof   Condition I defines a positive definite matrix, and our first step will be to show its equivalence to condition II. Therefore we suppose that I is satisfied, and deduce that each eigenvalue λ_i must be positive. The argument is simple. Suppose x_i is the corresponding unit eigenvector; then

$$A x_i = \lambda_i x_i, \quad \text{so} \quad x_i^T A x_i = x_i^T \lambda_i x_i = \lambda_i,$$

since $x_i^T x_i = 1$. Since condition I holds for all x, it will hold in particular for the eigenvector x_i, and the quantity $x_i^T A x_i = \lambda_i$ must be positive. ***A positive definite matrix has positive eigenvalues***.

Now we go in the other direction, assuming that all $\lambda_i > 0$ and deducing that $x^T A x > 0$. (This has to be proved for every vector x, not just the eigenvectors.) Since symmetric matrices have a full set of orthonormal eigenvectors (p. 224), we can write any x as a linear combination $c_1 x_1 + \cdots + c_n x_n$. Then

$$A x = c_1 A x_1 + \cdots + c_n A x_n = c_1 \lambda_1 x_1 + \cdots + c_n \lambda_n x_n.$$

Because of the orthogonality, and the normalization $x_i^T x_i = 1$,

$$x^T A x = (c_1 x_1^T + \cdots + c_n x_n^T)(c_1 \lambda_1 x_1 + \cdots + c_n \lambda_n x_n)$$
$$= c_1^2 \lambda_1 + \cdots + c_n^2 \lambda_n. \tag{6}$$

If every $\lambda_i > 0$, then $x^T A x > 0$, so condition II implies condition I.

We turn to III and IV, whose equivalence to I will be proved in three steps:

If I holds, so does III: First, the determinant of any matrix is the product of its eigenvalues. And if I holds, we already know that these eigenvalues are positive:

$$\det A = \lambda_1 \lambda_2 \cdots \lambda_n > 0.$$

To prove the same result for all the submatrices A_k, we check that if A is positive definite, so is every A_k. The trick is to look at the vectors whose last $n - k$ components are zero: if $x = \begin{bmatrix} x_k \\ 0 \end{bmatrix}$, then

$$x^T A x = \begin{bmatrix} x_k^T & 0 \end{bmatrix} \begin{bmatrix} A_k & * \\ * & * \end{bmatrix} \begin{bmatrix} x_k \\ 0 \end{bmatrix} = x_k^T A_k x_k.$$

If $x^T A x > 0$ for all nonzero x, then in particular $x_k^T A_k x_k > 0$ for all nonzero x_k. Thus Condition I holds for A_k, and the submatrix permits the same argument that worked for A itself. Its eigenvalues (which are not the same λ_i !) must be positive, and its determinant is the product of those eigenvalues.

If III *holds, so does* IV: This is easy to prove because there is a direct relation between the numbers $\det A_k$ and the pivots. According to p. 167, the kth pivot d_k is exactly the ratio of $\det A_k$ to $\det A_{k-1}$. Therefore if all the determinants are positive, so are all the pivots—and no row exchanges are needed for positive definite matrices.

If IV *holds, so does* I: We are given that the pivots are positive, and must deduce that $x^T A x > 0$. This is what we did in the 2 by 2 case, by completing the square. To prove this same result for matrices of any size, here is the essential fact: *In Gaussian elimination of a symmetric matrix, the upper triangular U is the transpose of the lower triangular L* (p. 38). *Therefore $A = LDU$ becomes $A = LDL^T$.*

EXAMPLE

$$A = \begin{bmatrix} 2 & -1 & 0 \\ -1 & 2 & -1 \\ 0 & -1 & 2 \end{bmatrix} = \begin{bmatrix} 1 & 0 & 0 \\ -\frac{1}{2} & 1 & 0 \\ 0 & -\frac{2}{3} & 1 \end{bmatrix} \begin{bmatrix} 2 & & \\ & \frac{3}{2} & \\ & & \frac{4}{3} \end{bmatrix} \begin{bmatrix} 1 & -\frac{1}{2} & 0 \\ 0 & 1 & -\frac{2}{3} \\ 0 & 0 & 1 \end{bmatrix} = LDL^T.$$

Multiplying on the left by x^T and on the right by x, we get a sum of squares in which the pivots are the coefficients:

$$x^T A x = \begin{bmatrix} u & v & w \end{bmatrix} \begin{bmatrix} 1 & 0 & 0 \\ -\frac{1}{2} & 1 & 0 \\ 0 & -\frac{2}{3} & 1 \end{bmatrix} \begin{bmatrix} 2 & & \\ & \frac{3}{2} & \\ & & \frac{4}{3} \end{bmatrix} \begin{bmatrix} 1 & -\frac{1}{2} & 0 \\ 0 & 1 & -\frac{2}{3} \\ 0 & 0 & 1 \end{bmatrix} \begin{bmatrix} u \\ v \\ w \end{bmatrix}$$

$$= 2(u - \tfrac{1}{2}v)^2 + \tfrac{3}{2}(v - \tfrac{2}{3}w)^2 + \tfrac{4}{3}(w)^2.$$

Positive pivots lead to a positive $x^T A x$, and condition IV implies condition I, which is the last step in the proof.† The theorem would be exactly the same in the complex case, for Hermitian matrices $A = A^H$.

It would be wrong to leave the impression in condition III that the upper submatrices A_k are extremely special. We could equally well test the determi-

---

† Our proof of this step was only "by example," and the real proof is coming in two more pages: IV $\Rightarrow$ V $\Rightarrow$ I.

nants of the lower right submatrices. Or we could use any chain of principal submatrices, starting with some diagonal entry a_{ii} as the first submatrix, and adding a new row and column pair each time. In particular, a *necessary condition* for positive definiteness is that every diagonal entry a_{ii} be positive; this is just the coefficient of x_i^2. As we know from the examples, however, it is far from sufficient to look only at the diagonal entries.

The pivots d_i are not to be confused with the eigenvalues λ_i ; for a typical positive definite matrix, they are two completely different sets of positive numbers. In our 3 by 3 example, probably the determinant test is the easiest:

$$\det A_1 = 2, \qquad \det A_2 = 3, \qquad \det A_3 = \det A = 4.$$

For a large matrix, it is simpler just to watch the pivots in elimination, which are the ratios $d_1 = 2$, $d_2 = \frac{3}{2}$, $d_3 = \frac{4}{3}$. Ordinarily the eigenvalue test is the longest, but for this example we know they are all positive,

$$\lambda_1 = 2 - \sqrt{2}, \qquad \lambda_2 = 2, \qquad \lambda_3 = 2 + \sqrt{2}.$$

Even though it is the hardest to apply to a single matrix A, this is actually the most useful test for theoretical purposes.

EXERCISE 6.2.1   Decide on the positive definiteness of

$$A = \begin{bmatrix} 2 & -1 & -1 \\ -1 & 2 & -1 \\ -1 & -1 & 2 \end{bmatrix}, \qquad B = \begin{bmatrix} 1 & 0 & 1 \\ 0 & 1 & 0 \\ 1 & 0 & 1 \end{bmatrix}, \qquad \text{and} \qquad C = \begin{bmatrix} 1 & 1 & 1 \\ 1 & 2 & 2 \\ 1 & 2 & 3 \end{bmatrix}.$$

Use the decomposition $B = LDL^T$ to write $x^T Bx = u^2 + v^2 + w^2 + 2uw$ as a sum of squares.

EXERCISE 6.2.2   Construct a matrix that has its largest entries on the main diagonal, say

$$A = \begin{bmatrix} 1 & \alpha & \beta \\ \alpha & 1 & \alpha \\ \beta & \alpha & 1 \end{bmatrix} \qquad \text{with} \qquad |\alpha| < 1, \quad |\beta| < 1,$$

and is not even positive semidefinite. (Compute $\det A$.)

EXERCISE 6.2.3   Show that if A is positive definite, so is A^2 and so is A^{-1}.

EXERCISE 6.2.4   Show that if A and B are positive definite, so is $A + B$. Which of the conditions I–IV is useful this time?

I hope you will allow one more test for positive definiteness. We are already very close to it, and it ought to give a better understanding even of the old tests:

6C A is positive definite if and only if it satisfies condition

(V)   There exists a nonsingular matrix W such that $A = W^TW$.

If this new condition holds, then

$$x^TAx = x^TW^TWx = \|Wx\|^2. \tag{7}$$

This is certainly positive, or at least nonnegative; the crucial question is whether $Wx = 0$. Since W is nonsingular, this can happen only if $x = 0$. Therefore condition V implies condition I: $x^TAx > 0$ for all nonzero vectors x. It is just like the weighted least squares on p. 142; the most general measure of length is $\|Wx\|$, and the square of this length brings in a positive definite $A = W^TW$.

It remains to show that if A satisfies conditions I–IV, then it also satisfies V. We have to find W, and in fact we have almost done it twice already:

(i)   In the last step of the main theorem, Gaussian elimination factored A into LDL^T. The pivots are positive, and if $\sqrt{D}$ has their square roots along the diagonal, then $A = L\sqrt{D}\sqrt{D}L^T$.† Therefore one choice for W is the upper triangular matrix $\sqrt{D}L^T$, which proves that IV implies I.

(ii)   Another approach, which yields a different W, is to use II instead of IV. The eigenvalues are positive, and the eigenvectors go into an orthogonal Q that diagonalizes A:

$$A = Q\Lambda Q^T = Q\sqrt{\Lambda}\sqrt{\Lambda}Q^T = (\sqrt{\Lambda}Q^T)^T(\sqrt{\Lambda}Q^T).$$

W is now the same $\sqrt{\Lambda}Q^T$ that psychologists use in principal component analysis.

EXERCISE 6.2.5   Construct these two choices of W if

$$A = \begin{bmatrix} 2 & -1 \\ -1 & 2 \end{bmatrix}.$$

EXERCISE 6.2.6   Compute the square root $R = Q\sqrt{\Lambda}Q^T$ of this same matrix A. Verify both from the numbers and from the formula that R is symmetric and $R^2 = A$. The eigenvalues of R are the positive square roots $\sqrt{\lambda_i}$; it is another possible choice for W; and *it is the only symmetric positive definite square root of A*.

EXERCISE 6.2.7   If A is positive definite and C is nonsingular, prove that $B = C^TAC$ is also positive definite.

---

† This is called the **Cholesky decomposition**, with the pivots split evenly between the upper and lower triangular pieces. It differs very little from the Gauss $A = LDL^T$, and I usually avoid computing those square roots.

EXERCISE 6.2.8   We might call a nonsymmetric matrix "positive definite" if its Hermitian part $(A^H + A)/2$ is positive definite. In this case, even though $x^H A x$ may be complex, its real part $\frac{1}{2}(x^H A x + x^H A^H x)$ will be positive.

(a)   Show that also the real part of every eigenvalue of A is positive. (If $Ax = (\alpha + i\beta)x$, take the real part of the inner product with x^H.)

(b)   Show by a 2 by 2 triangular example that the eigenvalues could be positive even without a positive definite $A^H + A$. (In the stability theorem 5K for $u' = Au$, a negative definite $A^H + A$ meant decay at every instant—but the equation could be stable without it.)

EXERCISE 6.2.9   Write down the conditions for a matrix to be *negative definite*, with special attention to condition III: How is $\det(-A)$ related to $\det A$?

Ellipsoids in n Dimensions

To understand the geometry of a positive definite matrix, the key equation is $x^T A x = 1$. The vectors x that satisfy this equation lie on a surface in n-dimensional space, and we want to show that it is an ellipse. Or more correctly, it is an n-dimensional ellipsoid centered at the origin.

EXAMPLE $A = \begin{bmatrix} 2 & -1 \\ -1 & 2 \end{bmatrix}$, $x = \begin{bmatrix} u \\ v \end{bmatrix}$, and $x^T A x = 2u^2 - 2uv + 2v^2 = 1$. The best plan is to rewrite this equation as a sum of squares, with the eigenvalues $\lambda_1 = 1$ and $\lambda_2 = 3$ as the coefficients:

$$2u^2 - 2uv + 2v^2 = 1\left(\frac{u}{\sqrt{2}} + \frac{v}{\sqrt{2}}\right)^2 + 3\left(\frac{u}{\sqrt{2}} - \frac{v}{\sqrt{2}}\right)^2 = 1. \tag{8}$$

Suppose the two expressions in parentheses are called w and z. Then the equation is $w^2 + 3z^2 = 1$, and its graph is identified as an ellipse. The longest axis ends at the point $w = 1$, $z = 0$, and the shortest axis ends at $w = 0$, $z = 1/\sqrt{3}$. In other words, the major axis has half-length $1/\sqrt{\lambda_1}$, and the minor axis has half-length $1/\sqrt{\lambda_2}$ (Fig. 6.2). Furthermore, *these axes point toward the eigenvectors* $x_1 = (1/\sqrt{2}, 1/\sqrt{2})$ *and* $x_2 = (1/\sqrt{2}, -1/\sqrt{2})$. Therefore the geometry is completely tied up with the eigenvalues and eigenvectors.

We can do the same thing for any other surface $x^T A x = 1$, as soon as we understand what happened in Eq. (8). As always, the key step was to diagonalize A: $Q^{-1}AQ = \Lambda$, or $A = Q\Lambda Q^T$, with the unit eigenvectors x_j in the columns of Q. This produces a sum of squares:

$$x^T A x = \begin{bmatrix} & x^T & \end{bmatrix} \begin{bmatrix} & & \\ x_1 & \cdots & x_n \\ & & \end{bmatrix} \begin{bmatrix} \lambda_1 & & \\ & \ddots & \\ & & \lambda_n \end{bmatrix} \begin{bmatrix} & x_1^T & \\ & \vdots & \\ & x_n^T & \end{bmatrix} \begin{bmatrix} & \\ x \\ & \end{bmatrix}$$

$$= \lambda_1(x_1^T x)^2 + \cdots + \lambda_n(x_n^T x)^2.$$

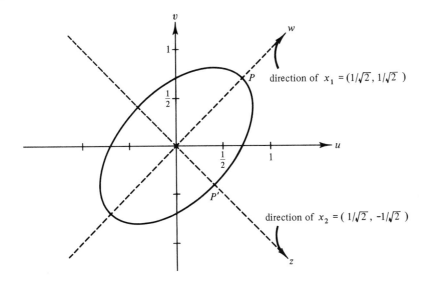

Fig. 6.2. The ellipse $2u^2 - 2uv + 2v^2 = 1$.

Now we introduce the new variables $y_1 = x_1^T x, \ldots, y_n = x_n^T x$—in matrix notation this is just $y = Q^T x$—and the equation for the surface simplifies to

$$x^T A x = \lambda_1 y_1^2 + \cdots + \lambda_n y_n^2 = 1. \tag{9}$$

This is the general case, of which Eq. (8) was a specific example. Even in this general case, it is easy to see that one point on the surface is $y_1 = 1/\sqrt{\lambda_1}$, $y_2 = \cdots = y_n = 0$. It lies at the end of the major axis, and it is the point farthest from the origin. Furthermore, it is just a multiple of the first eigenvector, $x = x_1/\sqrt{\lambda_1}$. It is marked P in Fig. 6.2. The point closest to the origin is at the end of the minor axis, with $y_n = 1/\sqrt{\lambda_n}$; this is the point P' along the last eigenvector x_n. The intermediate eigenvectors give the intermediate axes.

6D Suppose A is positive definite, and its unit eigenvectors are the columns of $Q: A = Q\Lambda Q^T$. Then the rotation $y = Q^T x$ produces the sum of squares

$$x^T A x = x^T Q\Lambda Q^T x = y^T \Lambda y = \lambda_1 y_1^2 + \cdots + \lambda_n y_n^2. \tag{10}$$

The equation $x^T A x = 1$ describes an ellipsoid whose axes end at the points where $\lambda_j y_j^2 = 1$ and where the remaining y components are zero. Undoing the rotation, these points are in the directions of the eigenvectors, and the axes have half length $1/\sqrt{\lambda_j}$.

EXERCISE 6.2.10 The ellipse $u^2 + 4v^2 = 1$ corresponds to $A = \begin{bmatrix} 1 & 0 \\ 0 & 4 \end{bmatrix}$. Write down the eigenvalues and eigenvectors, and sketch the ellipse.

EXERCISE 6.2.11 Reduce the equation $3u^2 - 2\sqrt{2}\, uv + 2v^2 = 1$ to a sum of squares by finding the eigenvalues of the corresponding A, and sketch the ellipse.

EXERCISE 6.2.12   In three dimensions, $\lambda_1 y_1^2 + \lambda_2 y_2^2 + \lambda_3 y_3^2 = 1$ represents an ellipsoid when all $\lambda_i > 0$. Describe all the different kinds of surfaces that appear in the positive semidefinite case when one or more of the eigenvalues is zero.

6.3 ■ SEMIDEFINITE AND INDEFINITE MATRICES; $Ax = \lambda Bx$

The goal of this section, now that the tests for a positive definite matrix are fully established, is to cast our net a little wider. We have three problems in mind:

   (1)   tests for a positive semidefinite matrix;
   (2)   the connection between the eigenvalues and the pivots of any symmetric matrix, definite or indefinite; and
   (3)   the generalized eigenvalue problem $Ax = \lambda Bx$.

The first will be very straightforward and quick; all the work has been done already. The second and third depend on the so-called "law of inertia" for symmetric matrices, which from a mathematical viewpoint is the main result of this section. One consequence of this law is that *the signs of the eigenvalues match the signs of the pivots*. Another is, as long as B is positive definite, that the problem $Ax = \lambda Bx$ has eigenvalues of the same sign as the conventional $Ax = \lambda x$. The eigenvectors are "orthogonal," but in a new way. This generalized eigenvalue problem is of special importance for applications, and we study one example in this section and another (called the finite element method) in Section 6.5.

For semidefinite matrices, the main point is to see the analogies with the positive definite case.

> **6E**   Each of the following tests is a necessary and sufficient condition for A to be *positive semidefinite*:
>
>   (I′) $x^T A x \geq 0$ for all vectors x.
>   (II′)   All the eigenvalues of A satisfy $\lambda_i \geq 0$.
>   (III′)   All the principal submatrices have nonnegative determinants.
>   (IV′)   All the pivots satisfy $d_i \geq 0$.
>   (V′)   There exists a matrix W, possibly singular, such that $A = W^T W$.
>
> If A has rank r, then $f = x^T A x$ is a sum of r perfect squares.

The connection between $x^T A x \geq 0$ and $\lambda_i \geq 0$, which is the most important, is exactly as before: The diagonalization $A = Q \Lambda Q^T$ leads to

$$x^T A x = x^T Q \Lambda Q^T x = y^T \Lambda y = \lambda_1 y_1^2 + \cdots + \lambda_n y_n^2, \qquad (11)$$

and this is nonnegative when the λ_i are nonnegative. If A has rank r, there are r nonzero eigenvalues and r perfect squares.

As for the determinant, it is the product of the λ_i and must also be non-negative. And since the principal submatrices are also semidefinite, their eigenvalues and their determinants are also ≥ 0; so we have deduced condition III′. (A principal submatrix is formed by throwing away rows and columns in matching pairs—say the first and fourth row and column, which retains the symmetry of A itself. It also retains semidefiniteness: If $x^T A x \geq 0$ for all x, then this property still holds when the first and fourth components of x are zero.) The novelty is that III′ applies to all the principal submatrices, and not only to those in the upper left corner. Otherwise, we could not distinguish between two matrices whose upper left determinants are all zero:

$$\begin{bmatrix} 0 & 0 \\ 0 & 1 \end{bmatrix} \text{ is positive semidefinite, and } \begin{bmatrix} 0 & 0 \\ 0 & -1 \end{bmatrix} \text{ is negative semidefinite.}$$

The pivots are again the ratios of the determinants and therefore nonnegative; the only problem is the possibility that row exchanges might be needed. In order to preserve symmetry, *we exchange the corresponding columns at the same time.* This means that "symmetric elimination" is carried out not on PA but on PAP^T. Since this exchange only moves around the principal submatrices, leaving their determinants unchanged, III′ safely implies IV′: after r stages of elimination the remaining submatrix in the lower right corner is identically zero, and $d_{r+1} = 0, \ldots, d_n = 0$.

Finally, IV′ means that $PAP^T = LDL^T$ (or $A = P^T LDL^T P$) with a non-negative pivot matrix D. The choice $W = \sqrt{D}\, L^T P$ produces $A = W^T W$, and condition V′ is established. Then V′ leads immediately back to I′, because $x^T A x = x^T W^T W x = \| Wx \|^2 \geq 0$, which completes the circle.

EXAMPLE

$$\begin{bmatrix} 2 & -1 & -1 \\ -1 & 2 & -1 \\ -1 & -1 & 2 \end{bmatrix}$$

is semidefinite, by the following tests:

(II′)   The eigenvalues are $\lambda_1 = 0$, $\lambda_2 = \lambda_3 = 3$.

(III′)   The principal submatrices have determinant 2 if they are 1 by 1; determinant 3 if they are 2 by 2; and det $A = 0$.

(IV′)

$$\begin{bmatrix} 2 & -1 & -1 \\ -1 & 2 & -1 \\ -1 & -1 & 2 \end{bmatrix} \rightarrow \begin{bmatrix} 2 & 0 & 0 \\ 0 & \frac{3}{2} & -\frac{3}{2} \\ 0 & -\frac{3}{2} & \frac{3}{2} \end{bmatrix} \rightarrow \begin{bmatrix} 2 & 0 & 0 \\ 0 & \frac{3}{2} & 0 \\ 0 & 0 & 0 \end{bmatrix},$$

and the pivots are $d_1 = 2$, $d_2 = \frac{3}{2}$, $d_3 = 0$.

EXERCISE 6.3.1   For the example above, find W and write $x^T A x = \| Wx \|^2$ as a sum of two squares. Describe the surface $x^T A x = 1$ in three-dimensional space.

EXERCISE 6.3.2   Apply any three tests to

$$A = \begin{bmatrix} 1 & 1 & 1 \\ 1 & 1 & 1 \\ 1 & 1 & 0 \end{bmatrix}.$$

Remark   The conditions for semidefiniteness could also be deduced from the original conditions I–V by the following trick: Add a small multiple of the identity, giving a positive definite matrix $A + \epsilon I$, and let ϵ approach zero. Since the determinants and eigenvalues depend continuously on ϵ, they will be positive until the last moment; at $\epsilon = 0$ they must still be nonnegative.

Congruence Transformations and the Law of Inertia

Earlier in this book, for elimination and for eigenvalues, we emphasized the elementary operations that make the matrix simpler. In each case, the essential thing was to know which properties of the matrix stayed unchanged. When a multiple of one row was subtracted from another, the list of "invariants" was pretty long: The nullspace and row space and rank and determinant all remain the same. In the case of eigenvalues, the basic operation was a similarity transformation $A \to S^{-1}AS$ (or $A \to M^{-1}AM$), and it is the eigenvalues themselves that are unchanged (and also, according to Appendix B, the Jordan form). Now we ask the same question for symmetric matrices: What are the "elementary operations" and their invariants for $x^{\mathrm{T}}Ax$?

The basic operation on a quadratic form is to change variables. A new vector y is related to x by some nonsingular matrix, $x = Cy$, and the quadratic form becomes $y^{\mathrm{T}}C^{\mathrm{T}}ACy$. Therefore the matrix operation which is fundamental to the theory of quadratic forms is a ***congruence transformation***:

$$A \to C^{\mathrm{T}}AC \qquad \text{for some nonsingular} \quad C. \tag{12}$$

The symmetry of A is preserved, since $C^{\mathrm{T}}AC$ remains symmetric. The real question is, What other properties of the matrix A remain unchanged by a congruence transformation? The answer is given by Sylvester's ***law of inertia***.

6F   The matrix $C^{\mathrm{T}}AC$ has the same number of positive eigenvalues as A, the same number of negative eigenvalues, and the same number of zero eigenvalues.

In other words, the *signs* of the eigenvalues (and not the eigenvalues themselves) are preserved by a congruence transformation. In the proof, we will suppose for convenience that A is nonsingular. Then $C^{\mathrm{T}}AC$ is also nonsingular, and there are no zero eigenvalues to worry about. (Otherwise we can work with the nonsingular $A + \epsilon I$ and $A - \epsilon I$, and at the end let $\epsilon \to 0$.)

We want to borrow a trick from topology, or more exactly from homotopy theory. Suppose C is linked to an orthogonal matrix Q by a continuous chain of matrices $C(t)$, none of which are singular: At $t = 0$ and $t = 1$, $C(0) = C$ and $C(1) = Q$. Then the eigenvalues of $C(t)^{\mathrm{T}}AC(t)$ will change gradually, as t goes from 0 to 1, from the eigenvalues of $C^{\mathrm{T}}AC$ to the eigenvalues of $Q^{\mathrm{T}}AQ$. Because $C(t)$ is never singular, *none of these eigenvalues can touch zero* (not to mention cross over it!). Therefore the number of eigenvalues to the right of zero, and the number to the left, is the same for $C^{\mathrm{T}}AC$ as for $Q^{\mathrm{T}}AQ$. And this number is also the same for A—which has exactly the same eigenvalues as the similar matrix $Q^{-1}AQ = Q^{\mathrm{T}}AQ$.† That is the proof.

EXAMPLE 1   If $A = I$ then $C^{\mathrm{T}}AC = C^{\mathrm{T}}C$, and this matrix is positive definite; take $C = W$ in condition V. Both the identity and $C^{\mathrm{T}}C$ have n positive eigenvalues, confirming the law of inertia.

EXAMPLE 2   If $A = \begin{bmatrix} 1 & 0 \\ 0 & -1 \end{bmatrix}$, then $C^{\mathrm{T}}AC$ has a negative determinant:

$$\det C^{\mathrm{T}}AC = (\det C^{\mathrm{T}})(\det A)(\det C) = -(\det C)^2 < 0.$$

Since this determinant is the product of the eigenvalues, $C^{\mathrm{T}}AC$ must have one positive and one negative eigenvalue, as A has, and again the law is obeyed.

EXAMPLE 3   This application is the important one:

6G   For any symmetric matrix A, the signs of the pivots agree with the signs of the eigenvalues. The eigenvalue matrix Λ and the pivot matrix D have the same number of positive entries, negative entries, and zero entries.

We know that $A = Q\Lambda Q^{\mathrm{T}}$, and Cholesky's formula is

$$A = LDL^{\mathrm{T}}, \quad \text{or} \quad D = L^{-1}A(L^{\mathrm{T}})^{-1} = (L^{-1}Q)\Lambda(L^{-1}Q)^{\mathrm{T}}. \quad (13)$$

The place of C is taken by $(L^{-1}Q)^{\mathrm{T}}$. By the law of inertia, $D = C^{\mathrm{T}}\Lambda C$ has eigenvalues of the same sign as Λ, and those eigenvalues are precisely the pivots. If there were row and column exchanges in the (symmetric) elimination, the same argument would apply to PAP^{T}.

EXERCISE 6.3.3   Using $C = \begin{bmatrix} 2 & 0 \\ 0 & -1 \end{bmatrix}$ and $A = \begin{bmatrix} 1 & 1 \\ 1 & 1 \end{bmatrix}$, confirm that $C^{\mathrm{T}}AC$ has eigenvalues of the same signs as A. Can you construct a chain of nonsingular matrices $C(t)$ linking C to an orthogonal Q?

---

† It is here that we needed Q to be orthogonal. One good choice for Q is to apply Gram–Schmidt to the columns of C; then $C = QR$, and the chain of matrices gradually reshapes R into the identity matrix: $C(t) = tQ + (1 - t)QR$. The family $C(t)$ goes slowly through Gram–Schmidt, from QR to Q; it is invertible because Q is invertible, and the other factor $tI + (1 - t)R$ is triangular with positive diagonal.

EXERCISE 6.3.4   From the signs of the pivots, show that

$$A = \begin{bmatrix} 1 & 3 & 0 \\ 3 & 8 & 7 \\ 0 & 7 & 6 \end{bmatrix}$$

has one negative and two positive eigenvalues. Then find the pivots for $A + 2I$, and show that adding 2 to the eigenvalues of A made them all positive. Therefore the negative eigenvalue of A must have been in the range $-2 < \lambda < 0$.

This exercise suggests a way to compute the eigenvalues, and it is the first practical method we have introduced. (It was clearly dominant about 1960, and is still an excellent method.) The idea is to keep cutting up the interval of uncertainty; once we know that $-2 < \lambda < 0$, the next step is to find out whether λ is above or below -1. Working with $A + I$, which adds 1 to all the eigenvalues of A, we discover that one of its pivots is negative. Therefore the original λ was below -1. The next step would investigate $A + \frac{3}{2}I$, which again has a negative pivot, so $-2 < \lambda < -\frac{3}{2}$. One binary digit is gained at every step.

For a large symmetric matrix, this method is too expensive; every step requires $n^3/6$ operations to find the pivots of a full matrix. But if A can be made tridiagonal before the calculation begins, which is done in Section 7.3, then every bisection of the interval of uncertainty costs only $2n$ multiplications. Elimination becomes very fast, and the search for eigenvalues by successive refinement becomes extremely simple. It is known as *Givens' method*.

The Generalized Eigenvalue Problem

I am not sure about economics, but physics and engineering and statistics are usually kind enough to produce symmetric matrices in their eigenvalue problems. They may, however, produce *two matrices rather than one*. The problem $Ax = \lambda x$ may be replaced by $Ax = \lambda Bx$, and for an example we look at the motion of two unequal masses.

Newton's law is $F_1 = m_1 a_1$ for the first mass and $F_2 = m_2 a_2$ for the second. If we set up the same system of springs as on p. 205, the forces F_1 and F_2 will be the same. We can describe the physics in a sentence: The masses are displaced by v and w, as in Fig. 6.3, and the springs pull them back with strength

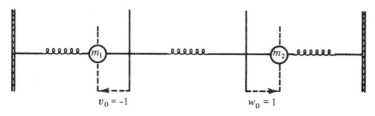

Fig. 6.3.   An oscillating system with unequal masses.

$F_1 = -2v + w$ and $F_2 = v - 2w$. The new and significant point comes on the left side of the equations, where the two masses appear:

$$m_1 \frac{d^2v}{dt^2} = -2v + w$$

or $\quad \begin{bmatrix} m_1 & 0 \\ 0 & m_2 \end{bmatrix} \frac{d^2u}{dt^2} = \begin{bmatrix} -2 & 1 \\ 1 & -2 \end{bmatrix} u.$

$$m_2 \frac{d^2w}{dt^2} = v - 2w$$

When the masses were equal, $m_1 = m_2 = 1$, this was the old system $u'' = Au$. Now it is $Bu'' = Au$, with a "mass matrix" B, and the eigenvalue problem arises when we look for exponential solutions $e^{i\omega t}x$:

$$Bu'' = Au \quad \text{becomes} \quad B(i\omega)^2 e^{i\omega t}x = Ae^{i\omega t}x.$$

Cancelling $e^{i\omega t}$, and writing λ for $(i\omega)^2$, this is

$$Ax = \lambda Bx, \quad \text{or} \quad \begin{bmatrix} -2 & 1 \\ 1 & -2 \end{bmatrix} x = \lambda \begin{bmatrix} m_1 & 0 \\ 0 & m_2 \end{bmatrix} x. \tag{14}$$

There is a solution only if the combination $A - \lambda B$ is singular, so λ must be a root of the polynomial equation $\det(A - \lambda B) = 0$. The special choice $B = I$ brings back the usual equation $\det(A - \lambda I) = 0$.

We shall work out an example, with $m_1 = 1$ and $m_2 = 2$:

$$\det(A - \lambda B) = \det \begin{bmatrix} -2 - \lambda & 1 \\ 1 & -2 - 2\lambda \end{bmatrix} = 2\lambda^2 + 6\lambda + 3, \quad \lambda = \frac{-3 \pm \sqrt{3}}{2}.$$

Both eigenvalues are negative, and the two natural frequencies are $\omega_i = \sqrt{-\lambda_i}$. The eigenvectors x_i are calculated in the usual way,

$$(A - \lambda_1 B)x_1 = \begin{bmatrix} \dfrac{-1 - \sqrt{3}}{2} & 1 \\ 1 & 1 - \sqrt{3} \end{bmatrix} x_1 = 0, \quad x_1 = \begin{bmatrix} \sqrt{3} - 1 \\ 1 \end{bmatrix},$$

$$(A - \lambda_2 B)x_2 = \begin{bmatrix} \dfrac{-1 + \sqrt{3}}{2} & 1 \\ 1 & 1 + \sqrt{3} \end{bmatrix} x_2 = 0, \quad x_2 = \begin{bmatrix} 1 + \sqrt{3} \\ -1 \end{bmatrix}.$$

These eigenvectors give the normal modes of oscillation. At the lower frequency ω_1, the two masses oscillate together—except that the first mass only moves as far as $\sqrt{3} - 1 \approx .73$, while the second (and larger) mass has for some reason the greater amplitude. In the faster mode, the components of x_2 have opposite signs and the masses move in opposite directions. This time the smaller mass goes much further.

The underlying theory is much easier to explain if B is split into W^TW. (B is assumed positive definite, as in the example.) Then the substitution $y = Wx$ changes

$$Ax = \lambda Bx = \lambda W^TWx \quad \text{into} \quad AW^{-1}y = \lambda W^Ty.$$

Writing C for W^{-1}, and multiplying through by $(W^T)^{-1} = C^T$, this becomes a standard eigenvalue problem for the *single* matrix $C^T A C$:

$$C^T A C y = \lambda y. \tag{15}$$

The eigenvalues λ_j are the same as for the original $Ax = \lambda Bx$, and the eigenvectors are related by $y_j = W x_j$.† The properties of the symmetric matrix $C^T A C$ lead directly to the corresponding properties of $Ax = \lambda Bx$:

(1)   The eigenvalues are real.

(2)   The eigenvalues have the same signs as the ordinary eigenvalues of A, by the law of inertia.

(3)   The eigenvectors y_j are orthonormal, so the original eigenvectors x_j are "B-orthonormal":

$$x_i{}^T B x_j = x_i{}^T W^T W x_j = y_i{}^T y_j = \begin{cases} 1 & \text{if } i = j \\ 0 & \text{if } i \neq j \end{cases}. \tag{16}$$

If the x_j are the columns of S, (16) means $S^T B S = I$. Similarly,

$$x_i{}^T A x_j = x_i{}^T \lambda_j B x_j = \begin{cases} \lambda_j & \text{if } i = j \\ 0 & \text{if } i \neq j. \end{cases} \tag{17}$$

In matrix terms that says $S^T A S = \Lambda$, and it expresses a remarkable fact: *The matrices A and B are simultaneously diagonalized by the congruence transformation S.*

Geometrically, this has a meaning which we do not understand very well. In the positive definite case, the two surfaces $x^T A x = 1$ and $x^T B x = 1$ are ellipsoids. Apparently $x = Sz$ gives a new choice of coordinates—not a pure rotation, because S is not an orthogonal matrix—such that these two ellipsoids become correctly aligned. They are

$$x^T A x = z^T S^T A S z = \lambda_1 z_1{}^2 + \cdots + \lambda_n z_n{}^2 = 1$$
$$x^T B x = z^T S^T B S z = z_1{}^2 + \cdots + z_n{}^2 = 1. \tag{18}$$

In fact, the second ellipsoid is a sphere! Certainly the main point of the theory is less obscure, and it is easy to summarize: As long as B is positive definite, the generalized eigenvalue problem $Ax = \lambda Bx$ behaves exactly as $Ax = \lambda x$.

EXERCISE 6.3.5   In the worked example with $m_1 = 1$ and $m_2 = 2$, verify that the normal modes are B-orthogonal: $x_1{}^T B x_2 = 0$. If both masses are displaced a unit distance and then released, with $v_0 = -1$, $w_0 = 1$, find the coefficients a_i in the resulting motion $u = a_1 \cos \omega_1 t\, x_1 + a_2 \cos \omega_2 t\, x_2$. What is the maximum distance from equilibrium achievable by the first mass (when the two cosines are $+1$ and -1)?

EXERCISE 6.3.6   Find the eigenvalues and eigenvectors of

$$\begin{bmatrix} 6 & -3 \\ -3 & 6 \end{bmatrix} x = \frac{\lambda}{18} \begin{bmatrix} 4 & 1 \\ 1 & 4 \end{bmatrix} x.$$

† A quicker way to produce a single matrix would have been $B^{-1}Ax = \lambda x$, but $B^{-1}A$ is not symmetric; we preferred to put half of B^{-1} on each side of A, giving a symmetric $C^T A C$.

EXERCISE 6.3.7 If the 2 by 2 symmetric matrices A and B are indefinite, then $Ax = \lambda Bx$ may have complex eigenvalues. Construct an example.

MINIMUM PRINCIPLES AND RAYLEIGH'S QUOTIENT ■ 6.4

In this section, as the end of the book approaches, we shall escape for the first time from linear equations. The unknown x will not be given as the solution to $Ax = b$ or $Ax = \lambda x$. Instead, it will be determined by a minimum principle.

It is astonishing how many natural laws can be expressed as minimum principles. Just the fact that heavy liquids sink to the bottom is a consequence of minimizing their potential energy. And when you sit on a chair or lie on a bed, the springs adjust themselves so that again the energy is minimized. Certainly there are more highbrow examples: A building is like a very complicated chair, carrying its own weight, and the fundamental principle of structural engineering is the minimization of total energy. In physics, there are "Lagrangians" and "action integrals"; a straw in a glass of water looks bent at the surface because light takes the path that reaches your eye as quickly as possible.†

We have to say immediately that these "energies" are nothing but *positive definite quadratic forms*. And obviously the derivative of a quadratic is linear. Therefore minimization will lead us back to the familiar linear equations, when we set all the first derivatives to zero. Our first goal in this section is *to find the minimum principle that is equivalent to $Ax = b$, and the one equivalent to $Ax = \lambda x$*. We will be doing in finite dimensions exactly what the calculus of variations does in a continuous problem, where the vanishing of the first derivatives gives a differential equation (Euler's equation). In every case, we are free to search either for a solution to the linear equation or for a minimum to the quadratic—and in many problems, as the next section will illustrate, the latter possibility ought not to be ignored.

The first step is very straightforward: We want to find the "parabola" P whose minimum occurs at the point where $Ax = b$. If A is just a scalar, that is easy to do:

$$P(x) = \tfrac{1}{2}Ax^2 - bx, \qquad \text{and} \qquad \frac{dP}{dx} = Ax - b.$$

This will be a minimum only if A is positive so the parabola opens upward, and then the vanishing of the first derivative gives $Ax = b$ (Fig. 6.4). In several dimensions this parabola turns into a paraboloid, but there is still the same formula for P; and to assure a minimum there is still a condition of positivity.

† I am convinced that plants and people also develop in accordance with minimum principles; perhaps civilization itself is based on a law of least action. The discovery of such laws is the fundamental step in passing from observations to explanations, and there must be some still to be found in the social sciences and in biology.

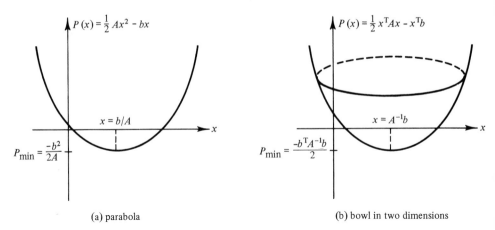

(a) parabola

(b) bowl in two dimensions

Fig. 6.4.   The minimum of a quadratic function $P(x)$.

6H   If A is symmetric positive definite, then $P(x) = \frac{1}{2}x^{T}Ax - x^{T}b$ assumes its minimum at the point where $Ax = b$.

Proof   Suppose x is the solution of $Ax = b$. For any vector y, we expand

$$
\begin{aligned}
P(y) - P(x) &= \tfrac{1}{2}y^{T}Ay - y^{T}b - \tfrac{1}{2}x^{T}Ax + x^{T}b \\
&= \tfrac{1}{2}y^{T}Ay - y^{T}Ax + \tfrac{1}{2}x^{T}Ax \qquad\qquad (19) \\
&= \tfrac{1}{2}(y - x)^{T}A(y - x).
\end{aligned}
$$

Since A is positive definite, this quantity can never be negative—and it is zero only if $y - x = 0$. At all other points $P(y)$ is larger than $P(x)$, so the minimum occurs at x.

EXERCISE 6.4.1   Consider the system $Ax = b$ given by

$$
\begin{bmatrix} 2 & -1 & 0 \\ -1 & 2 & -1 \\ 0 & -1 & 2 \end{bmatrix} \begin{bmatrix} x_1 \\ x_2 \\ x_3 \end{bmatrix} = \begin{bmatrix} 4 \\ 0 \\ 4 \end{bmatrix}.
$$

Construct the corresponding quadratic $P(x_1, x_2, x_3)$, compute its partial derivatives $\partial P/\partial x_i$, and verify that they vanish exactly at the desired solution.

EXERCISE 6.4.2   Find the minimum, if there is one, of $P_1 = \frac{1}{2}x^2 + xy + y^2 - 3y$ and $P_2 = \frac{1}{2}x^2 - 3y$. What matrix A is associated with P_2?

EXERCISE 6.4.3   Another quadratic that certainly has its minimum at $Ax = b$ is

$$
Q(x) = \tfrac{1}{2}\| Ax - b \|^2 = \tfrac{1}{2}x^{T}A^{T}Ax - x^{T}A^{T}b + \tfrac{1}{2}b^{T}b.
$$

Comparing Q with P, and ignoring the constant $\frac{1}{2}b^{T}b$, what system of equations do we get at the minimum? What are these equations called in the theory of least squares?

Our second goal was to find a minimization problem equivalent to $Ax = \lambda x$. That is not so easy. The function we choose cannot be simply a quadratic, or its differentiation would lead to a linear equation—and the eigenvalue problem is fundamentally nonlinear. The trick is to look at a *ratio* of quadratics, and the one we need is known as **Rayleigh's quotient**:

$$R(x) = \frac{x^{\mathrm{T}} A x}{x^{\mathrm{T}} x}.$$

We go directly to the main theorem:

61   Rayleigh's principle: The quotient $R(x)$ is minimized by the first eigenvector $x = x_1$, and its minimum value is the smallest eigenvalue λ_1 :

$$R(x_1) = \frac{x_1^{\mathrm{T}} A x_1}{x_1^{\mathrm{T}} x_1} = \frac{x_1^{\mathrm{T}} \lambda_1 x_1}{x_1^{\mathrm{T}} x_1} = \lambda_1. \tag{20}$$

Geometrically, imagine that we fix the numerator at 1, and make the denominator as large as possible. The numerator $x^{\mathrm{T}} A x = 1$ defines an ellipsoid, at least if A is positive definite. The denominator is $x^{\mathrm{T}} x = \| x \|^2$, so we are looking for the point on the ellipsoid farthest from the origin—in other words the vector x of greatest length. From our earlier description of the ellipsoid, its major axis points in the direction of the first eigenvector.

Algebraically, this is easy to see (without any requirement of positive definiteness) if we diagonalize A by an orthogonal matrix Q: $Q^{-1} A Q = \Lambda$ and $Q^{\mathrm{T}} = Q^{-1}$. With $x = Qy$, the quotient becomes much simpler:

$$R(x) = \frac{(Qy)^{\mathrm{T}} A (Qy)}{(Qy)^{\mathrm{T}} (Qy)} = \frac{y^{\mathrm{T}} \Lambda y}{y^{\mathrm{T}} y} = \frac{\lambda_1 y_1^2 + \cdots + \lambda_n y_n^2}{y_1^2 + \cdots + y_n^2}.$$

The minimum of this quotient certainly occurs at the point where $y_1 = 1$ and $y_2 = y_3 = \cdots = y_n = 0$. At that point the ratio equals λ_1. At any other point the ratio is larger, since λ_1 is the smallest of the λ's:

$$\lambda_1 (y_1^2 + y_2^2 + \cdots + y_n^2) \le (\lambda_1 y_1^2 + \lambda_2 y_2^2 + \cdots + \lambda_n y_n^2).$$

Dividing by the first expression in parentheses, $\lambda_1 \le R(x)$. The minimum occurs at the first eigenvector, or more accurately at any multiple $x = cx_1$, since the quotient is unchanged by the stretching factor c:

$$R(cx) = \frac{(cx)^{\mathrm{T}} A (cx)}{(cx)^{\mathrm{T}} (cx)} = \frac{c^2 \, x^{\mathrm{T}} A x}{c^2 \, x^{\mathrm{T}} x} = R(x).$$

Rayleigh's principle means that for every x, *the quotient $R(x)$ is an upper bound for λ_1*. We do not need to find the eigenvectors and change to y; this was useful in the proof, but in the applications we can choose any x whatsoever.

EXERCISE 6.4.4   For any matrix A, and for the special choice $x = (1, 0, \ldots, 0)$, what is the ratio $R(x)$? Show that λ_1 is less than or equal to every one of the diagonal entries a_{ii}.

EXERCISE 6.4.5   With $A = \begin{bmatrix} 2 & -1 \\ -1 & 2 \end{bmatrix}$, find a choice of x that gives a smaller $R(x)$ than the bound $\lambda_1 \leq 2$ which comes from the diagonal entries. What is the minimum value of $R(x)$?

EXERCISE 6.4.6   If B is positive definite, show from the Rayleigh quotient that the smallest eigenvalue of $A + B$ is larger than the smallest eigenvalue of A.

EXERCISE 6.4.7   If λ_1 is the smallest eigenvalue of A and μ_1 is the smallest eigenvalue of B, show that the smallest eigenvalue θ_1 of $A + B$ is at least as large as $\lambda_1 + \mu_1$. (Try the corresponding eigenvector x in the Rayleigh quotients.)

These two exercises are perhaps the most typical and most important results that come easily from Rayleigh's principle, but only with great difficulty from the eigenvalue equations themselves.

The applications of the Rayleigh quotient are not limited to the first eigenvalue and eigenvector, λ_1 and x_1. In fact, it is immediate that the *maximum* of the ratio

$$R(x) = \frac{\lambda_1 y_1^2 + \lambda_2 y_2^2 + \cdots + \lambda_n y_n^2}{y_1^2 + y_2^2 + \cdots + y_n^2} \tag{21}$$

occurs at the other end of the scale, where $y_n = 1$ and $y_1 = y_2 = \cdots = y_{n-1} = 0$. This point corresponds to the last eigenvector $x = x_n$, and the maximum value itself is λ_n. Therefore every value $R(x)$ is not only an upper bound for λ_1, it is also a lower bound for λ_n.[†] Finally, the intermediate eigenvectors $x_2, \ldots, x_{n-1}$ are all *saddle points* of the Rayleigh quotient.

EXERCISE 6.4.8   Find the partial derivatives of the quotient (21), and show that there is a stationary point at $y_2 = 1$, $y_1 = y_3 = \cdots = y_n = 0$. Find the maximum of

$$R(x) = (x_1^2 - x_1 x_2 + x_2^2)/(x_1^2 + x_2^2).$$

Minimax Principles for the Eigenvalues

The difficulty with saddle points is that for a typical vector x, we have no idea whether $R(x)$ is above or below the intermediate values $\lambda_2, \ldots, \lambda_{n-1}$. For the applications, it is an extremum principle (a minimum or a maximum) that is really useful. Therefore we look for such a principle, intending that the minimum or the maximum shall be attained at the jth eigenvector x_j.[‡] The

[†] By the reasoning of Exercise 6.4.4, every diagonal entry is below λ_n.
[‡] This topic is somewhat special, and it is actually the minimum principles we have already found on which the Rayleigh–Ritz finite element method is based. There is no difficulty in going directly to Section 6.5.

key idea is provided by a basic property of symmetric matrices: x_j is orthogonal to the other eigenvectors. Therefore suppose we constrain the vectors x in the minimum principle to be orthogonal to the first eigenvectors $x_1, \ldots, x_{j-1}$; this forces $y_1 = y_2 = \cdots = y_{j-1} = 0$. The remaining parameters $y_j, \ldots, y_n$ are unconstrained, and the Rayleigh quotient reduces to

$$\frac{\lambda_j y_j^2 + \lambda_{j+1} y_{j+1}^2 + \cdots + \lambda_n y_n^2}{y_j^2 + y_{j+1}^2 + \cdots + y_n^2}.$$

Its minimum is now λ_j, when $y_j = 1$.

6J Under the constraint that x is orthogonal to the eigenvectors $x_1, \ldots, x_{j-1}$, the quotient $R(x)$ is minimized by the next eigenvector x_j, and its minimum value is λ_j:

$$\lambda_j = \min R(x), \qquad \text{given} \qquad x^T x_1 = 0, \ldots, x^T x_{j-1} = 0. \tag{22}$$

Similarly, x_j maximizes $R(x)$ with x perpendicular to $x_{j+1}, \ldots, x_n$.

With this principle, it is theoretically possible to determine the eigenvalues in increasing order. The first, λ_1, is the absolute minimum of $R(x)$, and is attained at x_1. Then λ_2 will be the minimum under the constraint that x is orthogonal to x_1. And any such x, other than the eigenvector x_2, will give an $R(x)$ greater than λ_2.

EXAMPLE

$$A = \begin{bmatrix} 1 & -1 & & \\ -1 & 2 & -1 & \\ & -1 & 2 & -1 \\ & & -1 & 1 \end{bmatrix}$$

is positive semidefinite, and $x_1 = (1, 1, 1, 1)$ is a nullvector; it is the eigenvector corresponding to $\lambda_1 = 0$. To find an upper bound for λ_2, we can choose any trial vector x orthogonal to x_1, and compute $R(x)$:

$$x = (1, 1, -1, -1) \qquad \text{and} \qquad \frac{x^T A x}{x^T x} = 1, \qquad \text{so} \quad \lambda_2 \leq 1.$$

The only problem is that if we had not known x_1 exactly (which is rare), then we would not have known when x was orthogonal to it. In a problem of constrained minimization, the constraint has be to known!

We now ask whether there is an extremum principle for λ_2 and x_2 *that does not require knowledge of x_1*. The answer is subtle. Suppose that, not knowing x_1, we simply impose the constraint that all trial vectors x shall be perpendicular to some arbitrary vector z. If $z = x_1$, then the minimum value of $R(x)$ will be λ_2. In the much more likely case that z is different from x_1, we can still find out something about the constrained minimum: *It is not greater than* λ_2 . For any choice of z,

$$\min_{x^T z=0} R(x) \leq \lambda_2 . \tag{23}$$

Proof Some nonzero combination of the first two eigenvectors, $x = \alpha x_1 + \beta x_2$, will be orthogonal to z; this only imposes a single condition on α and β. For any such combination of the first two eigenvectors,

$$R(x) = \frac{(\alpha x_1 + \beta x_2)^T A (\alpha x_1 + \beta x_2)}{(\alpha x_1 + \beta x_2)^T (\alpha x_1 + \beta x_2)} = \frac{\lambda_1 \alpha^2 + \lambda_2 \beta^2}{\alpha^2 + \beta^2} \leq \lambda_2 .$$

Since we have found a candidate x for which $R(x) \leq \lambda_2$, the minimum is certainly less than λ_2. This leads directly to a "maximin principle":

6K If we minimize $R(x)$ subject to any one constraint $x^T z = 0$, and then choose the z that maximizes that minimum, the result is λ_2 :

$$\lambda_2 = \max_z \left[\min_{x^T z=0} R(x) \right]. \tag{24}$$

The quantity in brackets is $\leq \lambda_2$, according to (23), and for the particular choice $z = x_1$ that quantity *equals* λ_2 .

Geometrically, the maximin principle has a natural interpretation. Suppose an ellipsoid is cut by a plane through the origin. Then the cross section is again an ellipsoid, of one lower dimension. The inequality (23) simply means that the major axis of this cross section is at least as long as the second axis of the original ellipsoid ($\mu_1 \leq \lambda_2$ in Fig. 6.5). For a special choice of the cutting plane,

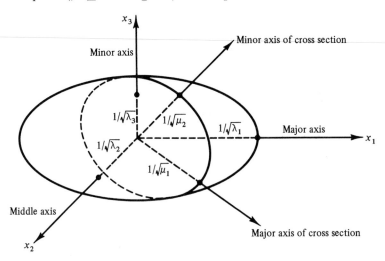

Fig. 6.5. The maximin and minimax principles: $\lambda_1 \leq \mu_1 \leq \lambda_2 \leq \mu_2 \leq \lambda_3$.

the major axis of the cross section will be exactly the second axis of the original ellipse; this choice is the plane perpendicular to the direction $z = x_1$ of the original major axis. In this case $\mu_1 = \lambda_2$.

This geometrical picture can be translated directly into matrix algebra:

6L   Given a symmetric matrix A with eigenvalues $\lambda_1 \leq \lambda_2 \leq \cdots \leq \lambda_n$, suppose the last row and column are deleted to form the submatrix A_{n-1}. Then if μ_1 is the smallest eigenvalue of this submatrix, it lies in the interval $\lambda_1 \leq \mu_1 \leq \lambda_2$.

To start, λ_1 is the absolute minimum of the quotient $R(x)$. The trick is to see that μ_1 is also the minimum of $R(x)$, under the constraint that the last component of x must be zero. This constrained minimum cannot be below the absolute minimum: $\mu_1 \geq \lambda_1$. On the other hand, the condition that the last component of x must vanish is the constraint $x^{\mathrm{T}}z = 0$, with the particular choice $z = (0, \ldots, 0, 1)$. Therefore $\mu_1 \leq \lambda_2$ according to (23).

EXAMPLE 1   The eigenvalues of $A = \begin{bmatrix} 2 & -1 \\ -1 & 2 \end{bmatrix}$ are 1 and 3, whereas the eigenvalue without the last row and column is 2.

EXAMPLE 2   The eigenvalues of

$$A = \begin{bmatrix} 2 & -1 & 0 \\ -1 & 2 & -1 \\ 0 & -1 & 2 \end{bmatrix}$$

are $2 - \sqrt{2}, 2, 2 + \sqrt{2}$. When the last row and column are deleted, the smallest eigenvalue is increased to 1, but this does not go beyond λ_2: $2 - \sqrt{2} < 1 < 2$.

EXAMPLE 3   If B is positive definite, then by Rayleigh quotients

$$\lambda_1(A + B) = \min \frac{x^{\mathrm{T}}(A + B)x}{x^{\mathrm{T}}x} > \min \frac{x^{\mathrm{T}}Ax}{x^{\mathrm{T}}x} = \lambda_1(A).$$

This was Exercise 6.4.6. Now the maximin principle (24) gives the same thing for the *second* eigenvalues, $\lambda_2(A + B) > \lambda_2(A)$, since again $R(x)$ is increased when B is included.

EXAMPLE 4   Given an oscillating system of masses and springs, suppose one of the masses is forced to remain at equilibrium. Then the lowest frequency of the system is increased, but not above the original λ_2.

EXERCISE 6.4.9   Find the eigenvalues of

$$A = \begin{bmatrix} 2 & -1 & 0 \\ -1 & 2 & 0 \\ 0 & 0 & 1 \end{bmatrix}$$

and show from $\lambda_1 \leq \mu_1 \leq \lambda_2$ that no matter which row–column pair is deleted, the new smallest eigenvalue must be $\mu_1 = 1$. What game is played with the ellipsoid $x^{\mathrm{T}}Ax = 1$?

EXERCISE 6.4.10   Generalize the maximin principle to allow j different constraints, with the same idea as for $j = 1$:

$$\max_{z_1, \cdots, z_j} \left[\min_{x^{\mathrm{T}}z_1 = 0, \cdots, x^{\mathrm{T}}z_j = 0} R(x) \right] = \lambda_{j+1}. \tag{25}$$

Taking $j = 2$, give an inequality on the smallest eigenvalue when the last two rows and columns of A are thrown away.

It is time to turn these theorems around and arrive at a ***minimax principle***. That means we intend first to maximize the Rayleigh quotient, and then to identify the minimum possible value of this maximum.

There are several ways to proceed, depending on which eigenvalue is wanted. To find the largest eigenvalue λ_n, we just maximize $R(x)$ and there we have it. But suppose we want to stay with λ_2. Since the Rayleigh quotient equals

$$\frac{\lambda_1 y_1{}^2 + \lambda_2 y_2{}^2 + \cdots + \lambda_n y_n{}^2}{y_1{}^2 + y_2{}^2 + \cdots + y_n{}^2},$$

the obvious way to produce λ_2 as a maximum is to require that $y_3 = y_4 = \cdots = y_n = 0$. These $n - 2$ constraints leave only a two-dimensional subspace, namely the subspace spanned by the first two eigenvectors x_1 and x_2. Over this particular two-dimensional subspace, the maximum value of $R(x)$ is λ_2—but the eigenvectors x_1 and x_2 are part of the problem, and they are unknown.

When this situation arose for the maximin principle, the key idea was to minimize with an arbitrary constraint $x^{\mathrm{T}}z = 0$. Therefore we copy that idea, and *maximize $R(x)$ over an arbitrary two-dimensional subspace*. We have no way to know whether this subspace contains any eigenvectors; it is very unlikely. Nevertheless, there is still an inequality on the maximum. Previously the minimum with $x^{\mathrm{T}}z = 0$ was below λ_2; now the maximum over the subspace will be above λ_2. If S_2 *is any two-dimensional subspace of* $\mathbf{R}^n$, then

$$\max_{x \text{ in } S_2} R(x) \geq \lambda_2. \tag{26}$$

Proof   Some x in the subspace is orthogonal to the first eigenvector x_1, and for that particular x we know that $R(x) \geq \lambda_2$. The minimax principle follows immediately:

6M   If we maximize $R(x)$ over all possible two-dimensional subspaces S_2, the minimum possible value for that maximum is λ_2:

$$\lambda_2 = \min_{S_2} \left[\max_{x \text{ in } S_2} R(x) \right]. \tag{27}$$

The quantity in brackets is always $\geq \lambda_2$, and for the particular subspace spanned by the first two eigenvectors, it equals λ_2.

EXERCISE 6.4.11   In Example 2 above, find the maximal eigenvalue μ_2 of the 2 by 2 submatrix in the upper left corner, and compare it to λ_2 and λ_3.

We close this section with two remarks. I hope that, even without detailed proofs, your intuition says they are correct.

Remark 1   The minimax principle extends from two-dimensional subspaces to j-dimensional subspaces, and produces λ_j :

$$\lambda_j = \min_{S_j} \left[\max_{x \text{ in } S_j} R(x) \right]. \tag{28}$$

Remark 2   For the generalized eigenvalue problem $Ax = \lambda Bx$, all the same principles hold if the denominator in the Rayleigh quotient is changed from x^Tx to x^TBx. Substituting $x = Sz$, with the eigenvectors of $B^{-1}A$ in the columns of S, $R(x)$ simplifies to

$$R(x) = \frac{x^TAx}{x^TBx} = \frac{\lambda_1 z_1^2 + \cdots + \lambda_n z_n^2}{z_1^2 + \cdots + z_n^2}. \tag{29}$$

This was the point of simultaneously diagonalizing A and B on p. 250; both quadratics become sums of perfect squares, and $\lambda_1 = \min R(x)$ as well as $\lambda_n = \max R(x)$. Even for *unequal* masses in an oscillating system, holding one mass at equilibrium will raise the lowest frequency and lower the highest.

EXERCISE 6.4.12   Which particular j-dimensional subspace in (28) is the one that gives the minimum? In other words, over which S_j is the maximum of $R(x)$ equal to λ_j ?

EXERCISE 6.4.13   The minimax principle (28) can be written in terms of constraints rather than subspaces:

$$\lambda_j = \min_{z_1, \dots, z_{n-j}} \left[\max_{x^Tz_1 = \cdots = x^Tz_{n-j} = 0} R(x) \right].$$

What is the connection between the z's and the subspaces S_j ?

EXERCISE 6.4.14   Show that the smallest eigenvalue λ_1 of $Ax = \lambda Bx$ is no bigger than the ratio a_{11}/b_{11} of the diagonal entries.

EXERCISE 6.4.15   (difficult)   Show that if the last row and column are removed from a symmetric matrix A, then the eigenvalues μ_i of the submatrix satisfy

$$\lambda_1 \leq \mu_1 \leq \lambda_2 \leq \mu_2 \leq \cdots \leq \mu_{n-1} \leq \lambda_n. \tag{30}$$

Remember that discarding the row and column amounts to a single constraint; the last component is $x_n = 0$, so x is orthogonal to $z = (0, \dots, 0, 1)$. Deduce $\mu_j \geq \lambda_j$ from the minimax principle, and $\mu_j \leq \lambda_{j+1}$ from maximin.

6.5 ■ THE RAYLEIGH–RITZ PRINCIPLE AND FINITE ELEMENTS

There were two main ideas in the previous section:

(i)   Solving $Ax = b$ is equivalent to minimizing $P(x) = \frac{1}{2}x^{\mathrm{T}}Ax - x^{\mathrm{T}}b$.

(ii)   Solving $Ax = \lambda_1 x$ is equivalent to minimizing $R(x) = x^{\mathrm{T}}Ax/x^{\mathrm{T}}x$.

Now we try to explain how these ideas can be applied.

The story is a long one, because these equivalences have been known for more than a century. In some classical problems of engineering and physics, such as the bending of a plate or the ground states (eigenvalues and eigenfunctions) of an atom, these minimizations were used to get rough approximations to the true solution. In a certain sense, these approximations *had* to be rough; the only tools were pencil and paper or a little machine. The mathematical principles were there, but they could not be implemented.

Obviously the digital computer was going to bring a revolution, but the first step in that revolution was to throw out the minimum principles; they were too old and too slow. It was the method of finite differences that jumped ahead, because it was easy to see how to "discretize" a differential equation. We saw it already, in Section 1.6; every derivative is replaced by a difference quotient. The physical region is covered by a lattice, or a "mesh," and at each mesh point there is an equation such as $-u_{j+1} + 2u_j - u_{j-1} = h^2 f_j$. Mathematically, the problem is expressed as $Au = f$—and numerical analysts were completely occupied in the 1950's with the development of quick ways to solve systems that are very large and very sparse.

What we did not fully recognize was that even finite differences become incredibly complicated for real engineering problems, like the stresses on an airplane or the natural frequencies of the human skull. *The real difficulty is not to solve the equations, but to set them up.* For an irregular region we need an irregular mesh, pieced together from triangles or quadrilaterals or tetrahedra, and then we need a systematic way of approximating the underlying physical laws. In other words, the computer has to help not only in the solution of the discrete problem, but in its formulation.

You can guess what has happened. The old methods have come back, but with a new idea and a new name. The new name is the *finite element method*, and the new idea has made it possible to use more of the power of the computer—in constructing a discrete approximation, solving it, and displaying the results—than any other technique in scientific computation.† The key is to keep the basic mathematical idea simple; then the applications can be complicated. The emphasis in these applications has moved away from airplane design to the safety of nuclear reactors, and at this writing there is a violent debate about its extension into fluid dynamics. For problems on this scale, however, the one undebatable point is their cost—I am afraid a billion dollars would be a conservative estimate of the expense so far. I hope some

† Please forgive this enthusiasm; I know the method may not be immortal.

readers will be interested enough, and vigorous enough, to master the finite element method and put it to good use.

To explain the method, we start with the classical Rayleigh–Ritz principle and then introduce the new idea of finite elements. We can work with the same differential equation $-u'' = f(x)$, and the same boundary conditions $u(0) = u(1) = 0$, which were studied earlier by finite differences. Admittedly this is an "infinite-dimensional" problem, with the vector b replaced by a function f and the matrix A by an operator $-d^2/dx^2$. But we can proceed by analogy, and write down the quadratic whose minimum is required, replacing inner products by integrals:

$$P(v) = \tfrac{1}{2}v^{\mathrm{T}}Av - v^{\mathrm{T}}b = \tfrac{1}{2}\int_0^1 v(x)\,(-v''(x))\,dx - \int_0^1 v(x)f(x)\,dx. \quad (31)$$

This is to be minimized over all functions v satisfying the boundary conditions, and *the function that gives the minimum will be the solution* u. The differential equation has been converted to a minimum principle, and it only remains to integrate one of the terms by parts:

$$\int_0^1 v(-v'')\,dx = \int_0^1 (v')^2 - [vv']_{x=0}^{x=1}, \quad \text{so} \quad P(v) = \tfrac{1}{2}\int_0^1 (v')^2 - \int_0^1 vf.$$

Now the quadratic term $\int (v')^2$ is symmetric, like $x^{\mathrm{T}}Ax$, and also it is *positive*; we are guaranteed a minimum.

How do we find this minimum? To compute it exactly is equivalent to solving the differential equation exactly, and that problem is infinite dimensional. *The Rayleigh–Ritz principle produces an n-dimensional problem by choosing only n trial functions* $v = V_1, v = V_2, \ldots, v = V_n$. It admits all linear combinations $V = y_1 V_1(x) + \cdots + y_n V_n(x)$, and computes the particular combination U that minimizes $P(V)$. To repeat: The idea is to minimize over a subspace instead of over all possible v, and the function that gives the minimum is U instead of u. We hope the two are close.

Substituting V for v, the quadratic turns into

$$P(V) = \tfrac{1}{2}\int_0^1 (y_1 V_1' + \cdots + y_n V_n')^2 - \int_0^1 (y_1 V_1 + \cdots + y_n V_n)f.$$

Remember that the V's are chosen in advance; the unknowns are the weights $y_1, \ldots, y_n$. If we put these weights into a vector y, then $P(V)$ is recognized as exactly one of the quadratics we are accustomed to:

$$P(V) = \tfrac{1}{2}y^{\mathrm{T}}Ay - y^{\mathrm{T}}b, \quad (32)$$

where $A_{ij} = \int V_i'V_j' =$ the coefficient of y_iy_j and $b_j = \int V_j f =$ the coefficient of y_j. We can certainly find the minimum of $\tfrac{1}{2}y^{\mathrm{T}}Ay - y^{\mathrm{T}}b$; it is equivalent to solving $Ay = b$. Therefore the steps in the Rayleigh–Ritz method are: (i) choose the trial functions V_j ; (ii) compute the coefficients A_{ij} and b_j ; (iii) solve $Ay = b$, and (iv) print out the approximate solution $U = y_1 V_1 + \cdots + y_n V_n$.

Everything depends on step (i). Unless the functions V_j are extremely simple, the other steps will be virtually impossible. And unless some combination of the V_j is close to the true solution u, those steps will be useless. The problem is to combine both computability and accuracy, and *the key idea that has made finite elements successful is the use of piecewise polynomials as the trial functions V.*

The simplest and most widely used element is piecewise linear. We begin by placing nodes at the points $x_1 = h$, $x_2 = 2h$, ..., $x_n = nh$, just as for finite differences; at the endpoints $x_0 = 0$ and $x_{n+1} = 1$, the boundary conditions require every V to be zero. Then V_j is the "roof function" which equals 1 at the node x_j, and zero at all the other nodes (Fig. 6.6a). It is concentrated in

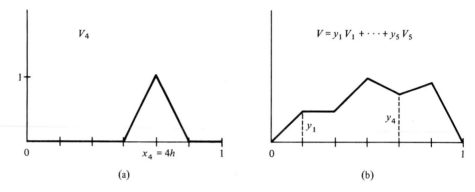

(a)   (b)

Fig. 6.6. Roof functions and their linear combinations.

a small interval around its node, and it is zero everywhere else. In fact any combination $y_1 V_1 + \cdots + y_n V_n$ must have the value y_j at that same node, because the other V's are zero, so the graph of this combination is easy to draw (Fig. 6.6b).

That completes step (i). Next we compute the coefficients $A_{ij} = \int V_i' V_j'$ in the "stiffness matrix" A. The slope V_j' equals $1/h$ in the small interval to the left of x_j, and $-1/h$ in the interval to the right. The same is true of V_i' near its node x_i, and if these "double intervals" do not overlap then the product $V_i' V_j'$ is identically zero. They do overlap only when

$$i = j \qquad \text{and} \qquad \int V_i' V_j' = \int \left(\frac{1}{h}\right)^2 + \int \left(-\frac{1}{h}\right)^2 = \frac{2}{h},$$

or

$$i = j \pm 1 \qquad \text{and} \qquad \int V_i' V_j' = \int \left(\frac{1}{h}\right)\left(\frac{-1}{h}\right) = \frac{-1}{h}.$$

Therefore the stiffness matrix is actually tridiagonal:

$$
A = \frac{1}{h}
\begin{bmatrix}
2 & -1 & & & \\
-1 & 2 & -1 & & \\
& -1 & 2 & -1 & \\
& & -1 & 2 & -1 \\
& & & -1 & 2
\end{bmatrix}.
$$

This looks just like finite differences, and it has led to a thousand discussions about the relation between these two methods. More complicated finite elements—polynomials of higher degree, which are defined on intervals for ordinary differential equations or on triangles or quadrilaterals for partial differential equations—also produce sparse matrices A. You could think of finite elements as a systematic way to construct accurate difference equations on irregular meshes, so that finite elements fall into the "intersection" of Rayleigh–Ritz methods and finite differences. The essential thing is the simplicity of these piecewise polynomials; on every subinterval their slopes are easy to find and to integrate.

The components b_j on the right side are completely new; instead of just the value of f at x_j, as for finite differences, they are now an average of f around that point: $b_j = \int V_j f$. Then in step (iii) we solve the tridiagonal system $Ay = b$, which gives the coefficients in the particular trial function $U = y_1 V_1 + \cdots + y_n V_n$ which is minimizing. And finally, connecting all these heights y_j by a broken line, we have a picture of the approximate solution U.

There is one property that is special to this particular problem: U is not just close to the true solution u, it is exactly equal to u at the nodes. In other words, U is the linear interpolate of u. This is too much to expect for a more complicated equation, and it does not happen. But the error $U - u$ is remarkably small even on a coarse mesh, and the underlying convergence theory is explained in the author's book "An Analysis of the Finite Element Method," written jointly with George Fix, (Prentice-Hall, Englewood Cliffs, New Jersey, 1973). There we also discuss partial differential equations, for which the method really comes into its own.

EXERCISE 6.5.1 With a uniform source term $f \equiv 1$, the solution of $-u'' = f$ is the parabola $u = (x - x^2)/2$. Compute the coefficients $b_j = \int V_j f$, and show that the exact values $y_j = u(x_j)$ do satisfy the finite element equations $Ay = b$.

EXERCISE 6.5.2 When $A = I$, the quadratic is $P(y) = \frac{1}{2} y^T y - y^T b$ and it is minimized at $y = b$. Prove that $P(y) - P(b) = \frac{1}{2} \| y - b \|^2$, and explain from this identity why the vector that minimizes $P(y)$ over a subspace of trial functions is also the vector closest to b. The Rayleigh–Ritz principle automatically picks out the *projection of the true solution onto the trial space.*

Eigenvalue Problems

The Rayleigh–Ritz idea—to minimize over a finite-dimensional family of V's in place of all admissible v's—is as useful for eigenvalue problems as for steady-state equations. This time it is the Rayleigh quotient that is minimized. Its true minimum is the fundamental frequency λ_1, and its approximate minimum Λ_1 will be increased when we restrict the class of trial functions from the v's to the V's. Again the discrete problem is manageable, and the principle can actually be applied, only when the functions V_j are easy to compute. Therefore the step which has been taken in the last 20 years was completely natural and inevitable: to apply the new finite element ideas to this long-established variational form of the eigenvalue problem.

The best example is the simplest one:

$$-u'' = \lambda u, \qquad \text{with} \qquad u(0) = u(1) = 0. \tag{33}$$

Its first eigenvector is $u = \sin \pi x$, with eigenvalue $\lambda_1 = \pi^2$. This combination satisfies Eq. (33), and it gives the minimum in the corresponding Rayleigh quotient

$$R(v) = \frac{v^{\mathrm{T}}[-d^2/dx^2]v}{v^{\mathrm{T}}v} = \frac{\int_0^1 v(-v'')}{\int_0^1 v^2} = \frac{\int_0^1 (v')^2}{\int_0^1 v^2}.$$

Physically this is a ratio of potential to kinetic energy, and they are in balance at the eigenvector. Normally this eigenvector would be unknown, and to approximate it we admit only the candidates $V = y_1 V_1 + \cdots + y_n V_n$:

$$R(V) = \frac{\int_0^1 (y_1 V_1' + \cdots + y_n V_n')^2}{\int_0^1 (y_1 V_1 + \cdots + y_n V_n)^2} = \frac{y^{\mathrm{T}} A y}{y^{\mathrm{T}} B y}.$$

Now we face the problem of minimizing $y^{\mathrm{T}}Ay/y^{\mathrm{T}}By$. If the matrix B were the identity, this would lead to the standard eigenvalue problem $Ay = \lambda y$. But our matrix B will be tridiagonal, and it is exactly this situation that brings in the *generalized eigenvalue problem* (p. 259). The minimum value Λ_1 will be the smallest eigenvalue of $Ay = \lambda By$, and the corresponding eigenvector y will give the approximation to the eigenfunction: $U = y_1 V_1 + \cdots + y_n V_n$.

As in the static problem, the method can be summarized in four steps: (i) choose the V_j; (ii) compute A and B; (iii) solve $Ay = \lambda By$; and (iv) print out Λ_1 and U. I don't know why that costs a billion dollars.

EXERCISE 6.5.3   For the piecewise linear functions V_1 and V_2, corresponding to the nodes $x = h = \frac{1}{3}$ and $x = 2h = \frac{2}{3}$, compute the 2 by 2 matrix B. Identify the discrete eigenvalue problem with Exercise 6.3.6, and compare Λ_1 with the true eigenvalue $\lambda_1 = \pi^2$.

EXERCISE 6.5.4   Explain why the minimax principle also guarantees that $\Lambda_2 \geq \lambda_2$.

$$\begin{bmatrix} 7 \end{bmatrix}$$

COMPUTATIONS WITH MATRICES

The aim of this book has been to explain some of the applicable parts of matrix theory. Compared to standard texts in abstract linear algebra, the underlying theory has not been radically changed; one of the best things about the subject is that the theory is really essential for the applications. What is different is the *change in emphasis* which comes with a new point of view. Gaussian elimination becomes more than just a way to find a basis for the row space, and the Gram–Schmidt process is not just a proof that every subspace has an orthonormal basis. Instead, we really *need* these algorithms. And we need a convenient description, $A = LU$ or $A = QR$, of what they do.

This chapter will take a few more steps in the same direction. I suppose these steps are governed by computational necessity, rather than by elegance, and I don't know whether to apologize for that; it makes them sound very superficial, and that is wrong. They deal with the oldest and most fundamental problems of the subject, $Ax = b$ and $Ax = \lambda x$—but virtually every one of them was conceived by the present generation of mathematicians. In numerical analysis there is a kind of survival of the fittest, and we want to describe some ideas that have survived so far. They fall into three groups:

1. *Techniques for solving* $Ax = b$. Gaussian elimination is a perfect algorithm, except perhaps if the particular problem has special properties—as almost every problem has. Our discussion will concentrate on the property

of *sparseness*, when most of the entries in A are zero, and on the development of *iterative rather than direct methods for solving $Ax = b$*. An iterative method is "self-correcting," and it repeats the corrections over and over; it will never reach the exact answer, but the object is to get close to it more quickly than elimination. In some problems that can be done; in many others, elimination is safer and faster if it takes advantage of the zeros. The competition is far from over, and our first goal is to identify the condition which guarantees convergence to the true solution $A^{-1}b$, and which governs its speed. Then we apply this condition to successive overrelaxation and to other rules of iteration; this is Section 7.4.

2.   *Techniques for solving $Ax = \lambda x$.*   The eigenvalue problem is one of the outstanding successes of numerical analysis. It is clearly defined, its importance is obvious, but until recently no one knew how to solve it. Dozens of algorithms have been suggested, and of course it is impossible to put them in an exact ordering from best to worst; everything depends on the size and the properties of A, and on the number of eigenvalues that are wanted. In other words, it is dangerous just to ask a computation center for an eigenvalue subroutine, without knowing anything about its contents. (Of course, I hope you do not need to check every Fortran statement.) The excellent book by Wilkinson and Reinsch† explains a whole range of algorithms, and we have chosen two or three ideas which have superseded almost all of their predecessors: the QR algorithm, the family of "power methods," and the preprocessing of a symmetric matrix to make it tridiagonal.

The first two methods are iterative, and the last is direct; it does its job in a finite number of steps, but it does not end up with the eigenvalues themselves, only with a much simpler matrix to use in the iterative steps.

3.   *The condition number of a matrix.*   Section 7.2 attempts to measure the sensitivity, or the "vulnerability," of a solution: If A and b are slightly changed, how great is the effect on $x = A^{-1}b$? Before starting on that question, we want to point out one obstacle (which is easy to overcome). There has to be a way to measure the change δA, and to estimate the size of A itself. The length of a vector is already defined, and now we need the *norm of a matrix*. Then the condition number, and with it the sensitivity of A, will follow directly from the norms of A and A^{-1}.

7.2 ■ THE NORM AND CONDITION NUMBER OF A MATRIX

An error and a blunder are very different things. An error is a small mistake, probably unavoidable even by a perfect mathematician or a perfect computer. A blunder is much more serious, and larger by at least an order of magnitude.

---

† J. M. Wilkinson and C. Reinsch, "Handbook for Automatic Computation, Linear Algebra," Springer, New York and Berlin, 1971.

When the computer rounds off a number to the eighth significant place, that is an error; but when a problem is so excruciatingly sensitive that this roundoff error completely changes the solution, then almost certainly someone has committed a blunder. Our goal in this section is to analyze the effect of errors, so that blunders can be avoided.

We are actually continuing a discussion that began in Chapter 1 with

$$A = \begin{bmatrix} 1 & 1 \\ 1 & 1.0001 \end{bmatrix} \quad \text{and} \quad A' = \begin{bmatrix} .0001 & 1 \\ 1 & 1 \end{bmatrix}.$$

We claimed that A' is well-conditioned, and not particularly sensitive to roundoff error—except that if Gaussian elimination is applied in a stupid way, then the matrix becomes completely vulnerable. It is a blunder to accept .0001 as the first pivot, and we must insist on a larger and safer choice by exchanging the rows of A'. When "partial pivoting" is built into the elimination algorithm, so that the computer automatically looks for the largest pivot, then the natural resistance to roundoff error is no longer compromised.

How do we measure this natural resistance, and decide whether a matrix is well-conditioned or ill-conditioned? If there is a small change in b or in A, how large a change does that produce in the solution x?

We begin with *a change in the right-hand side*, from b to $b + \delta b$. This error might come from experimental data or from roundoff; we may suppose that δb is small, but its direction is outside our control. The corresponding solution is changed from x to $x + \delta x$:

$$A(x + \delta x) = b + \delta b, \quad \text{so by subtraction} \quad A(\delta x) = \delta b.$$

This is a particularly simple case; we consider all perturbations δb, and estimate the resulting perturbation $\delta x = A^{-1}\delta b$. There will be a large change in the solution when A^{-1} is large—A is nearly singular—and it is especially large when δb points in the direction that is amplified most by A^{-1}.

To begin with, suppose that A is symmetric and that its eigenvalues are positive: $0 < \lambda_1 \leq \lambda_2 \leq \cdots \leq \lambda_n$. Any vector δb is a combination of the corresponding unit eigenvectors $x_1, \ldots, x_n$, and the worst perturbation is an error in the direction of the first eigenvector x_1:

$$\text{if} \quad \delta b = \epsilon x_1, \quad \text{then} \quad \delta x = \frac{\delta b}{\lambda_1}. \tag{1}$$

The error of size $\| \delta b \|$ is amplified by the factor $1/\lambda_1$, which is just the largest eigenvalue of A^{-1}. The amplification is greatest when λ_1 is close to zero, so that nearly singular matrices are the most sensitive.

There is only one drawback to this measure of sensitivity, and it is serious. Suppose we multiply all the entries of A by 1000; then λ_1 will be multiplied by 1000 and the matrix will look much less singular. This offends our sense of fair play; such a simple rescaling cannot make an ill-conditioned matrix well. It is true that δx will be smaller by a factor of 1000, but so will the solution

$x = A^{-1}b$, and the relative error $\| \delta x \|/\| x \|$ will be the same. The factor $\| x \|$ in the denominator normalizes the problem against a trivial change of scale. At the same time, there is a corresponding normalization for δb; our problem is to compare the *relative change* $\| \delta b \|/\| b \|$ with the *relative error* $\| \delta x \|/\| x \|$.

The worst case is when the numerator $\| \delta x \|$ is large—the perturbations lie in the direction of the eigenvector x_1—and when the denominator $\| x \|$ is small. The unperturbed solution x should be as small as possible compared to the unperturbed b. This means that *the original problem $Ax = b$ should be at the other extreme*, in the direction of the last eigenvector x_n:

$$\text{if} \quad b = x_n, \quad \text{then} \quad x = A^{-1}b = \frac{b}{\lambda_n}. \tag{2}$$

It is this combination, $b = x_n$ and $\delta b = \epsilon x_1$, that makes the relative error as large as possible. These are the extreme cases in the following inequalities:

7A For a positive definite matrix, the solution $x = A^{-1}b$ and the error $\delta x = A^{-1}\delta b$ always satisfy

$$\| x \| \geq \frac{\| b \|}{\lambda_n} \quad \text{and} \quad \| \delta x \| \leq \frac{\| \delta b \|}{\lambda_1}. \tag{3}$$

Therefore the relative error is bounded by

$$\frac{\| \delta x \|}{\| x \|} \leq \frac{\lambda_n}{\lambda_1} \frac{\| \delta b \|}{\| b \|}. \tag{4}$$

The quantity $c = \lambda_n/\lambda_1 = \lambda_{\max}/\lambda_{\min}$ is called the **condition number** of A.

EXAMPLE 1   The eigenvalues of

$$A = \begin{bmatrix} 1 & 1 \\ 1 & 1.0001 \end{bmatrix}$$

are approximately $\lambda_1 = 10^{-4}/2$ and $\lambda_2 = 2$. Therefore its condition number is about $c = 4 \cdot 10^4$, and we must expect a violent change in the solution from some very ordinary changes in the data. On p. 32 we actually compared the equations $Ax = b$ and $Ax' = b'$:

$$\begin{array}{ll} u + \quad v = 2 & u' + \quad v' = 2 \\ \multicolumn{2}{c}{\text{and}} \\ u + 1.0001v = 2.0001 & u' + 1.0001v' = 2.0002. \end{array}$$

The solution is changed from $x = (1, 1)$ to $x' = (0, 2)$:

$$\frac{\| x' - x \|}{\| x \|} = \frac{\|(-1, 1)\|}{\|(1, 1)\|} = 1, \quad \frac{\| b' - b \|}{\| b \|} = \frac{\|(0, .0001)\|}{\|(2, 2.0001)\|} \sim \frac{10^{-4}}{2\sqrt{2}}.$$

The relative amplification is $\| \delta x \|/\| x \| \sim 2\sqrt{2} \cdot 10^4 \| \delta b \|/\| b \|$. Without having made any special choice of the perturbation—our δb makes a $45°$ angle with the eigenvector x_1, which accounts for the missing $\sqrt{2}$ between our factor and the worst possible amplification $c = 4 \cdot 10^4$—we still found a tremendous change in the solution.

Notice that the condition number c is not directly affected by the size of the matrix; if $A = I$, or even $A = I/10$, the condition number is $c = \lambda_{max}/\lambda_{min} = 1$. By comparison, *the determinant is a terrible measure of ill-conditioning*, because it depends not only on the scaling but also on the order n; if $A = I/10$, then the determinant of A is 10^{-n}. In fact, this "nearly singular" matrix is as well-conditioned as possible.

EXAMPLE 2   Consider the n by n finite difference matrix

$$A = \begin{bmatrix} 2 & -1 & & & \\ -1 & 2 & -1 & & \\ & -1 & 2 & \cdot & \\ & & \cdot & \cdot & -1 \\ & & & -1 & 2 \end{bmatrix}.$$

Its largest eigenvalue is about $\lambda_n = 4$, and its smallest is about $\lambda_1 = \pi^2/n^2$. Therefore the condition number is approximately $c = \frac{1}{2}n^2$, and this time the dependence on the order n is genuine. The better we approximate $-u'' = f$, by increasing the number of unknowns, the harder it is to compute the approximation. It not only takes longer, but it is more affected by roundoff. At a certain crossover point, an increase in n will actually produce a poorer answer.

Fortunately for the engineer, this crossover occurs where the accuracy is already pretty good. Working in single precision, a typical computer might make roundoff errors of order 10^{-9}. If the approximation uses $n = 100$ unknowns, so that $c = 5000$, then such an error is amplified at most to be of order 10^{-5}—which is still more accurate than any ordinary measurements. But there will be trouble with 10,000 unknowns, or with a finite difference approximation to a higher order equation like $d^4u/dx^4 = f(x)$—for which the condition number grows as n^4.†

Our analysis so far has applied exclusively to symmetric matrices with positive eigenvalues. We could easily drop the positivity assumption, and use the absolute values of the λ_i; the condition number would become $c = \max |\lambda_i|/\min |\lambda_i|$. But to avoid the assumption of symmetry, as we certainly want to do, there will have to be a major change. This is easy to see for the matrices

$$A = \begin{bmatrix} 1 & 100 \\ 0 & 1 \end{bmatrix} \quad \text{and} \quad A^{-1} = \begin{bmatrix} 1 & -100 \\ 0 & 1 \end{bmatrix}. \tag{5}$$

† The usual rule of thumb, experimentally verified, is that the computer can lose $\log c$ decimal places to the roundoff errors in Gaussian elimination.

The eigenvalues all equal one, but it is certainly not true that the relative change in x is bounded by the relative change in b; the condition number is not given by $\lambda_{max}/\lambda_{min} = 1$. Compare the solutions

$$x = \begin{bmatrix} 0 \\ 1 \end{bmatrix} \quad \text{when} \quad b = \begin{bmatrix} 100 \\ 1 \end{bmatrix}; \quad x' = \begin{bmatrix} 100 \\ 0 \end{bmatrix} \quad \text{when} \quad b' = \begin{bmatrix} 100 \\ 0 \end{bmatrix}.$$

A 1% change in b has produced a hundredfold change in x; the amplification factor is 100^2. Since c represents an upper bound for this amplification, it must be at least 10,000. The difficulty with these matrices is that a large off-diagonal entry in A means an equally large entry in A^{-1}—contradicting the intuitive expectation that A^{-1} should get smaller as A gets bigger.

To find a proper definition of the condition number, we have to look back at Eq. (3). The reason λ_n entered that equation is that it was the largest possible ratio of $\| b \|$ to $\| x \|$; we were trying to make x small and $b = Ax$ large. (The extreme case occurred at the eigenvector x_n, where the ratio of Ax to x is exactly λ_n.) The only thing different when A is no longer symmetric is that the maximum of $\| Ax \|/\| x \|$ may be found at a vector x that is not one of the eigenvectors. This maximum is still an excellent measure of the size of A; it is called the **norm** of the matrix, and denoted by $\| A \|$.

7B   The norm of A is the number defined by

$$\| A \| = \max_{x \neq 0} \frac{\| Ax \|}{\| x \|}. \tag{6}$$

In other words, $\| A \|$ bounds the "amplifying power" of the matrix:

$$\| Ax \| \leq \| A \| \, \| x \| \quad \text{for all vectors } x, \tag{7}$$

and equality holds for at least one nonzero x.

The matrices A and A^{-1} in Eq. (5) will have norms somewhere between 100 and 101. In a moment we will calculate them exactly, but first we want to complete the connection between norms and condition numbers. Because $b = Ax$ and $\delta x = A^{-1}\delta b$, we know immediately from the definition (7) that

$$\| b \| \leq \| A \| \, \| x \| \quad \text{and} \quad \| \delta x \| \leq \| A^{-1} \| \, \| \delta b \|. \tag{8}$$

This is the right replacement for (3), when A is not symmetric; in the symmetric case $\| A \|$ is the same as λ_n, and $\| A^{-1} \|$ is the same as $1/\lambda_1$. And the right replacement for λ_n/λ_1 is therefore the product $\| A \| \, \| A^{-1} \|$:

7C The condition number of A is $c = \| A \| \, \| A^{-1} \|$, and the relative error satisfies

$$\frac{\| \delta x \|}{\| x \|} \leq c \frac{\| \delta b \|}{\| b \|}. \tag{9}$$

If we perturb the matrix A instead of the right side b, then

$$\frac{\| \delta x \|}{\| x + \delta x \|} \leq c \frac{\| \delta A \|}{\| A \|}. \tag{10}$$

The inequality (9) holds for every b and every δb, and it is just the product of the two inequalities in (8). What is remarkable is that the same condition number appears in (10), when the matrix itself is perturbed: If $Ax = b$ and $(A + \delta A)(x + \delta x) = b$, then by subtraction

$$A \delta x + \delta A (x + \delta x) = 0, \quad \text{or} \quad \delta x = -A^{-1}(\delta A)(x + \delta x).$$

Multiplying by δA amplifies a vector by no more than the norm $\| \delta A \|$, and then multiplying by A^{-1} amplifies by no more than $\| A^{-1} \|$. Therefore

$$\| \delta x \| \leq \| A^{-1} \| \, \| \delta A \| \, \| x + \delta x \|,$$

or

$$\frac{\| \delta x \|}{\| x + \delta x \|} \leq \| A^{-1} \| \, \| \delta A \| = c \frac{\| \delta A \|}{\| A \|}.$$

These inequalities mean that roundoff error comes from two sources: first the natural sensitivity of the problem, which is measured by c; and then the actual errors δb or δA that are committed in solving it. This was the basis of Wilkinson's error analysis; since the elimination algorithm actually produces approximate factors L' and U', it solves the equation with the wrong matrix $A + \delta A = L'U'$ instead of the right matrix $A = LU$. He proved that partial pivoting is adequate to keep δA under control—see his "Rounding Errors in Algebraic Processes"—so *the whole burden of the roundoff error is carried by the condition number c.*

EXERCISE 7.2.1 If A is an orthogonal matrix Q, show that $\| Q \| = 1$ and also $c(Q) = 1$. Orthogonal matrices (and their multiples αQ) are the only perfectly conditioned matrices.

EXERCISE 7.2.2 Which "famous" inequality gives $\| (A + B)x \| \leq \| Ax \| + \| Bx \|$, and why does it follow from (6) that $\| A + B \| \leq \| A \| + \| B \|$?

EXERCISE 7.2.3 Explain why $\| ABx \| \leq \| A \| \, \| B \| \, \| x \|$, and deduce from (6) that $\| AB \| \leq \| A \| \, \| B \|$. Show that this also implies $c(AB) \leq c(A)c(B)$.

EXERCISE 7.2.4 For the positive definite $A = \begin{bmatrix} 2 & -1 \\ -1 & 2 \end{bmatrix}$, compute $\| A^{-1} \| = 1/\lambda_1$, $\| A \| = \lambda_2$, and $c(A) = \lambda_2/\lambda_1$. Find a right side b and a perturbation δb so that the error is worst possible, $\| \delta x \|/\| x \| = c \, \| \delta b \|/\| b \|$.

EXERCISE 7.2.5 Show that if λ is any eigenvalue of A, $Ax = \lambda x$, then $|\lambda| \leq \|A\|$.

A Formula for the Norm

The norm of A measures the largest amount by which any vector (eigenvector or not) is amplified by matrix multiplication: $\|A\| = \max(\|Ax\|/\|x\|)$. The norm of the identity matrix is 1. To compute this "amplification factor" in general, we square both sides:

$$\|A\|^2 = \max \frac{\|Ax\|^2}{\|x\|^2} = \max \frac{x^T A^T A x}{x^T x}. \tag{11}$$

This brings back a symmetric $A^T A$, and its "Rayleigh quotient."

7D The norm of A is the square root of the largest eigenvalue of $A^T A$: $\|A\|^2 = \lambda_{\max}(A^T A)$. In case A is symmetric, then $A^T A = A^2$ and the norm is the largest eigenvalue: $\|A\| = \max|\lambda_i|$. In every case, the vector that is amplified the most is the corresponding eigenvector of $A^T A$:

$$\frac{x^T A^T A x}{x^T x} = \frac{x^T(\lambda_{\max}x)}{x^T x} = \lambda_{\max} = \|A\|^2.$$

Note 1 The norm and condition number of A are not actually computed in practical problems, but only estimated. There is not time to solve an eigenvalue problem for $\lambda_{\max}(A^T A)$.

Note 2 The condition number explains why the normal equations $A^T A x = A^T b$ are so hard to solve in least squares problems: The condition number $c(A^T A)$ is the *square* of $c(A)$. Forming $A^T A$ can turn a healthy problem into a sick one, and it is much better (except for very small problems) to use either Gram–Schmidt or the singular value decomposition $A = Q_1 \Sigma Q_2^T$.

Note 3 The entries σ_i in the diagonal matrix Σ are the *singular values* of A, and by construction (p. 136) their squares are the eigenvalues of $A^T A$. Therefore another formula for the norm is $\|A\| = \sigma_{\max}$. The orthogonal Q_1 and Q_2 leave lengths unchanged in $\|Ax\| = \|Q_1 \Sigma Q_2^T x\|$, so the largest amplification factor is the largest σ.

Note 4 Roundoff error enters not only $Ax = b$, but also $Ax = \lambda x$. This raises a new question: What is the "condition number of the eigenvalue problem"? The obvious answer is wrong; it is not the condition number of A itself. Instead *it is the condition number of the diagonalizing S which measures the sensitivity of the eigenvalues.* If μ is an eigenvalue of $A + E$, then its distance from one of the eigenvalues of A is

$$|\mu - \lambda| \leq \|S\| \, \|S^{-1}\| \, \|E\| = c(S)\|E\|. \tag{12}$$

In case S is an orthogonal matrix Q, the eigenvalue problem is perfectly conditioned: $c(Q) = 1$, and the change $\mu - \lambda$ in the eigenvalues is no greater than the change E in the matrix A. This happens whenever the eigenvectors are orthonormal—they are the columns of S. Therefore the best case is when A is symmetric, or more generally when $AA^T = A^T A$. Then A is a normal matrix, its diagonalizing S is an orthogonal Q (Section 5.6), and its eigenvalues are perfectly conditioned. You can see the presence of S in the formula for perturbations in each eigenvalue separately: If x_k is the kth column of S and y_k is the kth row of S^{-1}, then

$$\mu_k - \lambda_k = y_k E x_k + \text{terms of order } \| E \|^2. \tag{13}$$

In practice, $y_k E x_k$ is a very realistic estimate of the change in the eigenvalue, and the idea in every good algorithm is to keep the error matrix E as small as possible—usually by insisting, as QR will do in the next section, on orthogonal matrices at each step of the iteration.

EXERCISE 7.2.6 Find the exact norms of the matrices in (5).

EXERCISE 7.2.7 Show from Exercise 5.6.3, comparing the eigenvalues of $A^T A$ and AA^T, that $\| A \| = \| A^T \|$.

EXERCISE 7.2.8 For a positive definite A, the Cholesky decomposition is $A = LDL^T = W^T W$, where $W = \sqrt{D} L^T$. Show directly from 7D that the condition number of W is the square root of the condition number of A. It follows that the Gauss algorithm needs no row exchanges for a positive definite matrix; the condition does not deteriorate, since $c(A) = c(W^T)c(W)$.

EXERCISE 7.2.9 Show that the largest eigenvalue is not a satisfactory norm, by finding 2 by 2 counterexamples to $\lambda_{\max}(A + B) \leq \lambda_{\max}(A) + \lambda_{\max}(B)$ and to $\lambda_{\max}(AB) \leq \lambda_{\max}(A)\lambda_{\max}(B)$.

EXERCISE 7.2.10 Suppose $\| x \|$ is changed from the Euclidean length $(x_1^2 + \cdots + x_n^2)^{1/2}$ to the "maximum norm" or "L_∞ norm": $\| x \|_\infty = \max | x_i |$. (Example: $\| (1, -2, 1) \| = 2$.) Compute the corresponding matrix norm

$$\| A \|_\infty = \max_{x \neq 0} \frac{\| Ax \|_\infty}{\| x \|_\infty} = \max_{x_i = \pm 1} \| Ax \|_\infty \quad \text{if} \quad A = \begin{bmatrix} 1 & 2 \\ 3 & -4 \end{bmatrix}.$$

THE COMPUTATION OF EIGENVALUES ■ 7.3

There is no one best way to find the eigenvalues of a matrix. But there are certainly some terrible ways, which should never be tried, and also some ideas that do deserve a permanent place. We begin by describing one very rough and ready approach, the *power method*, whose convergence properties are easy to understand. Then we move steadily toward a much more sophisticated algorithm, which starts by making a symmetric matrix tridiagonal and ends by making it virtually diagonal. Its last step is done by Gram–Schmidt, so the method is known as QR.

The ordinary power method operates exactly on the principle of a difference equation. It starts with an initial guess u_0 and then successively forms $u_1 = A u_0$, $u_2 = A u_1$, and in general $u_{k+1} = A u_k$. Each step is a matrix–vector multiplication, and after k steps it produces $u_k = A^k u_0$, although the matrix A^k will never appear. In fact the essential thing is that multiplication by A should be easy to do—if the matrix is large, it had better be sparse—because convergence to the eigenvector is often very slow. Assuming A has a full set of eigenvectors $x_1, \ldots, x_n$, the vector u_k will be given by the usual formula for a difference equation:

$$u_k = c_1 \lambda_1{}^k x_1 + \cdots + c_n \lambda_n{}^k x_n . \tag{14}$$

Imagine the eigenvalues numbered in increasing order, and suppose that the largest eigenvalue is all by itself; there is no other eigenvalue of the same magnitude, and λ_n is not repeated. Thus $| \lambda_1 | \leq \cdots \leq | \lambda_{n-1} | < | \lambda_n |$. Then as long as the initial guess u_0 contained *some* component of the eigenvector x_n, so that $c_n \neq 0$, this component will gradually become dominant:

$$\frac{u_k}{\lambda_n{}^k} = c_1 \left(\frac{\lambda_1}{\lambda_n} \right)^k x_1 + \cdots + c_{n-1} \left(\frac{\lambda_{n-1}}{\lambda_n} \right)^k x_{n-1} + c_n x_n . \tag{15}$$

The vectors u_k point more and more accurately toward the direction of x_n, and the convergence factor is the ratio $r = | \lambda_{n-1} | / | \lambda_n |$. It is just like convergence to a steady state, which we studied for Markov processes, except that now the largest eigenvalue λ_n may not equal 1. In fact, we do not know the scaling factor $\lambda_n{}^k$ in (15), but some scaling factor should be introduced; otherwise u_k can grow very large or very small, in case $| \lambda_n | > 1$ or $| \lambda_n | < 1$. Normally we can just divide each u_k by its first component α_k before taking the next step; with this simple scaling, the power method becomes $u_{k+1} = A u_k / \alpha_k$, and it converges to a multiple of x_n .†

EXAMPLE   (from California) with eigenvalue 1 and eigenvector $\begin{bmatrix} .667 \\ .333 \end{bmatrix}$:

$$A = \begin{bmatrix} .9 & .2 \\ .1 & .8 \end{bmatrix} \quad \text{was the matrix of population shifts;}$$

$$u_0 = \begin{bmatrix} 1 \\ 0 \end{bmatrix}, \quad u_1 = \begin{bmatrix} .9 \\ .1 \end{bmatrix}, \quad u_2 = \begin{bmatrix} .83 \\ .17 \end{bmatrix}, \quad u_3 = \begin{bmatrix} .781 \\ .219 \end{bmatrix}, \quad u_4 = \begin{bmatrix} .747 \\ .253 \end{bmatrix}.$$

The most serious limitation is clear from this example: If r is close to 1, then convergence is very slow. In many applications $r > .9$, which means that more than 20 iterations are needed to reduce $(\lambda_2 / \lambda_1)^k$ just by a factor of 10. (The example had $r = .7$, and it was still slow.) Of course if $r = 1$, which

---

† The scaling factors α_k will also converge; they approach λ_n .

means $|\lambda_{n-1}| = |\lambda_n|$, then convergence may not occur at all. There are several ways to get around this limitation, and we shall describe three of them:

(1)   The *block power method* works with several vectors at once, in place of a single u_k. If we start with p orthonormal vectors, multiply them all by A, and then apply Gram–Schmidt to orthogonalize them again—that is a single step of the method—then the effect is to reduce the convergence ratio to $r' = |\lambda_{n-p}|/|\lambda_n|$. Furthermore we will simultaneously obtain approximations to p different eigenvalues and their eigenvectors.

(2)   The *inverse power method* operates with A^{-1} instead of A. A single step of the difference equation is $v_{k+1} = A^{-1}v_k$, which means that we solve the linear system $Av_{k+1} = v_k$ (and save the factors L and U!). Now the theory guarantees convergence to the *smallest eigenvalue*, provided the convergence factor $r'' = |\lambda_1|/|\lambda_2|$ is less than one. Often it is the smallest eigenvalue that is wanted in the applications, and then inverse iteration is an automatic choice.

(3)   The *shifted inverse power method* is the best of all. Suppose that A is replaced by $A - \alpha I$. Then all the eigenvalues λ_i are shifted by the same amount α, and the convergence factor for the inverse method will change to $r''' = |\lambda_1 - \alpha|/|\lambda_2 - \alpha|$. Therefore if α is chosen to be a good approximation to λ_1, r''' will be very small and the convergence is enormously accelerated. Each step of the method solves the system $(A - \alpha I)w_{k+1} = w_k$, and this difference equation is satisfied by

$$w_k = \frac{c_1 x_1}{(\lambda_1 - \alpha)^k} + \frac{c_2 x_2}{(\lambda_2 - \alpha)^k} + \cdots + \frac{c_n x_n}{(\lambda_n - \alpha)^k}.$$

Provided α is close to λ_1, the first of these denominators is so near to zero that only one or two steps are needed to make that first term completely dominant. In particular, if λ_1 has already been computed by another algorithm (such as QR), then α is this computed value. The standard procedure is to factor $A - \alpha I$ into LU† and to solve $Ux_1 = (1, 1, \ldots, 1)^\mathrm{T}$ by back-substitution.

If λ_1 is not already approximated by an independent algorithm, then the shifted power method has to generate its own choice of α—or, since we can vary the shift at every step if we want to, it must choose the α_k that enters $(A - \alpha_k I)w_{k+1} = w_k$. The simplest possibility is just to work with the scaling factors that bring each w_k back to a reasonable size, but there are other and better ways. In the symmetric case $A = A^\mathrm{T}$, the most accurate choice seems to be the Rayleigh quotient

$$\alpha_k = R(u_k) = \frac{u_k{}^\mathrm{T} A u_k}{u_k{}^\mathrm{T} u_k}.$$

---

† This may look extremely ill-conditioned, since $A - \alpha I$ is as nearly singular as we can make it. Fortunately the error is largely confined to the direction of the eigenvector. Since any multiple of an eigenvector is still an eigenvector, it is only that direction which we are trying to compute.

We already know that this quotient has a minimum at the true eigenvector—
the derivatives of R are zero, and its graph is like the bottom of a parabola.
Therefore the error $\lambda - \alpha_k$ in the eigenvalue is roughly the square of the error
in the eigenvector. The convergence factors $r''' = |\lambda_1 - \alpha_k|/|\lambda_2 - \alpha_k|$ are
changing at every step, and in fact r''' itself is converging to zero. The final
result, with these Rayleigh quotient shifts, is *cubic convergence*† of α_k to λ_1.

EXERCISE 7.3.1   For the matrix $A = \begin{bmatrix} 2 & -1 \\ -1 & 2 \end{bmatrix}$, with eigenvalues $\lambda_1 = 1$ and $\lambda_2 = 3$,
apply the power method $u_{k+1} = Au_k$ three times to the initial guess $u_0 = \begin{bmatrix} 1 \\ 0 \end{bmatrix}$. What is
the limiting vector u_∞ ?

EXERCISE 7.3.2   For the same A and the initial guess $u_0 = \begin{bmatrix} 3 \\ 4 \end{bmatrix}$, compare the results of

(i)   three inverse power steps

$$u_{k+1} = A^{-1}u_k = \frac{1}{3}\begin{bmatrix} 2 & 1 \\ 1 & 2 \end{bmatrix} u_k \; ;$$

(ii)   a single shifted step $u_1 = (A - \alpha I)^{-1}u_0$, with $\alpha = u_0^T A u_0/u_0^T u_0$.

The limiting vector u_∞ is now a multiple of the other eigenvector $(1, 1)$.

Tridiagonal and Hessenberg Forms

The power method is recommended only for a matrix that is large and very
sparse; when most of the entries are nonzero, the method is a mistake. Therefore we ask whether there is any simple way to create zeros. That is the goal of
the following paragraphs.

It should be said at the outset that, after computing a similar matrix $U^{-1}AU$
with more zeros than A, we do not intend to go back to the power method.
There are much more sophisticated variants, and the best of them seems to
be the QR algorithm. (The shifted inverse power method has its place at the
very end, in finding the eigenvector.) In any case the first step is to produce
as many zeros as possible, and to do it as quickly as possible. Our only restriction on speed will be the use of unitary (or orthogonal) transformations, which
preserve symmetry and preserve lengths; if A is symmetric, then so is $U^{-1}AU$,
and no entry can become dangerously large.

To go from A to $U^{-1}AU$, there are at least two possibilities: Either we can
produce one zero at every step (as in Gaussian elimination), or we can work
with a whole column at once. For a single zero, it is enough to use a plane
rotation as illustrated in Eq. (21) below; it has $\cos\theta$ and $\sin\theta$ in a 2 by 2
block. Then we could cycle through all the entries below the diagonal, choosing
at each step a rotation θ that will produce a zero; this is the principle of *Jacobi's*

† Linear convergence means that every step multiplies the error by a fixed factor $r < 1$;
quadratic convergence means that the error is squared at every step, as in Newton's method
$x_{k+1} - x_k = -f(x_k)/f'(x_k)$ for solving $f(x) = 0$; cubic convergence means that the error is
cubed at every step, going from 10^{-1} to 10^{-3} to 10^{-9}.

method. Unfortunately, it fails to diagonalize A after a finite number of rotations, since the zeros achieved in the early steps will be destroyed when later zeros are created.

To preserve those zeros, we have to settle for less than a triangular form; we accept *one nonzero diagonal below the main diagonal.* This is the so-called Hessenberg form. If the matrix is symmetric, then the upper triangular part will copy the lower triangular part, and the matrix will be tridiagonal.

Both these forms used to be obtained by a series of rotations in the right planes, but Householder has found a new way to accomplish exactly the same thing. His idea gives the "preparation step" for the QR method.† A *Householder transformation,* or an elementary reflector, is a matrix of the form

$$H = I - 2\frac{vv^{\mathrm{T}}}{\|v\|^2}.$$

Often v is normalized to become a unit vector $u = v/\|v\|$, and then H becomes $I - 2uu^{\mathrm{T}}$. In either case H is both *symmetric* and *orthogonal*:

$$H^{\mathrm{T}}H = (I - 2uu^{\mathrm{T}})(I - 2uu^{\mathrm{T}}) = I - 4uu^{\mathrm{T}} + 4uu^{\mathrm{T}}uu^{\mathrm{T}} = I.$$

Thus $H = H^{\mathrm{T}} = H^{-1}$. In the complex case, the corresponding matrix $I - 2uu^{\mathrm{H}}$ is both Hermitian and unitary. Householder's plan was to produce zeros with these matrices, and its success depends on the following identity:

7E Suppose z is the column vector $(1, 0, \ldots, 0)^{\mathrm{T}}$, and $\sigma = \|x\|$, and $v = x + \sigma z$. Then $Hx = -\sigma z = (-\sigma, 0, \ldots, 0)^{\mathrm{T}}$.

Proof

$$Hx = x - \frac{2vv^{\mathrm{T}}x}{\|v\|^2} = x - (x + \sigma z)\frac{2(x + \sigma z)^{\mathrm{T}}x}{(x + \sigma z)^{\mathrm{T}}(x + \sigma z)}$$

$$= x - (x + \sigma z) \qquad (\text{because } x^{\mathrm{T}}x = \sigma^2)$$

$$= -\sigma z. \qquad (16)$$

This identity can be used **right away**. We start with the first column of A, and remember that the final $U^{-1}AU$ is to be in tridiagonal or Hessenberg form. Therefore *only the $n - 1$ entries below the diagonal will be involved*:

$$x = \begin{bmatrix} a_{21} \\ a_{31} \\ \vdots \\ a_{n1} \end{bmatrix}, \quad z = \begin{bmatrix} 1 \\ 0 \\ \vdots \\ 0 \end{bmatrix}, \quad Hx = \begin{bmatrix} -\sigma \\ 0 \\ \vdots \\ 0 \end{bmatrix}. \qquad (17)$$

† You may want to bypass this preparation, and go directly to the QR algorithm on p. 280. It is only in actual computations that you need to produce the zeros first.

At this point Householder's matrix H is only of order $n - 1$, so it is imbedded into the lower right corner of a full-size matrix U_1 :

$$
U_1 = \begin{bmatrix} 1 & 0 & 0 & 0 & 0 \\ 0 & & & & \\ 0 & & H & & \\ 0 & & & & \\ 0 & & & & \end{bmatrix} = U_1^{-1}, \quad \text{and} \quad U_1^{-1}AU_1 = \begin{bmatrix} a_{11} & * & * & * & * \\ -\sigma & * & * & * & * \\ 0 & * & * & * & * \\ 0 & * & * & * & * \\ 0 & * & * & * & * \end{bmatrix}.
$$

Because of the 1 in its upper left corner, the matrix U_1 leaves the entry a_{11} completely unchanged—and more important, it does not touch the zeros which appear in (17). Therefore the first stage is complete, and $U_1^{-1}AU_1$ has the required first column.

The second stage is similar: x consists of the last $n - 2$ entries in the second column, z is the unit coordinate vector of matching length, and H_2 is of order $n - 2$. When it is imbedded in U_2, it produces

$$
U_2 = \begin{bmatrix} 1 & 0 & 0 & 0 & 0 \\ 0 & 1 & 0 & 0 & 0 \\ 0 & 0 & & & \\ 0 & 0 & & H_2 & \\ 0 & 0 & & & \end{bmatrix} = U_2^{-1}, \quad U_2^{-1}(U_1^{-1}AU_1)U_2 = \begin{bmatrix} * & * & * & * & * \\ * & * & * & * & * \\ 0 & * & * & * & * \\ 0 & 0 & * & * & * \\ 0 & 0 & * & * & * \end{bmatrix}.
$$

Finally U_3 will take care of the third column, and for a 5 by 5 matrix the Hessenberg form is achieved. In general U is the product of all the matrices $U_1 U_2 \cdots U_{n-2}$, and the number of operations required to compute it is of order n^3.

EXAMPLE

$$
A = \begin{bmatrix} 1 & 0 & 1 \\ 0 & 1 & 1 \\ 1 & 1 & 0 \end{bmatrix}, \quad x = \begin{bmatrix} 0 \\ 1 \end{bmatrix}, \quad v = \begin{bmatrix} 1 \\ 1 \end{bmatrix}, \quad H = \begin{bmatrix} 0 & -1 \\ -1 & 0 \end{bmatrix}.
$$

Imbedding H into U, the result is tridiagonal:

$$
U = \begin{bmatrix} 1 & 0 & 0 \\ 0 & 0 & -1 \\ 0 & -1 & 0 \end{bmatrix}, \quad U^{-1}AU = \begin{bmatrix} 1 & -1 & 0 \\ -1 & 0 & 1 \\ 0 & 1 & 1 \end{bmatrix}.
$$

$U^{-1}AU$ is a matrix that is ready to reveal its eigenvalues—the QR algorithm is ready to begin—but we digress for a moment to mention two other applications of these same Householder transformations.

I. *The factorization $A = QR$.* This was a shorthand for the Gram–Schmidt process in Chapter 3; now it can be carried out more simply and more stably. Remember that R is to be upper triangular—we no longer have to accept an extra nonzero diagonal below the main one, since there will be no U's or H's multiplying on the right (as in $U^{-1}AU$) to spoil the zeros already created. Therefore the first step in constructing Q is to work with the whole first column of A:

$$x = \begin{bmatrix} a_{11} \\ a_{21} \\ \vdots \\ a_{n1} \end{bmatrix}, \qquad z = \begin{bmatrix} 1 \\ 0 \\ \vdots \\ 0 \end{bmatrix}, \qquad v = x + \| x \| z, \qquad H_1 = I - 2\frac{vv^T}{\| v \|^2}.$$

The first column of $H_1 A$ is exactly as desired; it equals $-\| x \| z$; it is zero below the main diagonal; and it is the first column of R. The second step works with the second column of $H_1 A$, from the pivot on down, and produces an $H_2 H_1 A$ which is zero below that pivot. (The whole algorithm is very much like Gaussian elimination, and in fact it is a slightly slower alternative.) The result of $n - 1$ steps is again an upper triangular R, but the matrix that records the steps is not a lower triangular L. Instead it is the product $Q = H_1 H_2 \cdots H_{n-1}$, which can be stored in this factored form and never computed explicitly. That completes Gram–Schmidt.

II. *The singular value decomposition $Q_1^T A Q_2 = \Sigma$.* On p. 135, this decomposition gave at once the optimal solution $\bar{x}$ to any problem in least squares. Remember that Σ is a diagonal matrix of the same shape as A, and that its entries (the singular values μ) are the eigenvalues of $A^T A$. Since Householder transformations can only *prepare* for the eigenvalue problem, and not solve it, we cannot expect them to produce the final Σ. Instead, they produce a *bidiagonal matrix*, with zeros everywhere except along the main diagonal and the one above. Of course this preprocessing is numerically stable, because the H's are orthogonal.

The first step is exactly as in QR above: x is the first column of A, and $H_1 x$ is zero below the pivot. The next step is to multiply on the right by an $H^{(1)}$ which will produce zeros as indicated along the first row:

$$A \rightarrow H_1 A = \begin{bmatrix} * & * & * & * \\ 0 & * & * & * \\ 0 & * & * & * \end{bmatrix} \rightarrow H_1 A H^{(1)} = \begin{bmatrix} * & * & 0 & 0 \\ 0 & * & * & * \\ 0 & * & * & * \end{bmatrix}. \qquad (18)$$

Then two final Householder transformations achieve

$$H_2 H_1 A H^{(1)} = \begin{bmatrix} * & * & 0 & 0 \\ 0 & * & * & * \\ 0 & 0 & * & * \end{bmatrix} \quad \text{and} \quad H_2 H_1 A H^{(1)} H^{(2)} = \begin{bmatrix} * & * & 0 & 0 \\ 0 & * & * & 0 \\ 0 & 0 & * & * \end{bmatrix}.$$

This is the bidiagonal form we wanted, and illustrates again the quick way in which zeros can be produced by Householder transformations.

EXERCISE 7.3.3   Show that for any two different vectors of the same length, $\| x \| = \| y \|$, the choice $v = x - y$ leads to a Householder transformation such that $Hx = y$ and $Hy = x$.

EXERCISE 7.3.4   For

$$x = \begin{bmatrix} 3 \\ 4 \end{bmatrix}, \quad \text{and} \quad z = \begin{bmatrix} 1 \\ 0 \end{bmatrix},$$

compute $\sigma = \| x \|$, $v = x + \sigma z$, and the corresponding Householder matrix H. Verify that $Hx = -\sigma z$.

EXERCISE 7.3.5   Using 7.3.4, find the tridiagonal $U^{-1}AU$ produced from

$$A = \begin{bmatrix} 1 & 3 & 4 \\ 3 & 1 & 0 \\ 4 & 0 & 0 \end{bmatrix}.$$

The QR **Algorithm**

The algorithm is almost magically simple. It starts with the matrix A_0, factors it by Gram–Schmidt into $Q_0 R_0$, and then reverses the factors: $A_1 = R_0 Q_0$. This new matrix is similar to the original one, $Q_0^{-1} A_0 Q_0 = Q_0^{-1}(Q_0 R_0) Q_0 = A_1$, and the process continues with no change in the eigenvalues:

$$A_k = Q_k R_k \quad \text{and then} \quad A_{k+1} = R_k Q_k. \tag{19}$$

This equation describes the *unshifted QR algorithm*, and under fairly general circumstances it converges: A_k approaches a triangular form, and therefore its diagonal entries approach its eigenvalues, which are also the eigenvalues of the original A_0.†

As it stands, the algorithm is good but not very good. To make it special, it needs two refinements: (a) We must allow shifts of origin; and (b) we must ensure that the QR factorization at each step is very quick.

(a)   *The shifted algorithm.*   If the number α_k is close to an eigenvalue, step (19) should be shifted immediately to

$$A_k - \alpha_k I = Q_k R_k \quad \text{and then} \quad A_{k+1} = R_k Q_k + \alpha_k I. \tag{20}$$

This is justified by the fact that A_{k+1} is similar to A_k:

$$Q_k^{-1} A_k Q_k = Q_k^{-1}(Q_k R_k + \alpha_k I) Q_k = A_{k+1}.$$

---

† A_0 refers to the matrix with which the QR algorithm begins. If there was already some processing by Householder matrices to obtain a tridiagonal form, then A_0 is connected to the absolutely original A by $U^{-1}AU = A_0$.

What happens in practice is that the (n, n) entry of A_k—the one in the lower right corner—is the first to approach an eigenvalue. Therefore this entry is the simplest and most popular choice for the shift α_k. Normally its effect is to produce quadratic convergence, and in the symmetric case even cubic convergence, to the smallest eigenvalue. After perhaps three or four steps of the shifted algorithm, the matrix A_k looks like

$$
A_k = \begin{bmatrix} * & * & * & * \\ * & * & * & * \\ 0 & * & * & * \\ 0 & 0 & \epsilon & \lambda_1' \end{bmatrix}, \quad \text{with} \quad \epsilon \ll 1.
$$

We accept the computed λ_1' as a very close approximation to the true λ_1. To find the next eigenvalue, the QR algorithm continues with the smaller matrix (3 by 3, in the illustration) in the upper left corner. Its subdiagonal elements will have been somewhat reduced by the first QR steps, and another two steps are sufficient to find λ_2. This gives a systematic procedure for finding all the eigenvalues. In fact, **the QR method is now completely described**. It only remains to catch up on the eigenvectors—that is just a single inverse power step—and to take advantage of the zeros that Householder created.

(b)   The object of the preparatory Householder transformations, which put A_0 into tridiagonal or Hessenberg form, was to make each QR step very fast. Normally the Gram–Schmidt process (which is QR) would take $O(n^3)$ operations, but for a Hessenberg matrix this becomes $O(n^2)$ and for a tridiagonal matrix it is $O(n)$. Without this improvement the algorithm would be impossibly slow, and unless each new A_k is again in Hessenberg or tridiagonal form, the improvement will apply only to the first step and will be fruitless.

Fortunately, this does not happen. To show that A_1 has the same form as A_0, look at

$$
Q_0 = A_0 R_0^{-1} = \begin{bmatrix} * & * & * & * \\ * & * & * & * \\ 0 & * & * & * \\ 0 & 0 & * & * \end{bmatrix} \begin{bmatrix} * & * & * & * \\ 0 & * & * & * \\ 0 & 0 & * & * \\ 0 & 0 & 0 & * \end{bmatrix}.
$$

You can easily check that this multiplication leaves Q_0 with the same three zeros as A_0; Q_0 is itself in Hessenberg form. Then A_1 is constructed by reversing the factors, so

$$
A_1 = R_0 Q_0 = \begin{bmatrix} * & * & * & * \\ 0 & * & * & * \\ 0 & 0 & * & * \\ 0 & 0 & 0 & * \end{bmatrix} \begin{bmatrix} * & * & * & * \\ * & * & * & * \\ 0 & * & * & * \\ 0 & 0 & * & * \end{bmatrix}.
$$

Again the same three zeros will appear in the product; A_1 *is a Hessenberg matrix whenever* A_0 *is*. The symmetric case is even better, since $A_1 = Q_0^{-1}A_0Q_0$ remains symmetric:

$$A_1^\mathrm{T} = Q_0^\mathrm{T}A_0^\mathrm{T}(Q_0^{-1})^\mathrm{T} = Q_0^{-1}A_0Q_0 = A_1.$$

By the reasoning just completed, A_1 is also Hessenberg. Then, because it is both symmetric and Hessenberg, A_1 *is tridiagonal*. The same argument applies to each of the succeeding matrices A_2, A_3, ..., so every QR step begins with a tridiagonal matrix.

The last point to explain is the factorization itself, producing Q_0 and R_0 from the original A_0 (and Q_k and R_k from each A_k, or really from $A_k - \alpha_k I$). We may use Householder again, but it is simpler to annihilate each subdiagonal element in turn by a "plane rotation." The first one is

$$P_{21}A_0 = \begin{bmatrix} \cos\theta & -\sin\theta & & \\ \sin\theta & \cos\theta & & \\ & & 1 & \\ & & & 1 \end{bmatrix} \begin{bmatrix} a_{11} & * & * & * \\ a_{21} & * & * & * \\ 0 & * & * & * \\ 0 & 0 & * & * \end{bmatrix}. \tag{21}$$

The $(2, 1)$ entry in this product is $a_{11}\sin\theta + a_{21}\cos\theta$, and we just choose the angle θ that makes this combination zero. Then P_{32} is chosen in a similar way, to remove the $(3, 2)$ entry of $P_{32}P_{21}A_0$. After $n - 1$ of these elementary rotations, the final result is the upper triangular factor R_0:

$$R_0 = P_{nn-1}\cdots P_{32}P_{21}A_0. \tag{22}$$

That is as much as we can say—there is a lot more in the books of Wilkinson and Stewart—about one of the most remarkable algorithms in numerical mathematics.

EXERCISE 7.3.6 Show that starting from $A_0 = \begin{bmatrix} 2 & -1 \\ -1 & 2 \end{bmatrix}$, the unshifted algorithm produces only the modest improvement $A_1 = \frac{1}{5}\begin{bmatrix} 14 & -3 \\ -3 & 6 \end{bmatrix}$.

EXERCISE 7.3.7 Apply a single QR step to

$$A = \begin{bmatrix} \cos\theta & \sin\theta \\ \sin\theta & 0 \end{bmatrix},$$

with the shift $\alpha = a_{22}$—which in this case means without shift, since $a_{22} = 0$. Show that the off-diagonal entries go from $\sin\theta$ to $-\sin^3\theta$, an instance of cubic convergence.

EXERCISE 7.3.8 Show that the tridiagonal $A = \begin{bmatrix} 0 & 1 \\ 1 & 0 \end{bmatrix}$ is left unchanged by every step of the QR algorithm, and is therefore one of the (rare) counterexamples to convergence. It is removed by introducing an arbitrary shift.

EXERCISE 7.3.9 Show by induction that without shifts, $(Q_0Q_1\cdots Q_k)(R_k\cdots R_1R_0)$ is exactly the QR factorization of A^{k+1}. This identity connects QR to the power method and leads to an explanation of its convergence; if $|\lambda_1| > |\lambda_2| > \cdots > |\lambda_n|$, then these eigenvalues will gradually appear in descending order on the main diagonal of A_k.

ITERATIVE METHODS FOR $Ax = b$ ■ **7.4**

In contrast to eigenvalues, where there was no choice, we do not absolutely need an iterative method to solve $Ax = b$. Gaussian elimination will stop at the solution x after a finite number of steps, and as long as that number is reasonable there is no problem. On the other hand, when $n^3/3$ is enormous, we may have to settle for an approximate x that can be obtained more quickly—and it is no use to go part way through elimination and then stop. Our goal is to describe methods that start from any initial guess x_0, that produce an improved approximation x_{k+1} from the previous approximation x_k, and that can be terminated at will.

Such a method is easy to invent, simply by **splitting the matrix** A. If $A = S - T$, then the equation $Ax = b$ is the same as $Sx = Tx + b$. Therefore we can try the iteration

$$Sx_{k+1} = Tx_k + b. \tag{23}$$

Of course there is no guarantee that this method is any good, and a successful splitting needs to satisfy two different requirements:

(i)   The new vector x_{k+1} should be easy to compute. Therefore S should be a simple (and invertible!) matrix; it may be diagonal or triangular.

(ii)   The sequence x_k should converge to the true solution x. If we subtract the iteration (23) from the true equation $Sx = Tx + b$, the result is a formula involving only the errors $e_k = x - x_k$:

$$Se_{k+1} = Te_k. \tag{24}$$

This is just a difference equation. It starts with the initial error e_0, and after k steps it produces the new error $e_k = (S^{-1}T)^k e_0$. The question of convergence is exactly the same as the question of stability: $x_k \to x$ exactly when $e_k \to 0$.

7F   The iterative method (23) is convergent if and only if every eigenvalue λ of $S^{-1}T$ satisfies $|\lambda| < 1$. Its rate of convergence depends on the maximum size of $|\lambda|$, which is known as the **spectral radius** of $S^{-1}T$:

$$\rho(S^{-1}T) = \max_i |\lambda_i|. \tag{25}$$

Remember that a typical solution to $e_{k+1} = S^{-1}Te_k$ is

$$e_k = c_1\lambda_1^k x_1 + \cdots + c_n\lambda_n^k x_n.$$

Obviously, the largest of the $|\lambda_i|$ will eventually be dominant, and will govern the rate at which e_k converges to zero.

The two requirements on the iteration are to a certain extent conflicting. At one extreme, we could achieve immediate convergence with the splitting $S = A$ and $T = 0$; the first and only step of the iteration would be $Ax_1 = b$.

In that case the error matrix $S^{-1}T$ is zero, its eigenvalues and spectral radius are zero, and the rate of convergence (usually defined as $-\log \rho$) is infinite. But of course $S = A$ may not be easy to invert; that was the original reason for a splitting. At the other extreme, we can choose S to be the diagonal part of A, and the iteration becomes the one known as *Jacobi's method*:

$$a_{11}(x_1)_{k+1} = (-a_{12}x_2 - a_{13}x_3 - \cdots - a_{1n}x_n)_k + b_1$$

$$\vdots \tag{26}$$

$$a_{nn}(x_n)_{k+1} = (-a_{n1}x_1 - a_{n2}x_2 - \cdots - a_{nn-1}x_{n-1})_k + b_n .$$

If the diagonal entries a_{ii} are all nonzero, and if A is sparse so that most of the terms on the right are absent, this step from x_k to x_{k+1} is easy to carry out. The important question is whether the iteration converges, and how quickly.

EXAMPLE 1

$$A = \begin{bmatrix} 2 & -1 \\ -1 & 2 \end{bmatrix}, \quad S = \begin{bmatrix} 2 & \\ & 2 \end{bmatrix}, \quad T = \begin{bmatrix} 0 & 1 \\ 1 & 0 \end{bmatrix}, \quad S^{-1}T = \begin{bmatrix} 0 & \frac{1}{2} \\ \frac{1}{2} & 0 \end{bmatrix}.$$

If the components of x are v and w, the Jacobi step $Sx_{k+1} = Tx_k + b$ is

$$\begin{matrix} 2v_{k+1} = w_k + b_1 \\ \\ 2w_{k+1} = v_k + b_2 , \end{matrix} \quad \text{or} \quad \begin{bmatrix} v \\ w \end{bmatrix}_{k+1} = \begin{bmatrix} 0 & \frac{1}{2} \\ \frac{1}{2} & 0 \end{bmatrix} \begin{bmatrix} v \\ w \end{bmatrix}_k + \begin{bmatrix} b_1/2 \\ b_2/2 \end{bmatrix}.$$

The decisive matrix $S^{-1}T$ has eigenvalues $\pm\frac{1}{2}$, which means that the error is cut in half (one more binary digit becomes correct) at every step. In this example, which is much too small to be typical, the convergence is fast.

If we try to imagine a larger matrix A, there is an immediate and very practical difficulty with the Jacobi iteration (26). *It requires us to keep all the components of x_k until the calculation of x_{k+1} is complete.* A much more natural idea, which requires only half as much storage, is to start using each component of the new vector x_{k+1} as soon as it is computed; x_{k+1} takes the place of x_k a component at a time, and therefore x_k can be destroyed as fast as x_{k+1} is created. This means that the first equation remains as before,

$$a_{11}(x_1)_{k+1} = (-a_{12}x_2 - a_{13}x_3 - \cdots - a_{1n}x_n)_k + b_1 .$$

The next equation operates immediately with this new value of x_1 ,

$$a_{22}(x_2)_{k+1} = -a_{21}(x_1)_{k+1} + (-a_{23}x_3 - \cdots - a_{2n}x_n)_k + b_2 .$$

And the last equation will use new values exclusively,

$$a_{nn}(x_n)_{k+1} = (-a_{n1}x_1 - a_{n2}x_2 - \cdots - a_{nn-1}x_{n-1})_{k+1} + b_n .$$

This is called the *Gauss–Seidel method*, even though it was apparently unknown to Gauss and not recommended by Seidel. That is a surprising bit of history, because it is not a bad method. Notice that, when all the terms in x_{k+1} are moved to the left side, *the matrix S is now the lower triangular part of A*. On the right side, the other matrix T in the splitting is strictly upper triangular.

EXAMPLE 2

$$A = \begin{bmatrix} 2 & -1 \\ -1 & 2 \end{bmatrix}, \quad S = \begin{bmatrix} 2 & 0 \\ -1 & 2 \end{bmatrix}, \quad T = \begin{bmatrix} 0 & 1 \\ 0 & 0 \end{bmatrix}, \quad S^{-1}T = \begin{bmatrix} 0 & \frac{1}{2} \\ 0 & \frac{1}{4} \end{bmatrix}.$$

A single Gauss–Seidel step takes the components v_k and w_k into

$$\begin{array}{c} 2v_{k+1} = w_k + b_1 \\ 2w_{k+1} = v_{k+1} + b_2, \end{array} \quad \text{or} \quad \begin{bmatrix} 2 & 0 \\ -1 & 2 \end{bmatrix} x_{k+1} = \begin{bmatrix} 0 & 1 \\ 0 & 0 \end{bmatrix} x_k + b.$$

The eigenvalues of $S^{-1}T$ are again decisive, and of course they are easy to find: They are $\frac{1}{4}$ and 0. The error is divided by 4 every time, so *a single Gauss–Seidel step is worth two Jacobi steps.*† Since both methods require the same number of operations—we just use the new value instead of the old, and actually save on storage—the Gauss–Seidel method is better.

There is a way to make it better still. It was discovered during the years of hand computation (probably by accident) that convergence is faster if we go beyond the Gauss–Seidel correction $x_{k+1} - x_k$. Roughly speaking, the ordinary method converges monotonically; the approximations x_k stay on the same side of the solution x. Therefore it is natural to try introducing an *overrelaxation factor* ω to move closer to the solution. With $\omega = 1$, we recover Gauss–Seidel; with $\omega > 1$, the method is known as **successive overrelaxation** (SOR). The optimal choice of ω depends on the problem, but it never exceeds 2; it is often in the neighborhood of 1.9.

To describe the method more explicitly, let D, L, and U be the diagonal, the strictly lower triangular, and the strictly upper triangular parts of A. (This splitting has nothing to do with the $A = LDU$ of Gaussian elimination, and in fact we now have $A = L + D + U$.) The Jacobi method has $S = D$ on the left side and $T = -L - U$ on the right side, whereas Gauss–Seidel chose the splitting $S = D + L$ and $T = -U$. Now, to accelerate the convergence, we move to

$$[D + \omega L]x_{k+1} = [(1 - \omega)D - \omega U]x_k + \omega b. \tag{27}$$

Notice that for $\omega = 1$, there is no acceleration and we are back to Gauss–Seidel. But regardless of ω, the matrix on the left is lower triangular and the one on the right is upper triangular. Therefore x_{k+1} can still replace x_k, component by component, as soon as it is computed; a typical step is

$$a_{ii}(x_i)_{k+1} = a_{ii}(x_i)_k + \omega[(-a_{i1}x_1 - \cdots - a_{ii-1}x_{i-1})_{k+1}$$

$$+ (-a_{ii}x_i - \cdots - a_{in}x_n)_k + b_i].$$

---

† This rule holds in a large class of applications, even though it is possible to construct other examples in which Jacobi converges and Gauss–Seidel fails (or conversely). The symmetric case is the most straightforward: If all $a_{ii} > 0$, then Gauss–Seidel converges if and only if A is positive definite.

If the old guess x_k happened to coincide with the true solution x, then the new guess x_{k+1} would stay the same, and the quantity in brackets would vanish.

EXAMPLE 3   For the same matrix $A = \begin{bmatrix} 2 & -1 \\ -1 & 2 \end{bmatrix}$, each SOR step is

$$\begin{bmatrix} 2 & 0 \\ -\omega & 2 \end{bmatrix} x_{k+1} = \begin{bmatrix} 2(1-\omega) & \omega \\ 0 & 2(1-\omega) \end{bmatrix} x_k + \omega b.$$

If we divide through by ω, these two matrices are the S and T in the splitting $A = S - T$; the iteration is back to $Sx_{k+1} = Tx_k + b$. Therefore the crucial matrix $S^{-1}T$, whose eigenvalues govern the rate of convergence, is

$$L_\omega = \begin{bmatrix} 2 & 0 \\ -\omega & 2 \end{bmatrix}^{-1} \begin{bmatrix} 2(1-\omega) & \omega \\ 0 & 2(1-\omega) \end{bmatrix} = \begin{bmatrix} 1-\omega & \tfrac{1}{2}\omega \\ \tfrac{1}{2}\omega(1-\omega) & 1-\omega+\tfrac{1}{4}\omega^2 \end{bmatrix}.$$

The optimal choice of ω is the one which makes the largest eigenvalue of L_ω (in other words, its spectral radius) as small as possible. The whole point of overrelaxation is to discover this optimal ω. To start, the product of the eigenvalues must equal the determinant—and if we look at the two triangular factors that multiply to give L_ω, the first has determinant $\tfrac{1}{4}$ (after inverting) and the second has determinant $4(1-\omega)^2$. Therefore

$$\lambda_1 \lambda_2 = \det L_\omega = (1-\omega)^2.$$

This is a general rule, that the first matrix $(D + \omega L)^{-1}$ contributes $\det D^{-1}$ because L lies below the diagonal, and the second matrix contributes $\det(1-\omega)D$ because U lies above the diagonal. Their product, in the n by n case, is $\det L_\omega = (1-\omega)^n$. (This already explains why we never go as far as $\omega = 2$; the product of the eigenvalues would be too large to allow all $|\lambda_i| < 1$, and the iteration could not converge.) We also get a clue to the behavior of the eigenvalues, which is this: At $\omega = 1$ the Gauss–Seidel eigenvalues are 0 and $\tfrac{1}{4}$, and as ω increases these eigenvalues approach one another. *At the optimal ω the two eigenvalues are equal, and at that moment they must both equal $\omega - 1$ in order for their product to match the determinant.*† This value of ω is easy to compute because the sum of the eigenvalues always agrees with the sum of the diagonal entries (the trace of L_ω). Therefore the best parameter ω_{opt} is determined by

$$\lambda_1 + \lambda_2 = (\omega_{\text{opt}} - 1) + (\omega_{\text{opt}} - 1) = 2 - 2\omega_{\text{opt}} + \tfrac{1}{4}\omega_{\text{opt}}^2. \tag{28}$$

This quadratic equation gives $\omega_{\text{opt}} = 4(2 - \sqrt{3}) \approx 1.07$. Therefore the two equal eigenvalues are approximately $\omega - 1 = .07$, which is a major reduction from the Gauss–Seidel value $\lambda = \tfrac{1}{4}$ at $\omega = 1$. In this example, the right choice of ω has again doubled the rate of convergence, because $(\tfrac{1}{4})^2 \approx .07$.

---

† If ω is further increased, the eigenvalues become a complex conjugate pair—both have $|\lambda| = \omega - 1$, so their product is still $(\omega - 1)^2$ and their modulus is now increasing with ω.

The discovery that such an improvement could be produced so easily, almost as if by magic, was the starting point for 20 years of enormous activity in numerical analysis. The first problem was to develop and extend the theory of overrelaxation, and Young's thesis in 1950 contained the solution—a simple formula for the optimal ω. The key step in his thesis was to find a connection between the eigenvalues λ of the matrix L_ω and the eigenvalues μ of the original Jacobi matrix $D^{-1}(-L-U)$; that connection is expressed by

$$(\lambda + \omega - 1)^2 = \lambda\omega^2\mu^2. \tag{29}$$

It is valid for a wide class of finite difference matrices, and if we take $\omega = 1$ (Gauss–Seidel) it yields $\lambda^2 = \lambda\mu^2$. Therefore $\lambda = 0$ and $\lambda = \mu^2$. This is confirmed by Examples 1 and 2, in which $\mu = \pm\frac{1}{2}$ and $\lambda = 0$, $\lambda = \frac{1}{4}$. In fact, this situation is completely typical of the relation between Jacobi and Gauss–Seidel: All the matrices in Young's class have eigenvalues μ that occur in plus–minus pairs, and then (29) shows that the corresponding λ are 0 and μ^2. By using the latest approximations to x, we double the rate of convergence.

The important problem is to do better still; we want to choose ω so that the largest eigenvalue λ will be minimized. Fortunately, this problem is already solved. Young's equation (29) is nothing but the characteristic equation for the 2 by 2 example L_ω, and the best ω was the one which made the two roots λ both equal to $\omega - 1$. Exactly as in (28), where $\mu^2 = \frac{1}{4}$, this leads to

$$(\omega - 1) + (\omega - 1) = 2 - 2\omega + \mu^2\omega^2, \qquad \text{or} \qquad \omega = \frac{2(1 - \sqrt{1 - \mu^2})}{\mu^2}.$$

The only difference is that, for a large matrix, this pattern will be repeated for a number of different pairs $\pm\mu_i$—and we can only make a single choice of ω. The largest of these pairs gives the largest Jacobi eigenvalue $\mu_{\max}$, and it also gives the largest value of ω and of $\lambda = \omega - 1$. Therefore, since our goal is to make $\lambda_{\max}$ as small as possible, it is that extremal pair which specifies the best choice ω_{opt}:

$$\omega_{\mathrm{opt}} = \frac{2(1 - \sqrt{1 - \mu_{\max}^2})}{\mu_{\max}^2} \qquad \text{and} \qquad \lambda_{\max} = \omega_{\mathrm{opt}} - 1. \tag{30}$$

This is Young's formula for the optimal overrelaxation factor.

For our finite difference matrix A, with entries -1, 2, -1 down the three main diagonals, we can compute the improvement brought by ω. In the examples this matrix was 2 by 2; now suppose it is n by n, corresponding to the mesh width $h = 1/(n + 1)$. The largest Jacobi eigenvalue, according to Exercise 7.4.3, is $\mu_{\max} = \cos \pi h$. Therefore the largest Gauss–Seidel eigenvalue is $\mu_{\max}^2 = \cos^2 \pi h$, and the largest SOR eigenvalue is found by substituting in (30):

$$\lambda_{\max} = \frac{2(1 - \sin \pi h)}{\cos^2 \pi h} - 1 = \frac{(1 - \sin \pi h)^2}{\cos^2 \pi h} = \frac{1 - \sin \pi h}{1 + \sin \pi h}.$$

This can only be appreciated by an example. Suppose A is of order 21, which is very moderate. Then $h = \frac{1}{22}$, $\cos \pi h = .99$, and the Jacobi method is slow; $\cos^2 \pi h = .98$ means that even Gauss–Seidel will require a great many iterations. But since $\sin \pi h = \sqrt{.02} = .14$, the optimal overrelaxation method will have the convergence factor

$$\lambda_{\max} = \frac{.86}{1.14} = .75, \quad \text{with} \quad \omega_{\text{opt}} = 1 + \lambda_{\max} = 1.75.$$

The error is reduced by 25% at every step, and *a single SOR step is the equivalent of* 30 *Jacobi steps*: $(.99)^{30} = .75$.

That is a very striking result from such a simple idea. Its real applications are not in one-dimensional problems (ordinary differential equations); a tridiagonal system $Ax = b$ is already easy to solve. It is in more dimensions, for partial differential equations, that overrelaxation is important. If we replace the unit interval $0 \leq x \leq 1$ by the unit square $0 \leq x,y \leq 1$, and change the equation $-u_{xx} = f$ to $-u_{xx} - u_{yy} = f$, then the natural finite difference analogue is the "five-point scheme." The coefficients $-1, 2, -1$ in the x direction combine with coefficients $-1, 2, -1$ in the y direction to give a main diagonal of $+4$ and four off-diagonal entries of -1. But *the matrix A does not have a bandwidth of 5*; there is no way to number the N^2 mesh points in a square so that each point stays close to all four of its neighbors. If the ordering goes a row at a time, then every point will have to wait through a whole row for the neighbor above it to turn up, and the "five-point matrix" has a bandwidth N:

$$A = \quad$$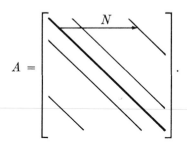

This matrix has probably had more attention, and been attacked in more different ways, then any other linear equation $Ax = b$. I think that the trend now is back to direct methods, based on an idea of Golub and Hockney; certain special matrices will fall apart when they are dropped the right way. (It is comparable to the so-called "fast Fourier transform.") Before that came the iterative methods of *alternating direction*, in which the splitting separated the tridiagonal matrix in the x direction from the one in the y direction. And before that came overrelaxation, because the Jacobi eigenvalue $\mu_{\max} = \cos \pi h$ is the same as in one dimension and so is the overrelaxation factor ω_{opt}. In every case the difficulty is to go from model problems to realistic problems, and each of these methods has its own possibilities for coping with equations more general than $-u_{xx} - u_{yy} = f$ and geometries more general than a square.

We cannot close without mentioning the method of *conjugate gradients*, which looked dead but is suddenly very much alive; it is direct rather than iterative, but unlike elimination, it can be stopped part way. And needless to say a completely new idea may still appear and win. But it seems fair to say that it was really the change from .99 to .75 that revolutionized the solution of $Ax = b$.

EXERCISE 7.4.1   For the matrix

$$A = \begin{bmatrix} 2 & -1 & 0 \\ -1 & 2 & -1 \\ 0 & -1 & 2 \end{bmatrix},$$

with eigenvalues $2 - \sqrt{2}$, 2, $2 + \sqrt{2}$, find the Jacobi matrix $D^{-1}(-L - U)$ and its eigenvalues, the Gauss–Seidel matrix $(D + L)^{-1}(-U)$ and its eigenvalues, and the numbers ω_{opt} and $\lambda_{\max}$ for SOR. You need not compute the matrix L_ω.

EXERCISE 7.4.2   For the n by n matrix

$$A = \begin{bmatrix} 2 & -1 & & \\ -1 & \cdot & & \cdot \\ & \cdot & & \cdot & -1 \\ & & -1 & 2 \end{bmatrix},$$

describe the Jacobi matrix $J = D^{-1}(-L - U)$. Show that the vector $x_1 = (\sin \pi h, \sin 2\pi h, \ldots, \sin n\pi h)$ is an eigenvector of J with eigenvalue $\lambda_1 = \cos \pi h = \cos \pi/(n + 1)$.

EXERCISE 7.4.3   For the same A, show that the other eigenvectors are $x_k = (\sin k\pi h, \sin 2k\pi h, \ldots, \sin nk\pi h)$ and the other eigenvalues are $\lambda_k = \cos k\pi h$, so that the λ's do occur in plus–minus pairs $(\lambda_n = -\lambda_1, \lambda_{n-1} = -\lambda_2, \ldots)$ and $\lambda_{\max}$ is $\cos \pi h$.

The following exercises require Gerschgorin's "circle theorem": *Every eigenvalue of A lies in at least one of the circles $C_1, \ldots, C_n$, where C_i has its center at the diagonal entry a_{ii} and its radius $r_i = \sum_{j \neq i} |a_{ij}|$ equal to the absolute sum along the rest of the row.*

Proof $Ax = \lambda x$ leads to

$$(\lambda - a_{ii})x_i = \sum_{j \neq i} a_{ij}x_j, \quad \text{or} \quad |\lambda - a_{ii}| \leq \sum_{j \neq i} |a_{ij}| \frac{|x_j|}{|x_i|}.$$

If the largest component of x is x_i, then these last ratios are ≤ 1, and λ lies in the ith circle: $|\lambda - a_{ii}| \leq r_i$.

EXERCISE 7.4.4   The matrix

$$A = \begin{bmatrix} 3 & 1 & 1 \\ 0 & 4 & 1 \\ 2 & 2 & 5 \end{bmatrix}$$

is called *diagonally dominant* because every $|a_{ii}| > r_i$. Show that zero cannot lie in any of the circles, and conclude that A is nonsingular.

EXERCISE 7.4.5   Write down the Jacobi matrix J for this diagonally dominant A, and find the three Gerschgorin circles for J. Show that all the radii satisfy $r_i < 1$, and explain why the Jacobi iteration will converge.

EXERCISE 7.4.6   The true solution to $Ax = b$ is slightly different from the elimination solution to $LUx_0 = b$; $A - LU$ misses zero because of roundoff. One possibility is to do everything in double precision, but a better and faster way is *iterative refinement*: Compute only one vector $r = b - Ax_0$ in double precision, solve $LUy = r$, and add the correction y to x_0. Problem: Multiply $x_1 = x_0 + y$ by LU, write the result as a splitting $Sx_1 = Tx_0 + b$, and explain why T is extremely small. This single step brings us almost exactly to x.

LINEAR
PROGRAMMING
AND GAME THEORY

Usually, the difference between algebra and analysis is more or less the difference between equations and inequalities. The line between the two has always seemed clear, but I have finally realized that linear programming is a counterexample: It is about inequalities, but it is unquestionably a part of linear algebra. The same is true of game theory, and there are three ways to approach these subjects: either intuitively through the geometry, or computationally through the simplex method, or algebraically through the theory of duality. For linear programming, these approaches are developed in this section and in Sections 8.2 and 8.3. Then Section 8.4 gives several examples of a matrix game (including poker), and explains the underlying minimax theorem.

The key to the chapter is to understand the geometric meaning of a linear inequality. We start in two dimensions, with the usual x-y plane. An equation like $x + 2y = 0$, or $x + 2y = 4$, has a simple interpretation: It represents a line. In fact these two equations represent two parallel lines, both of which are perpendicular to the vector $\begin{bmatrix} 1 & 2 \end{bmatrix}$; see Fig. 8.1. In matrix terms,

$$x + 2y = 0 \qquad \text{means} \qquad \begin{bmatrix} 1 & 2 \end{bmatrix} \begin{bmatrix} x \\ y \end{bmatrix} = \begin{bmatrix} 0 \end{bmatrix},$$

so the line is just the nullspace and $\begin{bmatrix} 1 & 2 \end{bmatrix}$ spans the row space; the equation itself says that their inner product is zero. As we change the right-hand side, we get a whole family of lines $x + 2y = b$, and the union of all these lines is the whole x-y plane. To use matrix terms again, each line $x + 2y = b$ is formed

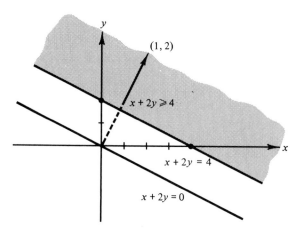

Fig. 8.1. Equations and inequalities.

by starting from any one particular solution of the equation—which gives a single point—and then adding to it the whole line of solutions to the homogeneous equation $x + 2y = 0$. That explains what is geometrically obvious: The lines $x + 2y = b$ are all parallel to the original line $x + 2y = 0$.

Now you can see the meaning of the inequality $x + 2y \geq 4$. The points that satisfy this constraint fill out a *halfspace*, lying above the boundary line $x + 2y = 4$. This halfspace is also a union of infinitely many lines $x + 2y = b$, but with the right side restricted by $b \geq 4$. (The opposite inequality $x + 2y \leq 4$ would describe the opposite halfspace, with the boundary line included again because equality is allowed.) In addition to inequalities of this kind, there is another constraint which is fundamental to linear programming: x and y are required to be nonnegative. Of course this requirement is itself a pair of linear inequalities, $x \geq 0$ and $y \geq 0$. Therefore we have two more halfspaces: The constraint $x \geq 0$ admits all points to the right of the vertical line $x = 0$, and $y \geq 0$ is the halfspace above $y = 0$.

The important step is to impose all three of these inequalities at once. Geometrically, they combine to give the shaded region in Fig. 8.2. You can recognize this region as the *intersection* of the three halfspaces $x + 2y \geq 4$, $x \geq 0$, $y \geq 0$. It is no longer a halfspace, but it is typical of what in linear programming is called a *feasible set*. To say it more directly, a feasible set is composed of the solutions to a family of linear inequalities. This set would be bounded rather than unbounded if we switch to the halfspace $x + 2y \leq 4$; keeping $x \geq 0$, $y \geq 0$, we get the triangle OAB. By imposing both of the inequalities $x + 2y \geq 4$ and $x + 2y \leq 4$, the feasible set can shrink to a line; the two opposing constraints combine to give the equation $x + 2y = 4$. And if we were to add a contradictory constraint like $x + 2y \leq -2$, the feasible set would become empty.†

The algebra of linear inequalities (or feasible sets) is one part of our subject. In linear programming, however, there is another essential ingredient: We

---

† The halfspace $x + 2y \leq -2$ has no intersection with the first quadrant $x \geq 0$, $y \geq 0$.

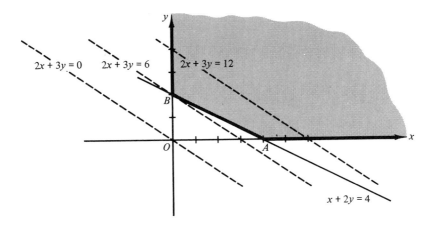

Fig. 8.2. The feasible set and the cost functions $2x + 3y$.

are interested *not in the set of all feasible points, but rather in the particular point that maximizes or minimizes a certain "cost function."* To the example $x + 2y \geq 4$, $x \geq 0$, $y \geq 0$, we add the cost function (or objective function) $2x + 3y$. Then the real problem in linear programming is to find the point x, y that lies in the feasible set and minimizes the cost.

The problem is illustrated by the geometry of Fig. 8.2. The family of costs $2x + 3y$ gives a family of parallel lines, and we have to find the minimum cost, in other words the first line to intersect the feasible set. That intersection clearly occurs at the point B, where $x^* = 0$ and $y^* = 2$; the minimum cost is $2x^* + 3y^* = 6$. The vector $(0, 2)$ is called *feasible* because it lies in the feasible set, it is *optimal* because it minimizes the cost function, and the minimum cost 6 is the *value* of the program. We will denote optimal vectors by an asterisk.

You can see that the optimal vector occurs at a corner of the feasible set. This is guaranteed by the geometry, because the lines that give the cost function (or the planes, when we get to more unknowns) are moved steadily up until they intersect the feasible set. The first contact must occur along its boundary. With a different cost function, however, the intersection might not be just a single point: If the cost happened to be $x + 2y$, then the whole edge between B and A would intersect at the same time, and there would be an infinity of optimal vectors along that edge (Fig. 8.3a). The value is still unique— $x^* + 2y^*$ equals 4 for all these optimal vectors—and therefore the minimum problem still has a definite answer. On the other hand, the maximum problem would have no solution! On our feasible set, the cost can go arbitrarily high and the maximum cost is infinite. Or another way to look at this possibility, staying with minimum problems, is to reverse the cost to $-x - 2y$. Then the minimum is $-\infty$, as in Fig. 8.3b, and again there is no solution. Every linear programming problem falls into one of three possible categories: Its feasible set is empty, or its cost function is unbounded on the feasible set, or there is a unique value for the linear program (with one or infinitely many optimal vectors). The first two should be very uncommon for a genuine problem in economics or engineering.

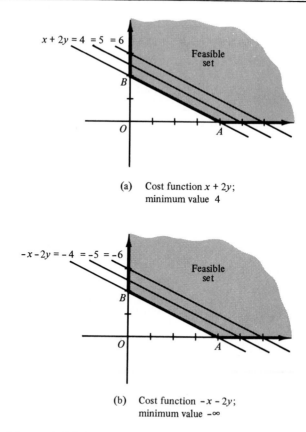

(a) Cost function $x + 2y$;
     minimum value 4

(b) Cost function $-x - 2y$;
     minimum value $-\infty$

Fig. 8.3. Special cases: infinitely many optimal vectors and none.

An Interpretation of the Problem and Its Dual

So much for the unusual cases. We want to return to our original example, with cost $2x + 3y$, and put it into words. It is an illustration of the "diet problem" in linear programming, with two sources of protein—say steak and peanut butter. Each pound of peanut butter gives a unit of protein, each pound of steak gives two units, and at least four units are required in the diet. Therefore a diet containing x pounds of peanut butter and y pounds of steak is constrained by $x + 2y \geq 4$, as well as by $x \geq 0$ and $y \geq 0$. (We cannot have negative steak or peanut butter.) This is the feasible set, and the problem is to minimize the cost. If a pound of peanut butter costs \$2 and a pound of steak is \$3, then the cost of the whole diet is $2x + 3y$. Fortunately, the optimal diet is steak: $x^* = 0$ and $y^* = 2$.

Every linear program, including this one, has a *dual*. If the original problem is a minimization, then its dual is a maximization, and the solution of one leads directly to the solution of the other; in fact the minimum in the given "primal problem" must equal the maximum in its dual. This is actually the central result in the theory of linear programming, and it will be explained in Section 8.3. Here we stay with the diet problem and try to interpret its dual.

In place of the shopper, who chooses between steak and peanut butter so as to get protein at minimal cost, the dual problem is the one faced by a druggist who sells *synthetic protein*. He intends to compete with steak and peanut butter, and immediately we meet the two ingredients of a typical linear programming problem: He wants to maximize the price p, but that price is subject to linear constraints. First, synthetic protein must not cost more than the protein in peanut butter (which was $2 a unit) or the protein in steak (which was $3 for two units). At the same time, the price must be nonnegative or the druggist will not sell. Since the diet requirement was four units of protein, the revenue to the druggist will be $4p$, and the dual problem is exactly this: *Maximize $4p$, subject to $p \leq 2$, $2p \leq 3$, and $p \geq 0$.* This is an example in which the dual is easier to solve than the primal; it has only one unknown. Obviously the constraint $2p \leq 3$ is the one that is really active, and the maximum price of synthetic protein is $1.50. Therefore the maximum revenue is $4p =$ $6. This was the minimal cost in the original problem, and the shopper ends up paying the same for both natural and synthetic protein. That is the meaning of the duality theorem.

One final remark: The constraints $p \leq 2$ and $2p \leq 3$ are easily converted from inequalities to equations, which is a useful step in a larger problem. We simply invent two new variables q and r, which represent the amount of "slack" in the inequalities: q is $2 - p$, r is $3 - 2p$, and both of these *slack variables* are forced to be nonnegative by the inequalities. In other words, the equations $p + q = 2$ and $2p + r = 3$ have the same effect as the original $p \leq 2$ and $2p \leq 3$ if we add the new constraints $q \geq 0$ and $r \geq 0$. Remembering the other constraint on the price, which was $p \geq 0$, the dual problem can be restated as a "canonical" maximum problem: Find the nonnegative p, q, and r that maximize $4p$, subject to the equations $p + q = 2$ and $2p + r = 3$.

We cannot end without rewriting this problem in matrix form: The unknown is a nonnegative column vector $y = [p \quad q \quad r]^{\mathrm{T}}$, the constraint equations are

$$\begin{bmatrix} 1 & 1 & 0 \\ 2 & 0 & 1 \end{bmatrix} \begin{bmatrix} p \\ q \\ r \end{bmatrix} = \begin{bmatrix} 2 \\ 3 \end{bmatrix}, \qquad \text{or} \qquad Ay = c,$$

and the revenue to be maximized is

$$y^{\mathrm{T}}b = [p \quad q \quad r] \begin{bmatrix} 4 \\ 0 \\ 0 \end{bmatrix} = 4p.$$

Therefore the problem is: *Maximize $y^{\mathrm{T}}b$, subject to $y \geq 0$ and $Ay = c$.*

EXERCISE 8.1.1   Sketch the feasible set with constraints $x + 2y \geq 6$, $2x + y \geq 6$, $x \geq 0$, $y \geq 0$. What points lie at the three "corners" of this set?

EXERCISE 8.1.2   (recommended)   On the preceding feasible set, what is the minimum value of the cost function $x + y$? Draw the line $x + y =$ constant that first touches the feasible set. How about the cost functions $3x + y$ and $x - y$?

EXERCISE 8.1.3   Convert this feasible set to one governed by equations rather than inequalities, by introducing nonnegative slack variables $u = x + 2y - 6$ and $v = 2x + y - 6$. The feasible set is now in four-dimensional space: Write down all six constraints, and find the x, y, u, v coordinates of the three original corner points $(0, 6)$, $(2, 2)$, $(6, 0)$.

EXERCISE 8.1.4   In the original diet problem, with cost $2x + 3y$ and constraint $x + 2y \geq 4$, let the slack variable z represent $4 - (x + 2y)$. For the vector $X = [x \quad y \quad z]^T$, choose a 1 by 3 matrix A, a 1-vector b, and a 3-vector c so the problem goes into matrix form: Minimize $c^T X$ subject to $AX = b$ and $X \geq 0$.

EXERCISE 8.1.5   Show that the feasible set constrained by $x \geq 0$, $y \geq 0$, $2x + 5y \leq 3$, $-3x + 8y \leq -5$, is empty.

EXERCISE 8.1.6   Show that the following problem is feasible but unbounded, so it has no optimal solution: Maximize $x + y$, subject to $x \geq 0$, $y \geq 0$, $-3x + 2y \leq -1$, $x - y \leq 2$.

EXERCISE 8.1.7   Add a single inequality constraint to $x \geq 0$, $y \geq 0$ in such a way that the feasible set contains only one point.

EXERCISE 8.1.8   What shape is the feasible set $x \geq 0$, $y \geq 0$, $z \geq 0$, $x + y + z = 1$, and what is the maximum of $x + 2y + 3z$?

Typical Applications

The next section will concentrate on the solution (and omit the formulation) of problems in linear programming. Therefore this is the time to describe some practical situations in which the underlying mathematical question can arise: *to minimize or maximize a linear cost function subject to linear constraints.* The list is very short, and the only exercise is to make it longer.

1. *The Transportation Problem.*   Suppose steel is produced at the plants $P_1, \ldots, P_r$ and required at the markets $M_1, \ldots, M_s$. If the total amount produced matches the total amount required, then a shipment is certainly feasible. We assume that the transportation costs are linear: Each unit shipped from P_i to M_j costs c_{ij}, and if x_{ij} units are shipped, then the total cost is $\sum \sum c_{ij} x_{ij}$. There are also constraints at both ends: The shipment out of each plant must equal its production, and the shipment into each market must equal its requirement. The problem is to minimize the cost subject to these constraints of supply and demand.

2. *The Traveling Salesman Problem.*   A salesman has to visit N different cities at minimum expense. If the travel costs from the ith city to the jth are c_{ij}, then again the problem is to minimize $\sum \sum c_{ij} x_{ij}$. The difference is that this time x_{ij} is the number of trips from i to j, and *this number is either 0 or 1.*

This is a problem in *integer linear programming*, just like the transportation problem for automobiles instead of steel. When the units are indivisible, the shipment has to be an integer.

3. *The Scheduling (or Inventory Control) Problem.* Suppose the demand for a product varies from month to month and that demand is known and has to be met. One possibility is to vary the production accordingly, but the fluctuations involve extra costs both for overtime and for employees who are unemployed. The alternative is to produce more than enough in the early months, and store the surplus. The scheduling problem, constrained by the given demand, is to minimize the sum of these two costs: fluctuation expense plus storage.

EXERCISE 8.1.9   Find another example.

THE SIMPLEX METHOD ■ 8.2

This section is about linear programming with n unknowns and m constraints. In the previous section we had $n = 2$ and $m = 1$; there were two nonnegative variables, and a single constraint $x + 2y \geq 4$. The more general case is not very hard to explain, and not very easy to solve.

The best way to describe the problem is to put it directly into matrix form. The n unknowns make up a nonnegative column vector $x = [x_1 \ \ x_2 \ \ \cdots \ \ x_n]^T$. To be "feasible," this vector has to satisfy m constraints in addition to $x \geq 0$; we write these constraints as $Ax \geq b$. The matrix A is m by n, and each of its rows gives one inequality (just as, in the rest of the book, every row of $Ax = b$ gave one equation). The matrix A and the vector b are given—for the constraint $x + 2y \geq 4$, we had $A = [1 \ \ 2]$ and $b = [4]$. The cost function is also given, and it is also linear; it equals $c_1x_1 + c_2x_2 + \cdots + c_nx_n$, or c^Tx. The problem is to find the feasible vector that minimizes the cost; that vector is optimal.

Standard minimum problem: Minimize c^Tx, subject to $x \geq 0$, $Ax \geq b$.

The geometric interpretation is straightforward. The first condition $x \geq 0$ restricts the vector to the "positive quadrant" in n-dimensional space—this is the region common to all the halfspaces $x_j \geq 0$. In two dimensions it is a quarter of the plane, and in three dimensions it is an eighth of the space; in general a vector has one chance in 2^n of being nonnegative. The other m constraints produce m additional halfspaces, and the feasible vectors are those that meet all of the $m + n$ conditions; they lie in the quadrant $x \geq 0$, and at the same time satisfy $Ax \geq b$. In other words, the feasible set is the intersection of $m + n$ halfspaces; it may be bounded (with flat sides!), or it may be unbounded, or it may be empty.

The cost function c^Tx brings to the problem a complete family of parallel planes. One member of the family, the one that goes through the origin, is the plane $c^Tx = 0$; if x satisfies this equation, then it is a vector with "zero cost." The other planes $c^Tx = $ constant give all other possible costs. As the cost varies, these planes sweep out the whole n-dimensional space, and the optimal vector x^* occurs at the point where they first touch the feasible set. That vector x^* is feasible, and its cost c^Tx^* is the minimum possible within the feasible set; it solves the standard minimum problem in linear programming.

Our aim in this section is to compute x^*. We could do it (in principle) by finding all the corners of the feasible set, and computing their costs; the one with the smallest cost would be optimal. In practice such a proposal is impossible; there are billions of corners, and we cannot compute them all. Instead we turn to the *simplex method*, which can fairly be called the most celebrated idea in recent computational mathematics. It was developed by Dantzig as a systematic way to solve problems in linear programming, and either by luck or genius it is an astonishing success. The steps of the method are summarized on p. 307, but first we will try to explain them as clearly as we can.

I think it is the geometric explanation that gives the method away. The first step simply locates a corner of the feasible set; this is "phase I," which we suppose to be complete. Then the heart of the method is in its second phase, *which goes from corner to corner along the edges of the feasible set.* At a typical corner there are n edges to choose from, some leading away from the optimal but unknown x^*, and others leading gradually toward it. Dantzig chose to go along an edge that is guaranteed to decrease the cost. That edge leads to a new corner with a lower cost, and there is no possibility of returning to anything that is more expensive. Eventually a special corner is reached, from which all the edges go the wrong way: The cost has been minimized. That corner is the optimal vector x^*, and the method stops.

The real problem is to turn this idea into linear algebra. First we interpret the words *corner* and *edge* in n dimensions. A corner is the meeting point of n different planes, each given by a single equation—just as three planes (or the front wall, the side wall, and the floor) produce a corner in three dimensions. Remember that the feasible set in linear programming is determined by m inequalities $Ax \geq b$ and n inequalities $x \geq 0$. Each corner of the set comes from turning n of these $n + m$ inequalities into equations, and finding the intersection of these n planes. In particular, one possibility is to choose the n equations $x_1 = 0, \ldots, x_n = 0$, and end up with the point at the origin. Like all the other possible choices, *this intersection point will only be a genuine corner if it also satisfies the m remaining constraints.* Otherwise it is not even in the feasible set, and is a complete fake. The example in Exercise 8.1.1 had $n = 2$ variables and $m = 2$ constraints; there are six possible intersections, illustrated

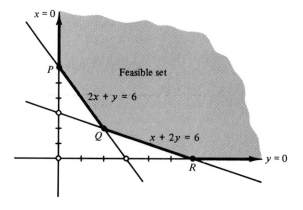

Fig. 8.4.   The corners and edges of the feasible set.

in Fig. 8.4, and three of them are actually corners of the feasible set. These three are marked P, Q, R; they are the vectors $(0, 6)$, $(2, 2)$, and $(6, 0)$, and one of them must be the optimal vector (unless the minimum cost is $-\infty$). The other three, including the origin, are fakes.

In general there are $(n + m)!/n!\,m!$ possible intersections, because that counts the number of ways to choose n planes out of $n + m$ candidates.† If the feasible set is empty, then of course none of these intersections will be genuine corners. It is the task of phase I either to find one genuine corner or to establish that the set is empty; we continue on the assumption that a corner has been found.

Now, for an edge: Suppose one of the n intersecting planes is removed, leaving only $n - 1$ equations and therefore one degree of freedom. The points that satisfy these $n - 1$ equations form an edge which comes out of the corner; geometrically, it is the intersection of the $n - 1$ planes. Again linear programming forces us to stay in the feasible set, and we have no choice of direction along the edge; only one direction remains feasible. But we do have a choice of n different edges, and phase II must make that choice.

To describe this phase, we have to rewrite $Ax \geq b$ in a form completely parallel to the n simple constraints $x_1 \geq 0, \ldots, x_n \geq 0$. This is the role of the slack variables $z = Ax - b$. The m constraints $Ax \geq b$ are translated into $z_1 \geq 0, \ldots, z_m \geq 0$, with one slack variable for every row of A. Then the equation $z = Ax - b$, or $Ax - z = b$, goes into matrix form as

$$[A \quad -I]\begin{bmatrix} x \\ z \end{bmatrix} = b.$$

The feasible set is governed by these m equations and the $n + m$ simple inequalities $x \geq 0$, $z \geq 0$—in other words, a standard problem has been turned into a canonical problem, with *equality constraints*.

To make the change complete, we want to leave no distinction whatsoever between the original x and the newly arrived z. The simplex method notices no difference, and it would be pointless to go on with the notations

$$[A \quad -I] \qquad \text{and} \qquad \begin{bmatrix} x \\ z \end{bmatrix}.$$

Therefore *we rename that larger matrix A, and we rename that longer vector x.* The equality constraints are then $Ax = b$, and the $n + m$ simple inequalities become just $x \geq 0$.‡ The original cost vector c needs to be extended by adding m more components, all zero; then the cost $c^\mathrm{T}x$ is the same for the new meaning of x and c as for the old. The only trace left of the slack variable z is in the fact that the new matrix A is m by $n + m$, and the new x has $n + m$ components.

---

† The size of that number explains why computing every corner is completely impractical for large n.

‡ Of course economics or engineering might have given us equality constraints in the first place, in which case the simplex method starts right out from $Ax = b$—with no need for slack variables.

We keep this much of the original notation, leaving m and n unchanged as a reminder of what happened.

EXAMPLE In the problem illustrated by Fig. 8.4, with the constraints $x + 2y \geq 6$, $2x + y \geq 6$, and the cost $x + y$, the new system has

$$A = \begin{bmatrix} 1 & 2 & -1 & 0 \\ 2 & 1 & 0 & -1 \end{bmatrix}, \qquad b = \begin{bmatrix} 6 \\ 6 \end{bmatrix}, \qquad \text{and} \qquad c = \begin{bmatrix} 1 \\ 1 \\ 0 \\ 0 \end{bmatrix}.$$

With this change to equality constraints, the simplex method can begin. Remember that phase I has already found a corner of the feasible set, which was a point where n planes met: n of the original inequalities $x \geq 0$ and $Ax \geq b$ (alias $z \geq 0$) were turned into equations. In other words, *a corner is a point where n components of the new vector x (the old $\begin{bmatrix} x \\ z \end{bmatrix}$) are zero.* In the system $Ax = b$, we think of these n components of x as the free variables and the remaining m components as the basic variables. Then, setting the n free variables to zero, the m equations $Ax = b$ determine the m basic variables. This solution x is called *basic*, to emphasize its complete dependence on those variables. It will be a genuine corner if, in addition to its n zero components, it is also nonnegative; then it belongs to the feasible set.

8A The corners of the feasible set are exactly the *basic feasible solutions* of $Ax = b$: A solution is basic when n of its $m + n$ components are zero, and it is feasible when it satisfies $x \geq 0$. Phase I of the simplex method finds one basic feasible solution, and phase II moves step by step to the optimal one.

The crucial decision is still to be made: Along which edge do we leave the corner? We are already at a basic feasible solution x, and for convenience we reorder its components so that the n zeros come last. At the same time we rearrange the columns of A in the same way: The first m columns multiply the basic variables and the last n columns correspond to the free variables:

$$x = \begin{bmatrix} x_B \\ 0 \end{bmatrix} \qquad \text{and} \qquad A = [B \quad F].$$

Thus the equation $Ax = b$ becomes $Bx_B = b$, and it is this system of m equations in m unknowns that determines the basic variables: $x_B = B^{-1}b$. We are assuming this to be positive, so the whole vector x is nonnegative, and we are at a genuine corner of the feasible set.

EXAMPLE The corner point P in Fig. 8.4 is the intersection of $x = 0$ with

$2x + y - 6 = 0$, so two of the components of x are zero:

$$Ax = \begin{bmatrix} 1 & 2 & -1 & 0 \\ 2 & 1 & 0 & -1 \end{bmatrix} \begin{bmatrix} 0 \\ * \\ * \\ 0 \end{bmatrix} = b.$$

The other two components of x happen to equal 6, indicating that this solution is feasible (nonnegative) as well as basic (two zero components). It is a genuine corner, as the figure confirms. The reordering exchanges the first and third components of x, and at the same time the first and third columns of A:

$$x \to \begin{bmatrix} x_B \\ 0 \end{bmatrix} = \begin{bmatrix} 6 \\ 6 \\ 0 \\ 0 \end{bmatrix}, \qquad A \to \begin{bmatrix} B & F \end{bmatrix} = \begin{bmatrix} -1 & 2 & 1 & 0 \\ 0 & 1 & 2 & -1 \end{bmatrix}.$$

The product Ax is unchanged, but now the free variables come last.

To move away from this corner, we let one of the zero components of x move away from zero. It has to become positive, since otherwise x will leave the feasible set. That matches the geometry: There is only one way to go along each edge, but there are n edges to choose from—just as there are n zero components that could be made nonzero. Along this edge, the bottom half of the vector x is changing from zero into a vector x_F with one positive component, and there are two questions to decide:

(i)   Which component (which edge) do we choose?
(ii)   When do we reach the next corner of the feasible set?

The answers come from solving $Ax = b$. The free variables are given by x_F, so—just as in back-substitution—the basic variables can be determined:

$$\begin{bmatrix} B & F \end{bmatrix} \begin{bmatrix} x_B \\ x_F \end{bmatrix} = b, \quad \text{or} \quad Bx_B + Fx_F = b, \quad \text{or} \quad x_B = B^{-1}b - B^{-1}Fx_F.$$

$$(1)$$

Now we take the two questions in turn.

(i)   The choice of edge depends on the cost vector c. When x is rearranged with the basic components x_B at the top, we make the same rearrangement of c. With matching components of x and c kept together, the inner product c^Tx is still the correct cost. Writing c_B and c_F for the first m components and the last n, the cost becomes

$$\begin{bmatrix} c_B \\ c_F \end{bmatrix}^T \begin{bmatrix} x_B \\ x_F \end{bmatrix} = c_B^Tx_B + c_F^Tx_F$$

$$= c_B^T(B^{-1}b - B^{-1}Fx_F) + c_F^Tx_F.$$

The key step is to rewrite this as

$$\text{cost} = c_B^T B^{-1} b + (c_F^T - c_B^T B^{-1} F)\, x_F . \tag{2}$$

At the corner itself, where $x_F = 0$, the cost equals the first term $c_B^T B^{-1} b$. It is the second term, and in particular the row vector $r = c_F^T - c_B^T B^{-1} F$, which decides whether or not this cost is minimal. Since x_F is to have one positive component, we want it to multiply a negative entry in r; then the second term in (2) reduces the cost. This leads to the **optimality condition**:

> **8B** If the vector $r = c_F^T - c_B^T B^{-1} F$ is nonnegative, then no reduction in cost can be achieved: The corner is already optimal and the minimum cost is $c_B^T B^{-1} b$. Therefore $r \geq 0$ is the stopping test for the simplex method. Assuming this test fails, any negative component of r corresponds to an edge along which the cost can be reduced, and the usual strategy is to choose the most negative component of r.

If this rule chooses the ith component of r, then the original corner was not optimal and the ith component (say α) of x_F will become positive. The remaining question is, How positive does α become? We have decided which component will move from free to basic, and now we have to decide which basic component is to become free (that is, zero).

(ii)   The top part of the vector x, according to (1), is

$$x_B = B^{-1} b - B^{-1} F x_F .$$

This vector has to be nonnegative to be feasible, and at the original corner it is positive: $x_F = 0$ and $B^{-1} b > 0$. As we increase x_F, some entries of x_B may begin to decrease, and *we reach the next corner when a component of x_B reaches zero.* It is that component (say the kth) which joins the free variables. At that point we have reached a new x which is both feasible and basic: It is feasible because we still have $x \geq 0$, and it is basic because we again have n zero components. The ith component of x_F, which went from zero to α, replaces the kth component of x_B, which dropped to zero. The other components of x_B will have moved around, but remain positive.

> **8C** Suppose u is the ith column of F, which was chosen in 8B to go from free to basic. Then its component at the new corner x is
>
> $$\alpha = \min \frac{(B^{-1} b)_j}{(B^{-1} u)_j} = \frac{(B^{-1} b)_k}{(B^{-1} u)_k} . \tag{3}$$
>
> This minimum is taken only over the positive components of $B^{-1} u$. If there are no positive components, then the next corner is infinitely far away, and the cost can be reduced forever; the minimal cost is $-\infty$. Otherwise, the old kth column of B is removed in favor of the new column u, and the simplex method begins again at the new corner.

This step of the method has changed the cost by αr_i, according to (2). It may now be minimal, or it may not; the new vector r will decide.

To understand the formula for α we look again at

$$x_B = B^{-1}b - B^{-1}Fx_F = B^{-1}b - \alpha B^{-1}u. \tag{4}$$

(Remember that x_F has only one nonzero component α, so the product Fx_F picks out the corresponding column u.) We can increase α until one of the components of x_B reaches zero; that signals the next corner. If $B^{-1}u \le 0$, this will never happen and the edge is infinitely long; you can see from (4) that increasing α only makes x_B more positive. In the opposite (and more normal) case, the vector $B^{-1}u$ has some positive components and the edge is finite; it ends when α reaches the smallest of the ratios in (3), since at that point the kth component of $B^{-1}b - \alpha B^{-1}u$ will drop to zero.

We have found the new corner, and the simplex step is finally complete. It is repeated, in a more organized way, in Table 8.1 (p. 307).

Remark on degeneracy Suppose that more than n of the variables are zero, either at the outset, or if two different components in the formula for α give the same minimum ratio of $B^{-1}b$ to $B^{-1}u$. We can choose either one of them to be made free, but the other will still be there, and the new corner will have $n + 1$ zero components instead of the usual n. Geometrically, there is an extra (and unwanted) plane intersecting at the new corner. This means trouble at the next step of the simplex method, because if $B^{-1}b$ is not strictly positive then the new ratio α may be zero, and the reduction in cost along the new edge would be $\alpha r_i = 0$. In other words, there is the possibility of cycling forever around the same set of corners, without moving toward the optimal x^*.

This possibility can be removed either by adjusting the simplex method to watch for it, or by assuming that cycling does not occur. Since the latter is in agreement with computational experience (and is much simpler), we shall make the standard nondegeneracy hypothesis that $B^{-1}b$ stays strictly positive.

End of the example We are still waiting to move from corner P in Fig. 8.4, corresponding to the basic feasible solution

$$x = \begin{bmatrix} x_B \\ 0 \end{bmatrix} = \begin{bmatrix} 6 \\ 6 \\ 0 \\ 0 \end{bmatrix}, \qquad A = \begin{bmatrix} B & F \end{bmatrix} = \begin{bmatrix} -1 & 2 & 1 & 0 \\ 0 & 1 & 2 & -1 \end{bmatrix}.$$

The first step in the simplex method is to compute r. Remember that the cost in Exercise 8.1.2 was $x + y$, which gives $c^T = (1, 1, 0, 0)$ after adding slack variables, and $c^T = (0, 1, 1, 0)$ after rearrangement. Therefore

$$r = c_F{}^T - c_B{}^T B^{-1}F = \begin{bmatrix} 1 & 0 \end{bmatrix} - \begin{bmatrix} 0 & 1 \end{bmatrix} \begin{bmatrix} -1 & 2 \\ 0 & 1 \end{bmatrix}^{-1} \begin{bmatrix} 1 & 0 \\ 2 & -1 \end{bmatrix} = \begin{bmatrix} -1 & 1 \end{bmatrix}.$$

The first component of r is negative, so the stopping test $r \ge 0$ is not passed and the corner P is not optimal. This component is the only negative one, so it is the first column of F that must be brought into the new basis. That column is named u. To decide which column of the old B to remove, we look at

$$u = \begin{bmatrix} 1 \\ 2 \end{bmatrix}, \qquad B^{-1}u = \begin{bmatrix} 3 \\ 2 \end{bmatrix}, \qquad \alpha = \min \frac{(B^{-1}b)_j}{(B^{-1}u)_j} = \text{smaller of } \frac{6}{3} \text{ and } \frac{6}{2}.$$

The first ratio is smaller, so the first column of B must go; the new A is

$$A^* = [B^* \;\; F^*] = \begin{bmatrix} 1 & 2 & -1 & 0 \\ 2 & 1 & 0 & -1 \end{bmatrix}.$$

It is an accident that this simplex step has put the original slack variables back at the end of the matrix; they happen to be zero at the new corner Q. The basic variables at Q are

$$(B^*)^{-1}b = \begin{bmatrix} 1 & 2 \\ 2 & 1 \end{bmatrix}^{-1} \begin{bmatrix} 6 \\ 6 \end{bmatrix} = \begin{bmatrix} 2 \\ 2 \end{bmatrix}, \qquad \text{so} \qquad x^* = \begin{bmatrix} 2 \\ 2 \\ 0 \\ 0 \end{bmatrix}.$$

The cost vector, after a matching rearrangement, is $c^T = (1, 1, 0, 0)$. Now we are ready for the second simplex step, which will be very brief; the new vector r is

$$r^* = c_F^* - c_B^*(B^*)^{-1}F^* = [0 \;\; 0] - [1 \;\; 1] \begin{bmatrix} 1 & 2 \\ 2 & 1 \end{bmatrix}^{-1} \begin{bmatrix} -1 & 0 \\ 0 & -1 \end{bmatrix} = [\tfrac{1}{3} \;\; \tfrac{1}{3}].$$

It has no negative components, the test $r \geq 0$ is passed, the simplex method stops, and the corner Q is optimal. The minimal cost is $c^T x^* = 4$.

EXERCISE 8.2.1   If the original cost function had been $3x + y$, so that after rearrangement $c^T = (0, 1, 3, 0)$ at the corner P, show that $r \geq 0$ and therefore that corner is optimal.

EXERCISE 8.2.2   If the original cost function had been $x - y$, so that after rearrangement $c^T = (0, -1, 1, 0)$ at the corner P, compute r and decide which column u should enter the basis. Then compute $B^{-1}u$ and show from its sign that you will never meet another corner. We are climbing the y axis in Fig. 8.4, and $x - y$ goes to $-\infty$.

EXERCISE 8.2.3   In the same example, change the cost to $x + 3y$. Verify that the simplex method takes you from P to Q to R, and that the corner R is optimal.

EXERCISE 8.2.4   Redraw Fig. 8.4 with the right sides of the constraints (the vector b) changed from 6 to -6. Show that the origin $x = y = 0$ is feasible, and is therefore a genuine corner.

EXERCISE 8.2.5   Phase I consists in finding a basic feasible solution to $Ax = b$. After changing signs to make $b \geq 0$, consider the auxiliary problem of minimizing $w_1 + w_2 +$

$\cdots+w_m$ subject to $x \geq 0$, $w \geq 0$, $Ax + w = b$. Whenever $Ax = b$ has a nonnegative solution, the minimum cost in this problem will be zero—with $w^* = 0$.

(a)   Show that, for this new problem, the corner $x = 0$, $w = b$ is both basic and feasible. Therefore *its* phase I is already set, and the simplex method can proceed to find the optimal pair x^*, w^*. If $w^* = 0$, then x^* is the required corner in the original problem.

(b)   With $A = [1 \ -1]$ and $b = [3]$, write out the auxiliary problem, its phase I vector from part (a), and its optimal vector. Find the corner of the feasible set $x_1 - x_2 = 3$, $x_1 \geq 0$, $x_2 \geq 0$, and draw a picture of this set.

EXERCISE 8.2.6   If we wanted to maximize instead of minimize the cost (still with $Ax = b$ and $x \geq 0$), what would be the stopping test on r and what rules would choose the column of F to make basic and the column of B to make free?

The Organization of a Simplex Step

So far we have achieved a transition from the geometry of the simplex method to the algebra—in other words, from the language of "corners" to "basic feasible solutions." That transition had to come before the computations could begin. Now we know that it is the vector r and the ratio α which are decisive, and we want to look once more at their calculation. This is the heart of the simplex method, and it can be organized in three different ways:

(1)   In a tableau.
(2)   By computing B^{-1}, and updating it when the new column u replaces the current kth column of B.
(3)   By computing either $B = LU$ or $B = QR$, and updating these factors instead of B^{-1}.

This list is really a brief history of the simplex method. In some ways the most fascinating stage was the first—the *tableau*—which dominated the subject for so many years. For most of us it brought an aura of mystery to linear programming, chiefly because it managed to avoid matrix notation almost completely (by the skillful device of writing out all matrices in full!). We can explain it in a few lines, for the sake of nostalgia, but you must understand that its day is absolutely (?) over.

The explanation begins with the Gauss–Jordan method for $Ax = b$, which creates zeros both above and below the pivots, and normalizes each pivot to unity. In other words, it transforms the first part of A into the identity:

$$[A \quad b] = [B \quad F \quad b] \to [I \quad B^{-1}F \quad B^{-1}b].$$

Certainly the matrix that transforms B into I has to be B^{-1}, and this explains the appearance of $B^{-1}F$ and $B^{-1}b$; each elimination step is applied to the whole row, so the whole matrix is premultiplied by B^{-1}. Next, to account for the cost function, we add one more row. It consists of the vector c^T, followed by a zero in the last column. Then we subtract multiples of the other rows from this one, as usual, to produce zeros below the pivots. Since the pivots are the 1's in the identity, this eliminates $c_B{}^T$—the part of c^T that lies at the front. In fact the elimination steps must subtract $c_B{}^T$ times the first m rows from the last row, and these steps produce

$$\begin{bmatrix} I & B^{-1}F & B^{-1}b \\ c_B{}^T & c_F{}^T & 0 \end{bmatrix} \rightarrow \begin{bmatrix} I & B^{-1}F & B^{-1}b \\ 0 & c_F{}^T - c_B{}^T B^{-1}F & -c_B{}^T B^{-1}b \end{bmatrix}.$$

This is the complete tableau. It now contains not only the solution $B^{-1}b$, but the crucial vector $r = c_F{}^T - c_B{}^T B^{-1}F$. A glance will show whether $r \geq 0$, in which case the method stops and the minimal cost $c_B{}^T B^{-1}b$ appears in the lower right corner (with a superfluous minus sign). Otherwise, the most negative component of r indicates which new column will enter B, and lying above that negative component is the vector $B^{-1}u$. Therefore we can find all the positive entries in $B^{-1}u$, and all their ratios with the corresponding entries of the last column $B^{-1}b$. If the smallest ratio of $B^{-1}b$ to $B^{-1}u$ occurs in component k, then it is the kth column of the present basis B that is to be replaced by u.

EXAMPLE   You can see these vectors fall into place if we return to the corner P in Fig. 8.4. Copying from p. 303, the starting tableau is

$$\begin{bmatrix} A & b \\ c^T & 0 \end{bmatrix} = \begin{bmatrix} B & F & b \\ c_B{}^T & c_F{}^T & 0 \end{bmatrix} = \begin{bmatrix} -1 & 2 & 1 & 0 & 6 \\ 0 & 1 & 2 & -1 & 6 \\ \hline 0 & 1 & 1 & 0 & 0 \end{bmatrix}.$$

Gauss–Jordan elimination multiplies the first row by -1, to give a unit pivot, and then it uses the second row to produce zeros in the second column. The final tableau is

$$\begin{bmatrix} 1 & 0 & 3 & -2 & 6 \\ 0 & 1 & 2 & -1 & 6 \\ \hline 0 & 0 & -1 & 1 & -6 \end{bmatrix} = \begin{bmatrix} I & B^{-1}F & B^{-1}b \\ \hline 0 & r & -\text{cost} \end{bmatrix}.$$

First we look at r. It has a negative entry, so the current cost of 6 is not optimal. The column above that negative entry is $B^{-1}u$, and its ratios with the last column are 6:3 and 6:2. Since the first ratio is the smaller, it is the first column of the matrix that will be pushed out of the basis.

That replacement step, which exchanges the third column and the first, is known as *pivoting*. It takes us to the new corner Q. The important point is to organize it properly, and everything depends on the following observation: *It is not necessary to return to the starting tableau, exchange the two columns, and start again.* Instead we can continue with the current tableau, leave the second column as is, and use row operations to turn the third column into what we want:

$$\begin{bmatrix} 1 & 0 & 3 & -2 & 6 \\ 0 & 1 & 2 & -1 & 6 \\ \hline 0 & 0 & -1 & 1 & -6 \end{bmatrix} \rightarrow \begin{bmatrix} \frac{1}{3} & 0 & 1 & -\frac{2}{3} & 2 \\ 0 & 1 & 2 & -1 & 6 \\ \hline 0 & 0 & -1 & 1 & -6 \end{bmatrix} \rightarrow \begin{bmatrix} \frac{1}{3} & 0 & 1 & -\frac{2}{3} & 2 \\ -\frac{2}{3} & 1 & 0 & \frac{1}{3} & 2 \\ \hline \frac{1}{3} & 0 & 0 & \frac{1}{3} & -4 \end{bmatrix}.$$

Now it is columns 2 and 3 that are basic, and columns 1 and 4 that contain the vector r in the bottom row. Because r is positive—it is the same $(\frac{1}{3}, \frac{1}{3})$ we computed earlier—the algorithm stops. The minimal cost, according to the lower right-hand corner, is 4.

That ends the example, to which we add two comments:

(i)   Although very neat, most of the calculations in the fourth column were completely wasted. The only vector really needed was r, in the last row, whose negative entry indicated the one important column $B^{-1}u$. In a larger problem, hundreds of columns would be wasted.

(ii)   Once $B^{-1}u$ is identified, the pivoting step is simple. The other basic columns are already set, like the second column in the example, and it remains only to turn $B^{-1}u$ into the kth column of the identity matrix (the column it is replacing). In other words, the pivoting step turns

$$
E = \begin{bmatrix} 1 & & & \vdots & & \\ & \ddots & & & & \\ & & \ddots & & & \\ & & & B^{-1}u & & \\ & & & \vdots & \ddots & \\ & & & & & 1 \end{bmatrix} \quad \text{into} \quad \begin{bmatrix} 1 & & & & & \\ & \ddots & & & & \\ & & \ddots & & & \\ & & & 1 & & \\ & & & & \ddots & \\ & & & & & 1 \end{bmatrix} = I.
$$

Of course the matrix that does this is E^{-1}. We conclude that after the initial B^{-1} has been computed at the first corner of the simplex method, *the B^{-1} for each succeeding corner is found by premultiplying the current B^{-1} by E^{-1}.†*

We are ready to leave behind the tableau. It is much quicker, and in the end much simpler, just to look at the simplex method and see what calculations are really necessary. Each step has to exchange a column of F for a column of B, and it has to decide (from r and α) which columns to choose. This step requires the following cycle of operations, beginning with the current basis matrix B and the current solution $x_B = B^{-1}b$:

Table 8.1   *A Step of the Simplex Method*

(1)   Compute the row vector $\lambda = c_B^\mathrm{T} B^{-1}$ and then $r = c_F^\mathrm{T} - \lambda F$.

(2)   If $r \geq 0$ stop; the current solution is optimal. Otherwise, if r_i is the most negative component, choose the ith column of F to enter the basis. Denote it by u.

(3)   Compute $v = B^{-1}u$.

(4)   Calculate the ratios of $B^{-1}b$ to $B^{-1}u$, admitting only the positive components of $B^{-1}u$. If there are no positive components, the minimal cost is $-\infty$; if the smallest ratio occurs at component k, then the kth column of the current B will be replaced.

(5)   Update B (or B^{-1}) and the solution $x_B = B^{-1}b$. Return to 1.

---

† This matrix is computed in Exercise 8.2.8. It is as simple as E.

This is sometimes called the *revised* simplex method, to distinguish it from the operations on a tableau. It is really the simplex method itself, boiled down.

The discussion is finished once we decide how to take steps 1, 3, and 5, which compute

$$\lambda = c_B^T B^{-1}, \qquad v = B^{-1}u, \qquad \text{and} \qquad x_B = B^{-1}b. \tag{5}$$

As this book is written, the most popular way is to work directly with B^{-1}, updating it by E^{-1} after each cycle. Instead of actually changing B^{-1}, it is enough to store the simple matrices E^{-1} and apply them in order to the given c_B^T, u, and b. This is called the "product form of the inverse." Periodically, excessive computing has to be avoided by directly reinverting the current B; then the history contained in the matrices E^{-1} can be erased.

A new way, and I believe a better way, is to use the ordinary methods of linear algebra. Just as the old-fashioned tableau was unrelated to the calculations that are actually necessary, so the explicit use of B^{-1} ignores both the sparsity of most large matrices and their numerical instability. The stability problem comes with the simple matrix E; its pivot is the kth component of $B^{-1}u$, and if that number is small we are in trouble. The alternative is to regard (5) as three equations sharing the same coefficient matrix B:

$$\lambda B = c_B^T, \qquad Bv = u, \qquad \text{and} \qquad Bx_B = b. \tag{6}$$

Then the standard decompositions $B = QR$ or $B = LU$ (or $PB = LU$ since we must allow row exchanges to improve on the stability!) lead directly to the three solutions.†

Just one more thing: *How many simplex steps do we have to take?* This is one question that is impossible to answer in advance. Experience shows that the method touches only about m different corners, which means an operation count of about m^2n—which is comparable to ordinary elimination for $Ax = b$, and is the reason for its success. But mathematics shows that the path length cannot always be bounded by any fixed multiple of m or any fixed power of m. It is now known that the worst feasible sets can force a path of more than $cm^{n/2}$ edges, where c is a number independent of m. For some reason, hidden in the geometry of many-dimensional polyhedra, the simplex method seems to be lucky.

EXERCISE 8.2.7   Form the tableau for the same corner P, with the cost function changed from $x + y$ to $x - y$. Show that the resulting r has a negative entry, but the column above it has no positive entries. This column is $B^{-1}u$, and therefore the minimum cost is $-\infty$.

---

† Luenberger's book explains how these factors can be updated rather than recomputed. See reference list.

EXERCISE 8.2.8   Verify that

$$
E^{-1} = \begin{bmatrix} 1 & & -v_1/v_k & & \\ & \ddots & \vdots & & \\ & & 1/v_k & & \\ & & \vdots & \ddots & \\ & & -v_n/v_k & & 1 \end{bmatrix} \quad \text{if} \quad E = \begin{bmatrix} 1 & & v_1 & & \\ & \ddots & \vdots & & \\ & & v_k & & \\ & & \vdots & \ddots & \\ & & v_n & & 1 \end{bmatrix}.
$$

EXERCISE 8.2.9   For any matrix B, show that the product BE is identical with B except that the kth column is changed to Bv (which is the desired u). Therefore the inverse of this new matrix is $(BE)^{-1} = E^{-1}B^{-1}$, and the factor E^{-1} updates the inverse correctly.

THE THEORY OF DUALITY ■ 8.3

Chapter 2 began by saying that although the elimination technique gives one approach to $Ax = b$, a different and deeper understanding is also possible. It is exactly the same for linear programming. The mechanics of the simplex method will solve a linear program, but it is really duality that belongs at the center of the underlying theory. It is an elegant idea, and at the same time fundamental for the applications; we shall explain as much as we understand.

It starts with the standard problem:

Minimize $c^T x$, subject to $x \geq 0$ and $Ax \geq b$.

But now, instead of creating an equivalent problem with equations in place of inequalities, the theory of duality creates an entirely different problem. *The dual problem starts from the same A, b, and c, and reverses everything.* In the primal problem c was in the cost function and b was in the constraint; in the dual these two vectors are switched, and the matrix A is transposed. Furthermore the inequality sign is changed, so the feasible vectors satisfy $A^T y \leq c$ rather than $Ax \geq b$. And finally we maximize the objective function $y^T b$, rather than minimizing $c^T x$. The only thing that stays the same is the requirement of nonnegativity; the unknown is a column vector y with m components, and it must satisfy $y \geq 0$. In short, the dual of a minimum problem is a maximum problem:

Maximize $y^T b$, subject to $y \geq 0$ and $A^T y \leq c$.

The dual of *this* problem is the original minimum problem.†

Obviously I have to give you some interpretation of all these reversals. They conceal a kind of competition between the minimizer and the maximizer, and the clearest explanation comes from returning to the diet problem; I hope you will follow it through once more. The minimum problem has n unknowns,

---

† There is complete symmetry between the two. We started with a minimization, but the simplex method applies equally well to a maximization—and anyway both problems get solved at once.

representing n different foods to be eaten in the (nonnegative) amounts $x_1, \ldots, x_n$. The m constraints represent m required vitamins, in place of the one earlier constraint of sufficient protein. The entry a_{ij} is the amount of the ith vitamin in the jth food, and the ith row of $Ax \geq b$ forces the diet to include that vitamin in at least the amount b_i. Finally, if c_j is the cost of the jth food, then $c_1 x_1 + \cdots + c_n x_n = c^T x$ is the cost of the diet. That cost is to be minimized; this is the primal problem.

In the dual, the druggist is selling vitamin pills rather than food. His prices y_i are adjustable as long as they are nonnegative. The key constraint, however, is that on each food he cannot charge more than the grocer; since the jth food contains vitamins in the amounts a_{ij}, the druggist's price for the equivalent in vitamins (which is $a_{1j} y_1 + \cdots + a_{mj} y_m$) cannot exceed the grocer's price c_j. That is the constraint $A^T y \leq c$. Working within this constraint, he can sell an amount b_i of each vitamin for a total income of $y_1 b_1 + \cdots + y_m b_m = y^T b$—which he maximizes.

You must recognize that the feasible sets for the two problems are completely different. The first is a subset of $\mathbf{R}^n$, marked out by the matrix A and the constraint vector b; the second is a subset of $\mathbf{R}^m$, determined by the transpose of A and the other vector c. Nevertheless, when the cost functions are included, the two problems do involve the same input A, b, and c. The whole theory of linear programming hinges on the relation between them, and we come directly to the fundamental result:

8D   Duality Theorem   If either the primal problem or the dual has an optimal vector, then so does the other, and their values are the same: *The minimum of $c^T x$ equals the maximum of $y^T b$*. Otherwise, if optimal vectors do not exist, there are two possibilities: Either both feasible sets are empty, or else one is empty and the other problem is unbounded (the maximum is $+\infty$ or the minimum is $-\infty$).

If neither feasible set is empty, we must be in the first and most important case: $c^T x^* = (y^*)^T b$.

Mathematically, this settles the competition between the grocer and the druggist: *The result is always a tie*. We will find a similar "minimax theorem" and a similar equilibrium in game theory. These theorems do not mean that the customer pays nothing for an optimal diet, or that the matrix game is completely fair to both players. They do mean that the customer has no economic reason to prefer vitamins over food, even through the druggist guaranteed to match the grocer on every food—and on expensive foods, like peanut butter, he will actually sell for less. We will show that expensive foods are kept out of the optimal diet, so the outcome can still be (and is) a tie.

This conclusion may seem like a total stalemate, but I hope you will not be fooled: It is the optimal vectors that contain the crucial information. In the primal problem, x^* tells the purchaser what to do; in the dual, y^* fixes the natural prices (or "shadow prices") at which the economy should run. Insofar as our linear model reflects the true economy, these vectors represent the actual decisions to be made; they still need to be found by the simplex method, and the duality theorem only tells us their most important property.

We want to start on the proof. In a sense it may seem obvious that the druggist can raise his prices to meet the grocer's, but it is not. Or rather, only the first part of it is: Since each individual food can be replaced by its vitamin equivalent, with no increase in cost, all adequate food diets must be at least as expensive as any price the druggist would charge. This is only a one-sided inequality, *druggist's price* ≤ *grocer's price*, but it is fundamental. It is called **weak duality**, and it is easy to prove for any linear program and its dual:

8E If x and y are any feasible vectors in the minimum and maximum problems, then $y^T b \leq c^T x$.

Proof Since the vectors are feasible, they satisfy

$$Ax \geq b \quad \text{and} \quad A^T y \leq c \quad (\text{or} \quad y^T A \leq c^T).$$

Furthermore, because feasibility also included $x \geq 0$ and $y \geq 0$, we can take inner products without spoiling the inequalities:

$$y^T Ax \geq y^T b \quad \text{and} \quad y^T Ax \leq c^T x. \tag{7}$$

But the two left sides are identical, which proves that $y^T b \leq c^T x$.

A point about the notation: Some authors prefer to avoid all transposes by turning the dual unknown y (as well as c) into a row vector. Then y is feasible when $yA \leq c$, $y \geq 0$, and the cost functions are cx for the primal and yb for the dual. We shall stick to column vectors, so you recognize that it really is A^T in the dual problem, but you can follow either convention without difficulty.

The one-sided inequality is easy to use. First of all, it prohibits the possibility that both problems are unbounded; if $y^T b$ is arbitrarily large, then there cannot even be a feasible x or we would contradict $y^T b \leq c^T x$. Similarly, if the primal problem is unbounded—the cost $c^T x$ can go down to $-\infty$—then the dual problem cannot admit a feasible y.

Second, and equally important, we can tell immediately that any vectors which achieve equality, $y^T b = c^T x$, must be optimal. At that point the grocer's price equals the druggist's price, and we recognize an optimal diet and optimal vitamin prices by the fact that the consumer has nothing to choose:

8F If the vectors x and y are feasible, and if $c^T x = y^T b$, then those vectors must be optimal.

Proof   According to 8E, no feasible y can make $y^T b$ larger than $c^T x$. Since our particular y achieves this value, it is optimal. Similarly no feasible x can bring $c^T x$ below the number $y^T b$, and any x that achieves this minimum must be optimal.

We give an example with two foods and two vitamins:

PRIMAL: Minimize $x_1 + 4x_2$ DUAL: Maximize $6y_1 + 7y_2$

subject to $x_1 \geq 0,\quad x_2 \geq 0,$ subject to $y_1 \geq 0,\quad y_2 \geq 0,$

$2x_1 + x_2 \geq 6$ $2y_1 + 5y_2 \leq 1$

$5x_1 + 3x_2 \geq 7.$ $y_1 + 3y_2 \leq 4.$

The choice $x_1 = 3$ and $x_2 = 0$ is feasible, with cost $x_1 + 4x_2 = 3$. In the dual problem $y_1 = \frac{1}{2}$ and $y_2 = 0$ give the same value $6y_1 + 7y_2 = 3$. Therefore these vectors must be optimal.

That seems almost too simple. Nevertheless it is worth a closer look, to find out what actually happens at the moment when $y^T b \leq c^T x$ becomes an equality. It is like calculus, where everybody knows the condition for a maximum or a minimum: The first derivatives are zero. On the other hand, everybody forgets that this condition is completely changed by the presence of constraints. The best example is a straight line sloping upward; its derivative is never zero, calculus is almost helpless, and the maximum is certain to occur at the right-hand end of the interval. That is exactly the situation that we face in linear programming! There are more variables, and an ordinary interval is replaced by a feasible set in several dimensions, but still the maximum will always be found at a corner of the feasible set. In the language of the simplex method, there is an optimal x which is *basic*: It has only m nonzero components.

The real problem in linear programming is to decide which corner it is. For this, we have to admit that calculus is not completely helpless. Far from it, because the device of "Lagrange multipliers" will bring back zero derivatives at the maximum and minimum, and in fact the dual variables y are exactly the Lagrange multipliers for the problem of minimizing $c^T x$. This device is also the key to nonlinear programming, but we do not intend to use it; we will stay with what we know. The conditions for a constrained minimum and maximum will be stated mathematically in Eq. (8), but first I want to express them in economic terms: *The diet x and the vitamin prices y are optimal when*

(i)   The grocer sells zero of any food that is overpriced compared to its vitamin equivalent.

(ii)   The druggist charges zero for any vitamin that is oversupplied in the diet.

In the example, $x_2 = 0$ because the second food is too expensive; its price exceeds the druggist's price, since $y_1 + 3y_2 \leq 4$ is a strict inequality $\frac{1}{2} + 0 < 4$. Similarly, $y_i = 0$ if the ith vitamin is oversupplied; it is a "free good," which means it is worthless. The example required seven units of the second vitamin, but the diet actually supplied $5x_1 + 3x_2 = 15$, so we found $y_2 = 0$. You can see how the duality has become complete; it is only when *both* of these conditions are satisfied that we have an optimal pair.

These conditions are easy to understand in matrix terms. We are comparing the vector Ax to the vector b (remember that feasibility requires $Ax \geq b$) and we look for any components in which equality fails. This corresponds to a vitamin that is oversupplied, so its price is $y_i = 0$. At the same time we compare A^Ty with c, and expect all strict inequalities (expensive foods) to correspond to $x_j = 0$ (omission from the diet). These are the "complementary slackness conditions" of linear programming, and the "Kuhn–Tucker conditions" of nonlinear programming:

8G *Equilibrium Theorem* Suppose the feasible vectors x and y satisfy the following complementary slackness conditions:

$$\text{if} \ (Ax)_i > b_i, \ \text{then} \ y_i = 0, \ \text{and if} \ (A^Ty)_j < c_j \ \text{then} \ x_j = 0. \quad (8)$$

Then x and y are optimal. Conversely, optimal vectors must satisfy (8).

Proof The key equations are

$$y^Tb = y^T(Ax) = (y^TA)x = c^Tx. \quad (9)$$

Normally only the middle equation is certain. In the first equation, we are sure that $y \geq 0$ and $Ax \geq b$, so we are sure of $y^Tb \leq y^T(Ax)$. Furthermore, there is only one way in which equality can hold: *Any time there is a discrepancy $b_i < (Ax)_i$, the factor y_i that multiplies these components must be zero.* Then this discrepancy makes no contribution to the inner products, and equality is saved.

The same is true for the remaining equation: Feasibility gives $x \geq 0$ and $A^Ty \leq c$ (or $y^TA \leq c^T$) and therefore $y^TAx \leq c^Tx$. We get equality only when the second slackness condition is fulfilled: If there is an overpricing $(A^Ty)_j < c_j$, it must be canceled through multiplication by $x_j = 0$. This leaves us with $y^Tb = c^Tx$ in the key equation (9), and it is this equality that guarantees (and is guaranteed by) the optimality of x and y.

So much for the one-sided inequality $y^Tb \leq c^Tx$. It was easy to prove, it gave a quick test for optimal vectors (they turn it into an equality), and now it has given a set of necessary and sufficient slackness conditions. The only thing it has not done is to show that the equality $y^Tb = c^Tx$ is really possible. Until the optimal vectors are actually produced, which cannot be done by a few simple manipulations, the duality theorem is not complete.

To produce them, we return to the simplex method—which has already computed the optimal x. Our problem is to identify at the same time the optimal y, showing that the method stopped in the right place for the dual problem (even though it was constructed to solve the primal). First we recall how it started. The m inequalities $Ax \geq b$ were changed to equations, by introducing the slack variables $z = Ax - b$ and rewriting feasibility as

$$[A \ -I]\begin{bmatrix} x \\ z \end{bmatrix} = b, \qquad \begin{bmatrix} x \\ z \end{bmatrix} \geq 0. \tag{10}$$

Then every stage of the method picked out m columns of the long matrix $[A \ -I]$ to be "basic columns," and used only the corresponding m components of the vector $\begin{bmatrix} x \\ z \end{bmatrix}$; the other n components were set to zero. For convenience we shifted these basic columns to the front, changing $[A \ -I]$ into $[B \ F]$, and moved the corresponding nonzero components of $\begin{bmatrix} x \\ z \end{bmatrix}$ to the top, which gave $\begin{bmatrix} B^{-1}b \\ 0 \end{bmatrix}$ because the free variables are zero. The cost vector c was first extended to $\begin{bmatrix} c \\ 0 \end{bmatrix}$ and then reshuffled into $\begin{bmatrix} c_B \\ c_F \end{bmatrix}$ by the same shifts that took B to the front and $B^{-1}b$ to the top. The stopping condition, which brought the method to an end, was $r \geq 0$—which is the same as $F^T(B^T)^{-1}c_B \leq c_F$.

We know that this condition was finally met, since the number of corners is finite. At that moment the cost function c^Tx was equal to

$$\begin{bmatrix} c_B \\ c_F \end{bmatrix}^T \begin{bmatrix} B^{-1}b \\ 0 \end{bmatrix} = c_B{}^T B^{-1}b, \qquad \text{the minimum cost.} \tag{11}$$

All that remains is to find a vector y in the dual giving this same value for y^Tb; then equality has been achieved. For this we go to the maximum problem and write down the constraints:

$$\begin{bmatrix} A^T \\ -I^T \end{bmatrix} y \leq \begin{bmatrix} c \\ 0 \end{bmatrix}. \tag{12}$$

Notice that the first components give $A^Ty \leq c$, and the last give $-y \leq 0$, or $y \geq 0$. Therefore (12) is the complete test for feasibility. When the simplex method reshuffles the matrix $[A \ -I]$, to put the basic variables first, this simply rearranges the constraints on y. The rows are exchanged, just as in elimination, and the feasibility condition (12) reappears as

$$\begin{bmatrix} B^T \\ F^T \end{bmatrix} y \leq \begin{bmatrix} c_B \\ c_F \end{bmatrix}. \tag{13}$$

In this position, we can recognize the y we want. It is the one that turns the first m rows into equalities: $B^Ty = c_B$, or $y^TB = c_B{}^T$, or $y^T = c_B{}^T B^{-1}$, a price vector which has the two properties that finally complete the whole argument:

(1)   It gives $y^Tb = c_B{}^T B^{-1}b$, which matches the minimum cost c^Tx.

(2)   It is feasible, because the first half of (13) is an equality and the second half is $F^Ty = F^T(B^T)^{-1}c_B \leq c_F$, which was exactly the stopping condition.

Therefore this choice of y is optimal, and the duality theorem has been proved. By locating the critical m by m matrix B, the simplex method has produced y^* as well as x^*.

EXERCISE 8.3.1 What is the dual of the following problem: Minimize $x_1 + x_2$, subject to $x_1 \geq 0$, $x_2 \geq 0$, $2x_1 \geq 4$, $x_1 + 3x_2 \geq 11$? Find the solution to both this problem and its dual, and find their common value.

EXERCISE 8.3.2 What is the dual of the following problem: Maximize y_2, subject to $y_1 \geq 0$, $y_2 \geq 0$, $y_1 + y_2 \leq 3$? Solve both this problem and its dual.

EXERCISE 8.3.3 Suppose A is the identity matrix (so that $m = n$) and the vectors b and c are nonnegative. Explain why $x^* = b$ is optimal in the minimum problem, find y^* in the maximum problem, and verify that the two values are the same.

EXERCISE 8.3.4 Construct a 1 by 1 example in which $Ax \geq b$, $x \geq 0$ is unfeasible, and the dual problem is unbounded.

EXERCISE 8.3.5 Starting with the 2 by 2 matrix $A = \begin{bmatrix} 1 & 0 \\ 0 & -1 \end{bmatrix}$, choose b and c so that both of the feasible sets $Ax \geq b$, $x \geq 0$ and $A^T y \leq c$, $y \geq 0$ are empty.

EXERCISE 8.3.6 If all the entries of A, b, and c are positive, show that both the primal and the dual are automatically feasible.

EXERCISE 8.3.7 Show that the vectors $x = (1, 1, 1, 0)^T$ and $y = (1, 1, 0, 1)^T$ are feasible in the standard dual problems, with

$$A = \begin{bmatrix} 0 & 0 & 1 & 0 \\ 0 & 1 & 0 & 0 \\ 1 & 1 & 1 & 1 \\ 1 & 0 & 0 & 1 \end{bmatrix}, \quad b = \begin{bmatrix} 1 \\ 1 \\ 1 \\ 1 \end{bmatrix}, \quad c = \begin{bmatrix} 1 \\ 1 \\ 1 \\ 3 \end{bmatrix}.$$

Then, after computing $c^T x$ and $y^T b$, explain how you know they are optimal.

EXERCISE 8.3.8 Verify that the vectors in the previous exercise satisfy the complementary slackness conditions (8), and find the one slack inequality in both the primal and the dual.

EXERCISE 8.3.9 Suppose that $A = \begin{bmatrix} 1 & 0 \\ 0 & 1 \end{bmatrix}$, $b = \begin{bmatrix} 1 \\ -1 \end{bmatrix}$, and $c = \begin{bmatrix} 1 \\ 1 \end{bmatrix}$. Find the optimal x and y, and verify the complementary slackness conditions (as well as $y^T b = c^T x$).

EXERCISE 8.3.10 If the primal problem is constrained by equations instead of inequalities—*minimize $c^T x$, subject to $Ax = b$ and $x \geq 0$*—then the requirement $y \geq 0$ is left out of the dual: *Maximize $y^T b$ subject to $A^T y \leq c$.* Show that the one-sided inequality $y^T b \leq c^T x$ still holds. Why was $y \geq 0$ needed in (7) but not here?

The Theory of Inequalities

There is more than one way to study duality. The approach we followed— to prove $y^T b \leq c^T x$, and then use the simplex method to get equality—was convenient because that method had already been established, but overall it was a long proof. Of course it was also a *constructive proof*; x^* and y^* were actually computed. Now we look briefly at a different approach, which leaves behind the mechanics of the simplex algorithm and looks more directly at the geometry. I think the key ideas will be just as clear (in fact, probably clearer) if we omit some of the details.

The best illustration of this approach came in the fundamental theorem of linear algebra. Remember that the problem in Chapter 2 was to solve $Ax = b$, in other words to find b in the column space of A. After elimination, and after the four fundamental subspaces, this solvability question was answered in a completely different way by Exercise 2.5.11:

8H   Either $Ax = b$ has a solution, or else there is a y such that $A^Ty = 0$, $y^Tb \neq 0$.

This is the theorem of the alternative, because it is impossible to find both x and y: If $Ax = b$, then $y^TAx = y^Tb \neq 0$; while the alternative requires $y^TAx = (A^Ty)^Tx = 0$. Both cannot happen. In the language of subspaces, either b is in the column space of A or else it has a nonzero component sticking into the perpendicular subspace, which is the left nullspace of A. That component is the required y.†

For inequalities, we want to find a theorem of exactly the same kind. The right place to start is with the same system $Ax = b$, but with the added constraint $x \geq 0$: When does there exist not just a solution to $Ax = b$, but a *nonnegative solution*? In other words, when is the feasible set nonempty in the problem with equality constraints?

To answer that question we look again at the attainable vectors b. In Chapter 2, when any x was allowed, Ax could be any combination of the columns of A; b was attainable if it was in the column space. Now we allow only non-negative combinations, and the attainable b's no longer fill out a subspace. Instead, they are represented by the cone-shaped region in Fig. 8.5. For an m by n matrix, there would be n columns in m-dimensional space, and the cone becomes a kind of open-ended pyramid. In the figure, there are four columns in two-dimensional space, and A is 2 by 4. If b lies in this cone, there is a nonnegative solution to $Ax = b$; otherwise there is not.

Our problem is to discover the alternative: *What happens if b lies outside the cone?* That possibility is illustrated in Fig. 8.6, and you can interpret the geometry at a glance. There is a "separating hyperplane," which goes through the origin and has the vector b on one side and the whole cone on the other side. (The prefix *hyper* is only to emphasize that the number of dimensions may be large; the plane consists, as always, of all vectors perpendicular to a fixed vector y.) The inner product between y and b is negative since they make an angle

---

† You see that this proof is not constructive! We only know that the component must be there, or b would have been in the column space.

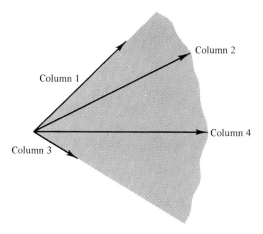

Fig. 8.5. The cone of nonnegative combinations of the columns.

greater than 90°, whereas the inner product between y and every column of A is positive. In matrix terms this means that $y^\mathrm{T}b < 0$ and $A^\mathrm{T}y \geq 0$, which represents the alternative we are looking for:

8I Either $Ax = b$ has a nonnegative solution, or else there is a y such that $A^\mathrm{T}y \geq 0$, $y^\mathrm{T}b < 0$.

This is the theorem of the separating hyperplane. It is fundamental to mathematical economics, and one reference for its proof is Gale's beautiful book on the theory of linear economic models.

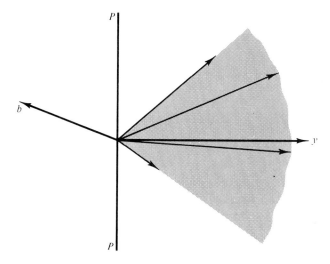

Fig. 8.6. b outside the cone with a separating hyperplane P.

EXAMPLE If $A = I$, then its cone is the positive quadrant. Every b in that quadrant is a nonnegative combination of the columns:

$$\text{if} \quad b = \begin{bmatrix} 2 \\ 3 \end{bmatrix}, \quad \text{then} \quad b = 2\begin{bmatrix} 1 \\ 0 \end{bmatrix} + 3\begin{bmatrix} 0 \\ 1 \end{bmatrix}.$$

For every b outside the quadrant, the second alternative must hold:

$$\text{if} \quad b = \begin{bmatrix} 2 \\ -3 \end{bmatrix}, \quad \text{then} \quad y = \begin{bmatrix} 0 \\ 1 \end{bmatrix} \quad \text{gives} \quad A^T y \geq 0 \quad \text{but} \quad y^T b = -3.$$

This theorem leads directly to a whole sequence of similar alternatives (see Gale). In fact you almost come to believe that whenever two alternatives are mutually exclusive, one or the other must be true. For example, it is impossible for a subspace S and its orthogonal complement $S^\perp$ both to contain positive vectors: Their inner product would be positive, whereas orthogonal vectors have inner product zero. On the other hand, it is not quite certain that either S or $S^\perp$ has to contain a positive vector: S might be the x axis and $S^\perp$ the y axis, in which case they contain only the "semipositive" vectors $\begin{bmatrix} 1 \\ 0 \end{bmatrix}$ and $\begin{bmatrix} 0 \\ 1 \end{bmatrix}$. What is remarkable is that this slightly weaker alternative does work: Either S contains a positive vector x, or $S^\perp$ contains a semipositive vector y. When S and $S^\perp$ are perpendicular lines in the plane, it is easy to see that one or the other must enter the first quadrant; but I do not see it very clearly in higher dimensions.

For linear programming, the most important pair of alternatives is when the constraints are inequalities rather than equations:

8J Either $Ax \geq b$ has a nonnegative solution, or else there is a y such that $A^T y \geq 0$, $y^T b < 0$, $y \leq 0$.

8J follows easily from 8I, using the slack variables $z = Ax - b$ to change the inequality into an equation:

$$[A \quad -I]\begin{bmatrix} x \\ z \end{bmatrix} = b.$$

If this has no nonnegative solution, then by 8I there must be a y such that

$$\begin{bmatrix} A^T \\ -I^T \end{bmatrix} y \geq \begin{bmatrix} 0 \\ 0 \end{bmatrix}, \quad y^T b < 0.$$

This is just the other alternative in 8J, and it is this result that leads to a "nonconstructive proof" of the duality theorem. But we promised to stick to the geometry and omit the algebraic details, so we keep that promise.

EXERCISE 8.3.11 If $A = \begin{bmatrix} 1 & 1 \\ 0 & 1 \end{bmatrix}$, describe the cone produced by its columns. If b lies inside that cone, say $b = (3, 2)^T$, what is the only feasible vector x? If b lies outside, say $b = (0, 1)^T$, what vector y will satisfy the alternative?

EXERCISE 8.3.12   In three dimensions, can you find a set of six vectors whose cone of nonnegative combinations fills the whole space?

EXERCISE 8.3.13   Use 8H to show that there is no solution (because the alternative holds) to

$$\begin{bmatrix} 2 & 2 \\ 4 & 4 \end{bmatrix} x = \begin{bmatrix} 1 \\ 1 \end{bmatrix}.$$

EXERCISE 8.3.14   Use 8I to show that there is no nonnegative solution (because the alternative holds) to

$$\begin{bmatrix} 1 & 3 & -5 \\ 1 & -4 & -7 \end{bmatrix} x = \begin{bmatrix} 2 \\ 3 \end{bmatrix}.$$

EXERCISE 8.3.15   Show that the alternatives in 8J cannot both hold.

GAME THEORY AND THE MINIMAX THEOREM ■ 8.4

The best way to explain a matrix game is to give an example. It has two players, and the rules are the same for every turn:

Player X holds up either one hand or two, and independently, so does player Y. If they make the same decision, Y wins $10. If they make opposite decisions, then X is the winner—$10 if he put up one hand, and $20 if he put up two. The net payoff to X is easy to record in the matrix

$$A = \begin{bmatrix} -10 & +10 \\ +20 & -10 \end{bmatrix} \quad \begin{array}{l} \text{one hand by X} \\ \text{two hands by X.} \end{array}$$

$$\begin{array}{cc} \text{one hand} & \text{two hands} \\ \text{by Y} & \text{by Y} \end{array}$$

If you think for a moment, you get a rough idea of the best strategy. It is obvious that X will not do the same thing every time, or Y would copy him and win everything. Similarly Y cannot stick to a single strategy, or X will do the opposite. Both players must use a *mixed strategy*, and furthermore the choice at every turn must be absolutely independent of the previous turns; otherwise, if there is some historical pattern for the opponent to notice, he can take advantage of it. This leaves the two players with the following calculation: X can decide that he will put up one hand with frequency x_1 (then he puts up both hands with frequency $x_2 = 1 - x_1$, and at every turn this decision is random) and Y can pick *his* probabilities y_1 and $y_2 = 1 - y_1$. We have seen that none of these probabilities should be 0 or 1; otherwise the opponent adjusts and wins. At the same time, it is not clear that they should all equal $\frac{1}{2}$, since Y would be losing $20 too often. (He would lose $20 a quarter of the time, $10 another quarter of the time, and win $10 half the time—an average loss of $2.50 per turn.) But the more Y moves toward a pure two-hand strategy, the more X will move toward one hand.

The fundamental problem is *to find an equilibrium*. Does there exist a mixed strategy y_1 and y_2 that, if used consistently by Y, offers no special advantage to X? And similarly, can X choose probabilities x_1 and x_2 that present Y with no reason to move his own strategy? At such an equilibrium, if it exists, the average payoff to X will have reached a *saddle point*: It is a maximum as far as X is concerned, and a minimum as far as Y is concerned. To find such a saddle point is to "solve" the game.

Suppose we begin from the point of view of X, who chooses frequencies x_1 and $1 - x_1$. When Y puts up one hand, X will lose $10 with frequency x_1 and win $20 with frequency $1 - x_1$; that is an average win of $-10x_1 + 20(1 - x_1)$. When Y puts up two hands, X will win with frequency x_1 (when he puts up one hand) and lose with frequency $1 - x_1$—an average win of $10x_1 - 10(1 - x_1)$. If Y has a mixed strategy, and X is to offer him no reason to change, these two averages should be equal:

$$-10x_1 + 20(1 - x_1) = 10x_1 - 10(1 - x_1).$$

When $x_1 = \frac{3}{5}$, both these payoffs are equal to $2. That is the average win for X, and it is a "maximin"—he has maximized his minimum expectation.

From the standpoint of Y, the calculation is different but similar. By adopting the probabilities y_1 and $1 - y_1$, he must expect to pay an average of $-10y_1 + 10(1 - y_1)$ when X puts up one hand, and $20y_1 - 10(1 - y_1)$ when X puts up two. The optimum occurs when X has no special inducement to go one way or the other, and then

$$-10y_1 + 10(1 - y_1) = 20y_1 - 10(1 - y_1).$$

Now we come to the essential point: *This optimal strategy for Y*, which has $y_1 = \frac{2}{5}$ and $y_2 = \frac{3}{5}$, *also produces an average payoff of* $-10 \cdot \frac{2}{5} + 10 \cdot \frac{3}{5} = \$2$. Such a payoff is not only the maximin that was optimal for X, but also the minimax that must be accepted by Y. Y has minimized his maximum loss, and it has led him to the same equilibrium ($2) for the value of the game.

This saddle point is remarkable, because it means that X plays his second strategy only $\frac{2}{5}$ of the time, even though it is this strategy that gives him a chance at $20. At the same time Y has been forced to adopt a losing strategy— he would like to match X, but instead he uses the opposite probabilities $\frac{2}{5}$ and $\frac{3}{5}$. You can check that X wins $10 with frequency $\frac{3}{5} \cdot \frac{3}{5} = \frac{9}{25}$, he wins $20 with frequency $\frac{2}{5} \cdot \frac{2}{5} = \frac{4}{25}$, and he loses $10 with the remaining frequency $\frac{12}{25}$. As expected, that gives him an average gain of $2.

EXERCISE 8.4.1 How will the optimal strategies be affected if the $20 is increased to $70, and what is the value (the average win for X) of this new game?

EXERCISE 8.4.2 Suppose Y is allowed a third strategy of putting up three hands and just paying out $1, regardless of the move by X. Explain why his optimal strategy is to do this every time; his optimal probabilities become $y_1 = 0$, $y_2 = 0$, and $y_3 = 1$, and the value of the game is \$1. X still cannot choose an arbitrary mixture and guarantee \$1 against any strategy for Y.

The Minimax Theorem

The most general "matrix game" is exactly like our simple example, with one important difference: X has m possible moves to choose from, and Y has n. Therefore the payoff matrix A, which stays the same for every repetition of the game, is allowed to have m rows and n columns. The entry a_{ij} represents the payment received by X when he chooses his ith strategy and Y chooses his jth; a negative entry simply means a negative payment, which is a win for Y. The result is still a *two-person zero-sum game*; whatever is lost by one player is won by the other. But the existence of a saddle point equilibrium is by no means obvious.

As in the example, player X is free to choose any mixed strategy $x = (x_1, \ldots, x_m)$. This mixture is always a probability vector; the x_i are nonnegative, and they add up to 1. These components give the frequencies for the m different pure strategies, and at every repetition of the game X will decide between them by some random device—the device being constructed to produce strategy i with frequency x_i. At the same time player Y is faced with a similar decision: He chooses a vector $y = (y_1, \ldots, y_n)$, also with $y_i \geq 0$ and $\sum y_i = 1$, which gives the frequencies in his own mixed strategy.

We cannot predict the result of a single play of the game; it is random. On the average, however, the combination of strategy i for X and strategy j for Y will turn up with probability $x_i y_j$—the product of the two separate probabilities. When it does come up, the payoff is a_{ij}. Therefore the expected payoff to X from this particular combination is $a_{ij} x_i y_j$, and *the total expected payoff from each play of the game is* $\sum \sum a_{ij} x_i y_j$. Again we emphasize that any or all of the entries a_{ij} may be negative; the rules are the same for X and Y, and it is the entries a_{ij} that decide who wins the game.

The expected payoff can be written more easily in matrix notation: The double sum $\sum \sum a_{ij} x_i y_j$ is just xAy, because of the matrix multiplication

$$xAy = \begin{bmatrix} x_1 & \cdots & x_m \end{bmatrix} \begin{bmatrix} a_{11} & a_{12} & \cdots & a_{1n} \\ \vdots & \vdots & & \vdots \\ a_{m1} & a_{m2} & \cdots & a_{mn} \end{bmatrix} \begin{bmatrix} y_1 \\ y_2 \\ \vdots \\ y_n \end{bmatrix} = a_{11} x_1 y_1 + \cdots + a_{mn} x_m y_n.$$

It is xAy that player X wants to maximize and player Y wants to minimize.†

EXAMPLE 2   Suppose A is the n by n identity matrix, $A = I$. Then the expected payoff becomes $xIy = x_1 y_1 + \cdots + x_n y_n$, and the idea of the game is not hard to explain: X is hoping to hit on the same choice as Y, in which case he receives the payoff $a_{ii} = \$1$. At the same time, Y is trying to evade X. If

† This time x is a row vector!

X chooses any of his strategies more often than any other, then Y can escape more often; therefore the optimal mixture is $x^* = (1/n,\ 1/n,\ \ldots,\ 1/n)$. Similarly Y cannot overemphasize any strategy or X will discover him, and therefore his optimal choice also has equal probabilities adding up to 1, $y^* = (1/n,\ 1/n,\ \ldots,\ 1/n)^T$. The probability that both will choose strategy i is $(1/n)^2$, and the sum over all such combinations is the expected payoff to X. The total value of the game is n times $(1/n)^2$, or $1/n$, as is confirmed by

$$x^*Ay^* = [1/n \quad \cdots \quad 1/n]\begin{bmatrix} 1 & & \\ & \ddots & \\ & & 1 \end{bmatrix}\begin{bmatrix} 1/n \\ \vdots \\ 1/n \end{bmatrix} = \left(\frac{1}{n}\right)^2 + \cdots + \left(\frac{1}{n}\right)^2 = \frac{1}{n}.$$

As n increases, Y has a better chance to escape.

Notice that the symmetric matrix $A = I$ did not guarantee that the game was fair. In fact, the true situation is exactly the opposite: It is a *skew-symmetric matrix*, $A^T = -A$, which means a completely fair game. Such a matrix faces the two players with identical decisions, since a choice of strategy i by X and j by Y wins a_{ij} for X, and a choice of i by Y and j by X wins the same amount a_{ij} for Y (because $a_{ji} = -a_{ij}$). The optimal strategies x^* and y^* must be the same, and the expected payoff must be $x^*Ay^* = 0$. The value of the game, when $A^T = -A$, is zero.

EXAMPLE 3

$$A = \begin{bmatrix} 0 & 1 & 1 \\ -1 & 0 & 1 \\ -1 & -1 & 0 \end{bmatrix}.$$

In words, X and Y both choose a number between 1 and 3, and the one with the smaller number wins $1. (If X chooses 2 and Y chooses 3, the payoff is $a_{23} = \$1$; if they choose the same number, we are on the main diagonal and nobody wins.) Evidently neither player can choose a strategy involving 2 or 3, or the other can get underneath him. Therefore the pure strategies $x^* = y^* = (1, 0, 0)$ are optimal—both players choose 1 every time—and the value is $x^*Ay^* = a_{11} = 0$.

It is worth remarking that the matrix that leaves all decisions unchanged is not the identity matrix, but the matrix E that has *every* entry e_{ij} equal to 1. Adding a multiple of E to the payoff matrix, $A \to A + \alpha E$, simply means that X wins an additional amount α at every turn. The value of the game is increased by α, but there is no reason to change x^* and y^*.

Now we return to the general theory, putting ourselves first in the place of X. Suppose he chooses the mixed strategy $x = (x_1, \ldots, x_m)$. Then Y will eventually recognize that strategy and choose his own strategy so as to minimize the payment xAy: X will receive $\min_y xAy$. An intelligent player X will select a vector x^* (it may not be unique) that *maximizes this minimum.* By this choice, X guarantees that he will win at least the amount

$$\min_{y} x^*Ay = \max_{x} \min_{y} xAy. \qquad (14)$$

He cannot expect to win more.

Player Y does the opposite. For any of his own mixed strategies y, he must expect X to discover the vector that will maximize xAy.† Therefore Y will choose the mixture y^* that minimizes this maximum and guarantees that he will lose no more than

$$\max_{x} xAy^* = \min_{y} \max_{x} xAy. \qquad (15)$$

Y cannot expect to do better.

I hope you see what the key result will be, if it is true. We want the amount (14) that X is guaranteed to win to coincide with the amount (15) that Y must be satisfied to lose. Then the mixtures x^* and y^* will yield a saddle point equilibrium, and the game will be solved: X can only lose by moving from x^* and Y can only lose by moving from y^*. The existence of this saddle point was proved by von Neumann, and it is known as the *minimax theorem*:

8K For any m by n matrix A, the minimax over all mixed strategies equals the maximin:

$$\max_{x} \min_{y} xAy = \min_{y} \max_{x} xAy. \qquad (16)$$

This quantity is the value of the game. If the maximum on the left is attained at x^*, and the minimum on the right is attained at y^*, then those strategies are optimal and they yield a saddle point from which nobody wants to move:

$$xAy^* \leq x^*Ay^* \leq x^*Ay \qquad \text{for all } x \text{ and } y. \qquad (17)$$

At this saddle point, x^* is at least as good as any other x (since $xAy^* \leq x^*Ay^*$). Similarly, the second player could only pay more by leaving y^*.

Just as in the duality theory, we can begin with a one-sided inequality in the direction maximin $\leq$ minimax. It is no more than a combination of the definition (14) of x^* and the definition (15) of y^*:

$$\max_{x} \min_{y} xAy = \min_{y} x^*Ay \leq x^*Ay^* \leq \max_{x} xAy^* = \min_{y} \max_{x} xAy. \qquad (18)$$

This only says that if X can guarantee to win at least α, and Y can guarantee to lose no more than β, then necessarily $\alpha \leq \beta$. The achievement of von Neumann was to prove that $\alpha = \beta$. That is the minimax theorem; it means

---

† This may not be x^*. If Y adopts a foolish strategy, then X could get more than he is guaranteed by (14). Game theory has to assume that the players are smart.

that equality must hold throughout (18), and the saddle point property (17) is deduced from it in Exercise 8.4.6.

For us, the most striking thing about the proof is that *it uses exactly the same mathematics as the theory of linear programming.* Intuitively, that is almost obvious; X and Y are playing "dual" roles, and they are both choosing strategies from the "feasible set" of probability vectors: $x_i \geq 0$, $\sum x_i = 1$, $y_i \geq 0$, $\sum y_i = 1$. What is amazing is that even von Neumann did not immediately recognize the two theories as the same. (He proved the minimax theorem in 1928, linear programming began before 1947, and Gale, Kuhn, and Tucker published the first proof of duality in 1951—based however on von Neumann's notes!) Their proof actually appeared in the same volume where Dantzig demonstrated the equivalence of linear programs and matrix games, so we are reversing the true historical sequence by deducing the minimax theorem from duality.

Briefly, the minimax theorem can be proved as follows. First, we make sure that the payoff matrix A is positive by adding the same large number α to all its entries. (This change to $A + \alpha E$, which is our new A, cannot affect the optimal strategies; every payoff goes up by α, and so do the minimax and maximin.) Then if $c = (1, 1, \ldots, 1)^{\mathsf{T}}$ is the column vector of m 1's, and $b = (1, \ldots, 1)$ is the row vector of n 1's, consider the dual linear programs

   (P)   minimize xc (D)   maximize by
          subject to $xA \geq b$, $x \geq 0$;   subject to $Ay \leq c$, $y \geq 0$.

Both are feasible: Take $y = 0$ and take x very large. By the duality theorem for linear programming, there exist optimal vectors x^* and y^* such that $x^*c = by^*$. Because of the ones in b and c, this means that $\sum x_i^* = \sum y_i^*$. If these sums equal θ, then division by θ changes the sums to one—and *the resulting mixed strategies x^*/θ and y^*/θ are optimal in game theory.* For any other mixed strategies x and y,

$$Ay^* \leq c \text{ implies } xAy^* \leq xc = 1, \quad \text{and} \quad x^*A \geq b \text{ implies } x^*Ay \geq by = 1.$$

The main point is that $xAy^* \leq 1 \leq x^*Ay$. Dividing by θ, this says that player X cannot win more than $1/\theta$ against the strategy y^*/θ, and player Y cannot lose less than $1/\theta$ against the strategy x^*/θ. Therefore, those are the strategies to be chosen, and maximin = minimax = $1/\theta$.

This completes the theory, but it leaves unanswered the most natural question of all: In the ordinary sense of the word, which games are actually equivalent to "matrix games"? Do chess and bridge and poker fit into the framework of von Neumann's theory?

It seems to me that chess does not fit very well, for two reasons. First, a strategy for the white pieces does not consist just of the opening move; it must include a decision on how to respond to the first reply of black, and then how to respond to his second reply, and so on to the end of the game. There are so many alternatives at every step that X has billions of pure strategies, and the same is true for his opponent. Therefore m and n are impossibly large. Furthermore, I do not see much of a role for chance. If white can find a winning strategy or if black can find a drawing strategy—neither has ever been found—

then that would effectively end the game of chess. Of course it could continue to be played, like tic-tac-toe, but the excitement would tend to go away.

Unlike chess, bridge does contain some elements of deception, for example, deciding what to do in a finesse. Therefore it counts as a matrix game. Unfortunately, m and n are again fantastically big, but perhaps one separate element of the game could be analyzed for an optimal strategy. The same is true in baseball, where the pitcher and batter try to outguess each other on the choice of pitch. (Or the catcher tries to guess when the runner will steal, by calling a pitchout; he cannot do it every time without walking the batter, so there must be an optimal frequency—depending on the base runner and on the situation.) Again a small part of the game could be isolated and analyzed.

On the other hand, blackjack is not a matrix game (at least in a casino) because the house follows fixed rules. When my friend Ed Thorp found and published a winning strategy—forcing a change in the rules at Las Vegas—his advice depended entirely on keeping track of the cards already seen. There was no element of chance, and therefore no mixed strategy x^*.

The perfect example of an ordinary game that is also a matrix game is *poker*. Bluffing is an essential part of good poker, and to be effective it has to be unpredictable. That means that the decision to bluff should be made completely at random, if you accept the fundamental assumption of game theory that your opponent is intelligent. (If he finds a pattern, he wins.) The probabilities for and against bluffing will depend on the cards that are known, and on the size of the bets—both those that have already been made and those that remain to be made. In fact, the number of alternatives again makes it impractical to find an absolutely optimal strategy x^*. Nevertheless a good poker player must come pretty close to x^*, and we can compute it exactly if we accept the following enormous simplification of the game:

Player X is dealt a jack or a king, with equal probability, whereas Y always gets a queen. After looking at his hand, X can either fold and concede his ante of $1, or he can bet an additional $2. If X bets, then Y can either fold and concede his own ante of $1, or decide to match the $2 and see whether X is bluffing. In this case the higher card wins the $3 that was bet by the opponent.

Even in this simple game, the possible "pure strategies" are not at all obvious, and it is instructive to write them down. Y has only two possibilities, since he can only react to X:

(1)   If X bets, Y folds.
(2)   If X bets, Y sees him and tries for $3.

X has four strategies, some reasonable and some foolish:

(1)   He bets the extra $2 on a king and folds on a jack.
(2)   He bets in either case (bluffing).
(3)   He folds in either case, and concedes $1.
(4)   He folds on a king and bets on a jack.

The payoff matrix needs a little patience to compute:

$a_{11} = 0$, since X loses \$1 half the time on a jack and wins on a king (Y folds).

$a_{12} = 1$, since X loses \$1 half the time and wins \$3 half the time (Y tries to beat him, even though X has a king).

$a_{21} = 1$, since X bets and Y folds (the bluff succeeds).

$a_{22} = 0$, since X wins \$3 with the king and loses \$3 with the jack (the bluff fails).

The third strategy always loses \$1, and the fourth is also unsuccessful:

$$A = \begin{bmatrix} 0 & 1 \\ 1 & 0 \\ -1 & -1 \\ 0 & -2 \end{bmatrix}.$$

The optimal strategy in this game is for X to bluff half the time, $x^* = (\frac{1}{2}, \frac{1}{2}, 0, 0)$, and the underdog Y must choose $y^* = (\frac{1}{2}, \frac{1}{2})^\mathrm{T}$. The value of the game is fifty cents.

That is a strange way to end this book, by teaching you how to play a watered down version of poker, but I guess even poker has its place within linear algebra and its applications.

EXERCISE 8.4.3   With payoff matrix $A = \begin{bmatrix} 1 & 2 \\ 3 & 4 \end{bmatrix}$, explain the calculation by X of his maximin and by Y of his minimax. What strategies x^* and y^* are optimal?

EXERCISE 8.4.4   If a_{ij} is the smallest entry in its row and the largest in its column, why will X always choose the pure strategy i and Y always choose j? Show that the previous exercise had such an entry, and then construct an A without one.

EXERCISE 8.4.5   Find x^*, y^*, and the value v for

$$A = \begin{bmatrix} 1 & 0 & 0 \\ 0 & 2 & 0 \\ 0 & 0 & 3 \end{bmatrix}.$$

EXERCISE 8.4.6   Explain each of the inequalities in (18). Then, once the minimax theorem has turned them into equalities, derive—again in words—the saddle point equations (17).

EXERCISE 8.4.7   Show that $x^* = (\frac{1}{2}, \frac{1}{2}, 0, 0)$ and $y^* = (\frac{1}{2}, \frac{1}{2})^\mathrm{T}$ are optimal strategies in poker, by computing x^*Ay and xAy^* and verifying the conditions (17) for a saddle point.

EXERCISE 8.4.8   Has it been proved that there cannot exist a chess strategy that always wins for black? I believe this is known only when the players are given two moves at a time; if black had a winning strategy, white could move a knight out and back and then follow that strategy, leading to the impossible conclusion that both would win.

LINEAR TRANSFORMATIONS, MATRICES, AND CHANGE OF BASIS

Suppose A is an m by n matrix, and x is an n-dimensional vector. The vector x is in $\mathbf{R}^n$, and the product Ax is in $\mathbf{R}^m$. This map from x to Ax describes a transformation of the space $\mathbf{R}^n$ into the space $\mathbf{R}^m$, and the rules of matrix multiplication lead immediately to its most important property: For any vectors x and y, and any scalars b and c,

$$A(bx + cy) = b(Ax) + c(Ay). \tag{1}$$

A transformation with property (1) is called *linear*, and thus every matrix A leads to a linear transformation.

Our goal in this appendix is to study other examples of linear transformations, and to try to understand all such transformations at once. This goal is the foundation of an approach to linear algebra, starting with a property like (1) and developing its consequences, which is much more abstract than the one taken in this book; we preferred to begin directly with matrices. But it is time now to look beyond simultaneous linear systems $Ax = b$, and their coefficient matrices, for other instances of linear transformations.

First, we need to remember that there are vector spaces other than $\mathbf{R}^n$ and $\mathbf{C}^n$. All a vector space requires is some reasonable way to form combinations $bx + cy$, and the assurance that every such combination lies within the space.

EXAMPLE 1 Let P_n be the space of all polynomials of degree n, $p = a_0 + a_1 t + \cdots + a_n t^n$. These polynomials are the "vectors," and the sum of two such vectors is certainly in P_n. The particular polynomials $p_0 = 1$, $p_1 = t, \ldots, p_n = t^n$ provide a basis for this space, and its dimension is $n + 1$.

EXAMPLE 2   Given a linear differential equation of order n with constant coefficients

$$\frac{d^n u}{dt^n} + c_1 \frac{d^{n-1}u}{dt^{n-1}} + \cdots + c_n u = 0, \tag{2}$$

the solutions $u(x)$ form a vector space S_n. For any two solutions u and v, the law of superposition assures that $bu + cv$ is also a solution; in other words, the solutions can be added together and multiplied by scalars. The simplest solutions are the pure exponentials $u = e^{\lambda t}$—or possibly also $te^{\lambda t}$ if λ happens to be a double root of the critical equation, $t^2 e^{\lambda t}$ if it is a triple root, and so on. There are n solutions of this special kind, and they form a basis for the space of all solutions.†

EXAMPLE 3   Let H be the Hilbert space described on p. 126. It contains all vectors $v = (v_1, v_2, v_3, \ldots)$ that have infinitely many components but a finite length:

$$\| v \|^2 = v_1^2 + v_2^2 + v_3^2 + \cdots < \infty.$$

There is certainly no finite basis—H is infinite dimensional—because we need infinitely many pieces of information to specify v.

With this list of vector spaces, it will be easier to illustrate the idea of a linear transformation. Remember that the essential property is linearity; the transformation A may map one vector space into any other (or into itself), but it must satisfy $A(bx + cy) = b(Ax) + c(Ay)$.

EXAMPLE 4   On the space of polynomials of degree n, let A be the operation of *differentiation*: $A = d/dt$. Thus for any polynomial p,

$$Ap = A(a_0 + a_1 t + \cdots + a_n t^n) = a_1 + 2a_2 t + \cdots + na_n t^{n-1}. \tag{3}$$

We may think of A as a linear transformation of P_n into the smaller space P_{n-1}, or of P_n into P_n. In the first case the transformation is *onto* P_{n-1}, and each polynomial in P_{n-1} can be produced by differentiating a suitable polynomial in P_n. This is no longer true in the second case, since there are polynomials in P_n (such as t^n) that never appear on the right side of (3): *The "differentiation matrix," once we find it, will not be invertible.*

EXAMPLE 5   Acting this time on the polynomial space P_{n-1}, let A be the operation of *integration*:

$$A(a_0 + a_1 t + \cdots + a_{n-1} t^{n-1}) = a_0 t + a_1 \frac{t^2}{2} + \cdots + a_{n-1} \frac{t^n}{n}. \tag{4}$$

This defines a transformation from P_{n-1} into the larger space P_n, and once

---

† When the coefficients c_i depend on t, S_n is still a vector space of dimension n. But a basis, or even one member of the space, is much harder to find.

more it is linear:

$$\int_0^t [bp(t) + cq(t)]\, dt = b \int_0^t p(t)\, dt + c \int_0^t q(t)\, dt.$$

Does integration produce all polynomials in P_n, so that the map is onto? No, but it is one-to-one; different polynomials have different integrals.

EXAMPLE 6 S_n is the space of all solutions to the differential equation in (2). Suppose x is any vector in $\mathbf{R}^n$, and its components are the initial values of the solution u: At $t = 0$,

$$u(0) = x_1, \qquad \frac{du}{dt}(0) = x_2, \ldots, \qquad \frac{d^{n-1}u}{dt^{n-1}}(0) = x_n.$$

By specifying these initial values, we have picked out one particular solution of the differential equation. This correspondence between x in $\mathbf{R}^n$ and u in S_n is a linear transformation between the two spaces; the solution to a linear equation depends linearly on its initial values. This transformation is both onto and one-to-one: Every solution can be produced from the right initial values, and different initial values must produce different solutions. Therefore the n by n matrix that describes the transformation will be invertible.

EXAMPLE 7 Let A be the *left shift* on the Hilbert space H:

$$Av = A(v_1, v_2, v_3, \ldots) = (v_2, v_3, v_4, \ldots).$$

This shift is linear and onto; we can produce any vector as output by choosing a suitable input. Since we are free to choose any v_1 whatsoever, the operator is not one-to-one; v_1 does not appear in the output. It is shifted completely away, and there is a one-dimensional nullspace: $A(v_1, 0, 0, \ldots) = (0, 0, \ldots)$.

In such examples, the property of linearity is not very interesting to verify; if it is there, it is practically impossible to miss. Nevertheless, it is the most useful property a transformation can have,† and it leads us naturally back to matrices. In fact, our main purpose in this appendix is *to show that every linear transformation can be represented by a matrix*. This representing matrix will depend on the bases we choose for the vector spaces, and it is constructed according to the following rule:

Suppose the space V has a basis $v_1, \ldots, v_n$ and the space W has a basis $w_1, \ldots, w_m$. Then every linear transformation A from V to W is represented by an m by n matrix $[A]$. Its entries a_{ij} are determined by applying A to each v_j, and expressing the result as a combination of the w's:

$$Av_j = \sum_{i=1}^{m} a_{ij} w_i, \qquad j = 1, \ldots, n. \tag{5}$$

† Invertibility is perhaps in second place.

In other words, the jth column of the matrix is determined by observing what the transformation does to the basis vector v_j.

EXAMPLE 8 Suppose A is the operation of integration, with the special polynomials $1, t, t^2, \ldots$ as a basis: v_j and w_j are given by t^{j-1}. To compute the matrix $[A]$, we look to see how A acts on the v_j:

$$Av_j = \int_0^t t^{j-1}\,dt = \frac{t^j}{j} = \frac{1}{j}\,w_{j+1}.$$

Down the jth column of the matrix, the only nonzero entry a_{ij} is $1/j$. If V has dimension $n = 4$, W will have dimension $m = 5$, and the 5 by 4 matrix is

$$[A] = \begin{bmatrix} 0 & 0 & 0 & 0 \\ 1 & 0 & 0 & 0 \\ 0 & \frac{1}{2} & 0 & 0 \\ 0 & 0 & \frac{1}{3} & 0 \\ 0 & 0 & 0 & \frac{1}{4} \end{bmatrix}.$$

Suppose we want to integrate $v(t) = 2t + 8t^3$. Then we express v as

$$2t + 8t^3 = 0v_1 + 2v_2 + 0v_3 + 8v_4,$$

and multiply these coefficients by A:

$$[A]\begin{bmatrix} 0 \\ 2 \\ 0 \\ 8 \end{bmatrix} = \begin{bmatrix} 0 \\ 0 \\ 1 \\ 0 \\ 2 \end{bmatrix}, \quad \text{or} \quad \int 2t + 8t^3 = w_3 + 2w_5 = t^2 + 2t^4.$$

EXAMPLE 9 Suppose A is the operation of differentiation, taking P_n into the smaller space P_{n-1}. With the same choice of basis, the action of A on v_j is just

$$Av_j = \frac{d}{dt} t^{j-1} = (j-1)t^{j-2} = (j-1)\,w_{j-1}.$$

Thus the jth column of $[A]$ again contains only a single entry, equal to $j - 1$ and appearing in row $i = j - 1$. The 4 by 5 case looks like

$$[A] = \begin{bmatrix} 0 & 1 & 0 & 0 & 0 \\ 0 & 0 & 2 & 0 & 0 \\ 0 & 0 & 0 & 3 & 0 \\ 0 & 0 & 0 & 0 & 4 \end{bmatrix}.$$

Remark   We think of differentiation and integration as inverse operations. Or at least integration followed by differentiation leads back to the original function; the second matrix is a *left-inverse* of the first. The product of the two matrices displayed above is the 4 by 4 identity matrix. But a rectangular matrix cannot have a two-sided inverse, and in fact the product in the other order is not the identity:

$$[A]_{\text{integ}}[A]_{\text{diff}} = \begin{bmatrix} 0 & 0 & 0 & 0 & 0 \\ 0 & 1 & 0 & 0 & 0 \\ 0 & 0 & 1 & 0 & 0 \\ 0 & 0 & 0 & 1 & 0 \\ 0 & 0 & 0 & 0 & 1 \end{bmatrix}.$$

The zero entry in the upper left corner corresponds to the fact that starting with a constant function, differentiation followed by integration produces zero.

EXAMPLE 10   The matrix that represents the left shift on Hilbert space is infinitely large, but otherwise it is nothing special:

$$[A] = \begin{bmatrix} 0 & 1 & 0 & 0 & \cdot \\ 0 & 0 & 1 & 0 & \cdot \\ 0 & 0 & 0 & 1 & \cdot \\ 0 & 0 & 0 & 0 & \cdot \\ \cdot & \cdot & \cdot & \cdot & \cdot \end{bmatrix}, \quad \text{and} \quad [A] \begin{bmatrix} v_1 \\ v_2 \\ v_3 \\ v_4 \\ \cdot \end{bmatrix} = \begin{bmatrix} v_2 \\ v_3 \\ v_4 \\ v_5 \\ \cdot \end{bmatrix}.$$

It remains to justify our general rule (5) for constructing the matrix, and to show that $[A]$ contains complete information about the linear transformation A that it represents. In other words, knowing the bases chosen for V and W, and knowing $[A]$, we should be able to reconstruct the effect of A on an arbitrary vector $v = \sum_1^n x_j v_j$ in the space V.

If the vector x gives the coefficients of v when it is expanded in terms of the basis $v_1, \ldots, v_n$, then the vector $y = [A]x$ gives the coefficients of Av when it is expanded in terms of the basis $w_1, \ldots, w_m$. Therefore the effect of A on any v is reconstructed by matrix multiplication:

$$Av = \sum_{i=1}^m y_i w_i = \sum_{i,j} a_{ij} x_j w_i. \tag{6}$$

In the proof, we must depend on the linearity of A: For any $v = \sum x_j v_j$,

$$Av = A\left(\sum_1^n x_j v_j\right) = \sum_1^n x_j Av_j.$$

Substituting the expression (5) for Av_j, we have verified formula (6). Given the action of A on each member of a basis, linearity is sufficient to determine the action of A on every other vector in the space.

One more point about the relation between linear transformations and matrices. If A and B are linear transformations, say from V to W and from W to Z, then their product BA is a transformation from V to Z; it starts with v in V, goes to Av in W, and finishes with $BAv = B(Av)$ in Z. This last equation is exactly like the rule (7) on p. 13, which was the foundation for matrix multiplication. Therefore it is completely natural that matrix multiplication correctly reflects this product of linear transformations:

If the matrices $[A]$ and $[B]$ represent the linear transformations A and B with respect to bases $\{v_i\}$ in V, $\{w_i\}$ in W, and $\{z_i\}$ in Z, then the product of these two matrices represents the composite transformation BA:

$$[BA] = [B][A]. \tag{7}$$

It hardly needs saying that this product BA is again a linear transformation—even though it looks like some kind of quadratic. It might be supposed that with $A(x) = x$ and $B(x) = x$, the product BA had something to do with x^2, but of course the square of the identity is the identity: $B(A(x))$ is still x.

EXERCISE A.1   Find the 3 by 5 matrix which represents the second derivative d^2/dt^2, taking P_4 into P_2 (polynomials of degree four into degree two).

EXERCISE A.2   For the left shift in Example 10, show that the right shift $[A]^T$ is only a one-sided inverse. The rows of $[A]$ are orthonormal, but the columns are not, which was impossible for n by n matrices.

EXERCISE A.3   Suppose that $V = W =$ the space of 2 by 2 matrices, with basis

$$v_1 = w_1 = \begin{bmatrix} 1 & 0 \\ 0 & 0 \end{bmatrix}, \qquad v_2 = w_2 = \begin{bmatrix} 0 & 1 \\ 0 & 0 \end{bmatrix},$$

$$v_3 = w_3 = \begin{bmatrix} 0 & 0 \\ 1 & 0 \end{bmatrix}, \qquad v_4 = w_4 = \begin{bmatrix} 0 & 0 \\ 0 & 1 \end{bmatrix}.$$

Consider the transformation A that takes every matrix into its transpose, and apply rule (5) to find the 4 by 4 matrix $[A]$. Why is $[A]^2 = I$?

EXERCISE A.4   At the end of Chapter 2, we discussed the transformation that premultiplied every member of the above space V by the fixed matrix $\begin{bmatrix} 1 & 0 \\ 2 & 0 \end{bmatrix}$. Find the 4 by 4 matrix that represents this transformation on V.

EXERCISE A.5   Suppose the differential equation (2) is simply $d^n u/dt^n = 0$. Describe the space S_n of all solutions, by specifying a basis.

EXERCISE A.6   Suppose that in the x-y plane the transformation A takes every vector into its mirror image in the x axis and the transformation B takes every vector into its image on the opposite side of the origin. Find the matrices $[A]$, $[B]$, and $[BA]$, using the standard basis vectors e_1 and e_2, and describe the effect on a vector of applying A and then B.

Change of Basis

Suppose $v_1, \ldots, v_n$ and $w_1, \ldots, w_n$ are both bases for the same vector space, and suppose that a single vector v is expressed first with respect to one basis and then with respect to the other:

$$v = \sum_1^n x_j v_j = \sum_1^n y_i w_i. \tag{8}$$

What is the relation between one set of coefficients and the other?

If the two bases are related by $v_j = \sum_{i=1}^n s_{ij} w_i$, then the x's and y's are related in a different way by the same coefficients s_{ij}:

$$y_i = \sum_{j=1}^n s_{ij} x_j. \tag{9}$$

The easiest check is just to verify that the new coefficients y_i satisfy $\sum x_j v_j = \sum y_i w_i$, so that they really are the coefficients in the expansion of v:

$$\sum_{j=1}^n x_j v_j = \sum_{j=1}^n \sum_{i=1}^n x_j s_{ij} w_i \quad \text{does equal} \quad \sum_{i=1}^n y_i w_i = \sum_{i=1}^n \sum_{j=1}^n s_{ij} x_j w_i.$$

A change of basis also enters matrix theory in a way which is slightly subtle but absolutely fundamental. (Often it is disguised as a change of variables, which amounts mathematically to the same thing.) Suppose we decide on a transformation A from some vector space to itself, and choose a basis $w_1, \ldots, w_n$—which will serve both for W and for $V = W$. Then Eq. (5) produces a corresponding matrix $[A]_w$ (with entries a_{ij}) from the rule $Aw_j = \sum a_{ij} w_i$. Now we ask, for the *same* transformation A, how its matrix would look if we had used a different basis $v_1, \ldots, v_n$ for the space, that is, for both V and W. To find this new representation of A, we let S be the matrix that describes the change of basis. As above, $v_j = \sum s_{ij} w_i$.

The new matrix $[A]_v$, representing A with respect to the basis of v's, is related to the original matrix $[A]_w$ constructed with respect to the w's by

$$[A]_v = S^{-1}[A]_w S. \tag{10}$$

In other words, a change of basis induces a similarity transformation of the representing matrix.

Starting with any vector $v = \sum x_j v_j$, Sx gives its coefficients with respect to the w's, then $[A]_w Sx$ gives the coefficients of Av with respect to the w's, and finally $S^{-1}[A]_w Sx$ gives the coefficients of Av with respect to the original v's. Since this is exactly the property that defines $[A]_v$, the two sides of (10) must be equal.

EXAMPLE 11   Suppose A is the transformation that takes any vector (x, y) across the 45° line into its mirror image (y, x). With respect to the standard basis $w_1 = (1, 0)$, $w_2 = (0, 1)$, the corresponding matrix is easy to compute; since A takes w_1 into w_2 and w_2 into w_1, its matrix is the permutation

$$[A]_w = \begin{bmatrix} 0 & 1 \\ 1 & 0 \end{bmatrix}.$$

Suppose we rotate the plane so that one of the new basis vectors is along the 45° line, $v_1 = (1, 1)$, and the other is perpendicular to it, $v_2 = (-1, 1)$. Then expressing v_1 as $w_1 + w_2$ and v_2 as $-w_1 + w_2$,

$$S = \begin{bmatrix} 1 & -1 \\ 1 & 1 \end{bmatrix}.$$

With respect to the new basis, the transformation A is particularly simple. Since v_1 lies right on the 45° line, it is its own mirror image, and $Av_1 = v_1$. The other basis vector $v_2 = (-1, 1)$ is exactly reversed, $Av_2 = -v_2$. Therefore

$$[A]_v = \begin{bmatrix} 1 & 0 \\ 0 & -1 \end{bmatrix}.$$

This matrix is diagonal, because the basis v_1, v_2 is formed from the eigenvectors of A. These eigenvectors appear in the columns of the "change of basis matrix" S, so **just as in Chapter 5 it diagonalizes A:**

$$\Lambda = [A]_v = S^{-1}[A]_w S.$$

EXERCISE A.7   If v_1 is perpendicular to v_2, and A represents projection onto the line through v_2, what is the matrix $[A]_v$? What are the eigenvalues of this (or any other) projection matrix?

EXERCISE A.8   Find a matrix whose eigenvectors are $v_1 = (1, 3)$ with eigenvalue 4, and $v_2 = (2, 0)$ with eigenvalue 0.

THE JORDAN FORM

Given a square matrix A, we want to choose M so that $M^{-1}AM$ is as nearly diagonal as possible. In the simplest case, A has a complete set of eigenvectors and they become the columns of M—otherwise known as S. The Jordan form is $J = M^{-1}AM = \Lambda$, it is constructed entirely from 1 by 1 blocks $J_i = \lambda_i$, and the goal of a diagonal matrix is completely achieved. In the more general and more difficult case, some eigenvectors are missing and a diagonal form is impossible. That case is now our main concern.

We repeat the theorem that is to be proved:

5S If a matrix has s linearly independent eigenvectors, then it is similar to a matrix which is in the special Jordan form

$$
J = M^{-1}AM = \begin{bmatrix} J_1 & & & \\ & \cdot & & \\ & & \cdot & \\ & & & \cdot \\ & & & & J_s \end{bmatrix},
$$

with blocks

$$
J_i = \begin{bmatrix} \lambda_i & 1 & & \\ & \cdot & \cdot & \\ & & \cdot & 1 \\ & & & \lambda_i \end{bmatrix}.
$$

An example of such a Jordan matrix is

$$J = \begin{bmatrix} 8 & 1 & 0 & 0 & 0 \\ 0 & 8 & 0 & 0 & 0 \\ 0 & 0 & 0 & 1 & 0 \\ 0 & 0 & 0 & 0 & 0 \\ 0 & 0 & 0 & 0 & 0 \end{bmatrix} = \begin{bmatrix} \begin{bmatrix} 8 & 1 \\ 0 & 8 \end{bmatrix} & & \\ & \begin{bmatrix} 0 & 1 \\ 0 & 0 \end{bmatrix} & \\ & & [0] \end{bmatrix} = \begin{bmatrix} J_1 & & \\ & J_2 & \\ & & J_3 \end{bmatrix}.$$

The double eigenvalue $\lambda = 8$ has only a single eigenvector, in the first coordinate direction $e_1 = (1, 0, 0, 0, 0)^T$; as a result $\lambda = 8$ appears only in a single block J_1. The triple eigenvalue $\lambda = 0$ has two eigenvectors, e_3 and e_5, which correspond to the two Jordan blocks J_2 and J_3.

The key question is this: *If A is some other 5 by 5 matrix, under what conditions will its Jordan form be this same matrix J? When will there exist an M such that $M^{-1}AM = J$?* As a first requirement, any similar matrix A must share the same eigenvalues 8, 8, 0, 0, 0. But this is far from sufficient—the diagonal matrix with these eigenvalues is not similar to J—and our question really concerns the eigenvectors.

To answer it, we rewrite the relationship $M^{-1}AM = J$ in the simpler form $AM = MJ$:

$$A \begin{bmatrix} x_1 & x_2 & \cdots & x_5 \end{bmatrix} = \begin{bmatrix} x_1 & x_2 & \cdots & x_5 \end{bmatrix} \begin{bmatrix} 8 & 1 & & & \\ 0 & 8 & & & \\ & & 0 & 1 & \\ & & 0 & 0 & \\ & & & & 0 \end{bmatrix}$$

Carrying out the multiplications a column at a time,

$$A x_1 = 8x_1 \quad \text{and} \quad A x_2 = 8x_2 + x_1 \tag{1}$$

$$A x_3 = 0x_3 \quad \text{and} \quad A x_4 = 0x_4 + x_3 ; \quad A x_5 = 0x_5 . \tag{2}$$

Now we can recognize the conditions on A. It must have three genuine eigenvectors, just as J has. The one with $\lambda = 8$ will go into the first column of M, exactly as it would have gone into the first column of S: $A x_1 = 8x_1$. The other two, which will be named x_3 and x_5, go into the third and fifth columns of M: $A x_3 = A x_5 = 0$. Finally there must be two other special vectors, the "generalized eigenvectors" x_2 and x_4. We think of x_2 as belonging to a ***string of vectors***, headed by x_1 and described by Eq. (1). In fact x_2 is the only other vector in the string, and the corresponding block J_1 is of order two. Equation (2) describes two different strings, one in which x_4 follows x_3, and another in which x_5 is alone; the blocks J_2 and J_3 are 2 by 2 and 1 by 1.

The search for the Jordan form of A becomes a search for these strings of vectors, each one headed by an eigenvector: For every i,

$$\text{either} \quad Ax_i = \lambda_i x_i \quad \text{or} \quad Ax_i = \lambda_i x_i + x_{i-1}. \quad (3)$$

The vectors x_i go into the columns of M, and each string produces a single block in J. Essentially, we have to show how these strings can be constructed for every matrix A. Then if the strings match the particular equations (1) and (2), our J will be the Jordan form of A.

I think Filippov's idea, published in Volume 26 of the Moscow University Vestnik, makes the construction as clear and simple as possible. It proceeds by mathematical induction, starting from the fact that every 1 by 1 matrix is already in its Jordan form. We may assume that the construction is achieved for all matrices of order less than n—this is the "induction hypothesis"—and then explain the steps for a matrix of order n. We hope to combine a general explanation for A with a particular one for J, and therefore we imagine from the start that they share the same five eigenvalues.

There are three steps in Filippov's algorithm.

Step 1 Since one of the eigenvalues is $\lambda = 0$, the matrix A is singular and its column space has dimension $r < n$. Looking only within this smaller space, the induction hypothesis guarantees that a Jordan form is possible—there must be r independent vectors w_i in the column space such that

$$\text{either} \quad Aw_i = \lambda_i w_i \quad \text{or} \quad Aw_i = \lambda_i w_i + w_{i-1}. \quad (4)$$

In the example $A = J$, the rank is $r = 3$ and the column space is spanned by e_1, e_2, and e_3; these will serve as the vectors w_i, producing

$$Jw_1 = 8w_1, \quad Jw_2 = 8w_2 + w_1, \quad Jw_3 = 0w_3. \quad (5)$$

Step 2 Suppose the nullspace and the column space of A have an intersection of dimension p. Of course every vector in the nullspace is an eigenvector corresponding to $\lambda = 0$. Therefore there must have been p strings in step 1 which started from this eigenvalue, and we are interested in the vectors w_i that come at the *end* of these strings. Each of these p vectors is in the column space, so each one is a combination of the columns of A: $w_i = Ay_i$ for some y_i.

In the example $A = J$, the nullspace contains e_3 and e_5, and its intersection with the column space is spanned by e_3; therefore $p = 1$. There is one string in Eq. (5) corresponding to $\lambda = 0$, e_3 comes at the end (as well as the beginning) of that string, and $e_3 = Je_4$. Therefore $y = e_4$.

Step 3 The nullspace always has dimension $n - r$. Therefore, independent from its p-dimensional intersection with the column space, it must contain $n - r - p$ additional basis vectors z_i lying *outside* that intersection. The example has $n - r - p = 5 - 3 - 1 = 1$, and the nullvector $z = e_5$ is independent of the column space; it will be this z that produces the 1 by 1 block $J_3 = (0)$.

Now we put these steps together, to give Jordan's theorem:

The r vectors w_i, the p vectors y_i, and the $n - r - p$ vectors z_i form Jordan strings for the matrix A, and these vectors are linearly independent. They go into the columns of M, and $J = M^{-1}AM$ is in Jordan form.

If we want to renumber these vectors as $x_1, \ldots, x_n$, and match them to the conditions of Eq. (3), then each y_i should be inserted immediately after the w_i it came from; it completes a string in which $\lambda_i = 0$. The z's come at the very end, each one alone in its own string; again the eigenvalue is zero, since the z's lie in the nullspace. The blocks with nonzero eigenvalues are already finished at step 1, the blocks with zero eigenvalues grow by one row and column at step 2, and step 3 contributes any 1 by 1 blocks $J_i = (0)$.

The only technical point is to verify the independence of the whole collection w_i, y_i, and z_i. Therefore we assume that some combination is zero:

$$\sum c_i w_i + \sum d_i y_i + \sum g_i z_i = 0. \tag{6}$$

Multiplying by A, and using Eqs. (4) for the w_i,

$$\sum c_i \begin{bmatrix} \lambda_i w_i \\ \text{or} \\ \lambda_i w_i + w_{i-1} \end{bmatrix} + \sum d_i A y_i = 0. \tag{7}$$

The $A y_i$ are the special w_i at the end of strings corresponding to $\lambda_i = 0$, so they cannot appear in the first sum. Since (7) is some kind of combination of the w_i, which were independent by the induction hypothesis—they supplied the Jordan form within the column space—we conclude that each d_i must be zero. Returning to (6), this leaves $\sum c_i w_i = -\sum g_i z_i$, and the left side is in the column space. Since the z's were independent of that space, each g_i must be zero. Finally $\sum c_i w_i = 0$, and the independence of the w_i produces $c_i = 0$.

If the original A had not been singular, the three steps would have been applied instead to $A' = A - cI$. (The constant c is chosen to make A' singular, and it can be any one of the eigenvalues of A.) The algorithm puts A' into its Jordan form $M^{-1}A'M = J'$, by producing the strings x_i from the w_i, y_i, and z_i. Then the Jordan form for A uses the same strings and the same M:

$$M^{-1}AM = M^{-1}A'M + M^{-1}cM = J' + cI = J.$$

This completes the proof that every A is similar to some Jordan matrix J. Except for a reordering of the blocks, *it is similar to only one such J*; there is a unique Jordan form for A. Thus the set of all matrices is split into a number of subsets, with the following property: *All the matrices in the same subset have the same Jordan form, and they are all connected to one another by similarity transformations, but no two matrices in different subsets are similar.* With this classification, we end the book.

EXAMPLE

$$A = \begin{bmatrix} 0 & 1 & 2 \\ 0 & 0 & 1 \\ 0 & 0 & 0 \end{bmatrix},$$

with triple eigenvalue $\lambda = 0$. This is the matrix on p. 228, with rank $r = 2$ and only one eigenvector. Within the column space there is a single string w_1, w_2, which happens to coincide with the last two columns:

$$A \begin{bmatrix} 1 \\ 0 \\ 0 \end{bmatrix} = 0 \quad \text{and} \quad A \begin{bmatrix} 2 \\ 1 \\ 0 \end{bmatrix} = \begin{bmatrix} 1 \\ 0 \\ 0 \end{bmatrix},$$

or

$$A w_1 = 0 \quad \text{and} \quad A w_2 = 0 w_2 + w_1.$$

The nullspace lies entirely within the column space, and it is spanned by w_1. Therefore $p = 1$ in step 2, and the vector y comes from the equation

$$A y = w_2 = \begin{bmatrix} 2 \\ 1 \\ 0 \end{bmatrix}, \quad \text{whose solution is} \quad y = \begin{bmatrix} 0 \\ 0 \\ 1 \end{bmatrix}.$$

Finally the string w_1, w_2, y goes into the matrix M:

$$M = \begin{bmatrix} 1 & 2 & 0 \\ 0 & 1 & 0 \\ 0 & 0 & 1 \end{bmatrix}, \quad \text{and} \quad M^{-1}AM = \begin{bmatrix} 0 & 1 & 0 \\ 0 & 0 & 1 \\ 0 & 0 & 0 \end{bmatrix} = J.$$

Application to $du/dt = Au$ As always, we simplify the problem by uncoupling the unknowns. This uncoupling is complete only when there is a full set of eigenvectors, and $u = Sv$; the best change of variables in the present case is $u = Mv$. This produces the new equation $M \, dv/dt = AMv$, or $dv/dt = Jv$, which is as simple as the circumstances allow. It is coupled only by the off-diagonal 1's within each Jordan block. In the example just above, which has a single block,

$$\frac{dv}{dt} = \begin{bmatrix} 0 & 1 & 0 \\ 0 & 0 & 1 \\ 0 & 0 & 0 \end{bmatrix} v \quad \text{or} \quad \begin{aligned} da/dt &= b \\ db/dt &= c \\ dc/dt &= 0 \end{aligned} \quad \text{or} \quad \begin{aligned} a &= a_0 + b_0 t + c_0 t^2/2 \\ b &= b_0 + c_0 t \\ c &= c_0. \end{aligned}$$

The system is solved by working upward from the last equation, and a new power of t enters at every step. (An l by l block has powers as high as t^{l-1}.)

The exponentials of J, in this case and in the earlier 5 by 5 example, are

$$
e^{Jt} = \begin{bmatrix} 1 & t & t^2/2 \\ 0 & 1 & t \\ 0 & 0 & 1 \end{bmatrix} \quad \text{and} \quad e^{Jt} = \begin{bmatrix} e^{8t} & te^{8t} & 0 & 0 & 0 \\ 0 & e^{8t} & 0 & 0 & 0 \\ 0 & 0 & 1 & t & 0 \\ 0 & 0 & 0 & 1 & 0 \\ 0 & 0 & 0 & 0 & 1 \end{bmatrix}.
$$

You can see how the coefficients of a, b, and c appear in the first exponential. And in the second example, you can identify all five of the "special solutions" to $du/dt = Au$. Three of them are the pure exponentials $u_1 = e^{8t}x_1$, $u_3 = e^{0t}x_3$, and $u_5 = e^{0t}x_5$, formed exactly as before from the three eigenvectors of A. The other two involve the generalized eigenvectors:

$$
u_2 = e^{8t}(tx_1 + x_2) \quad \text{and} \quad u_4 = e^{0t}(tx_3 + x_4). \tag{8}
$$

The most general solution to $du/dt = Au$ is a combination $c_1u_1 + \cdots + c_5u_5$, and the combination which matches u_0 at time $t = 0$ is again

$$
u_0 = c_1x_1 + \cdots + c_5x_5, \quad \text{or} \quad u_0 = Mc, \quad \text{or} \quad c = M^{-1}u_0.
$$

This only means that $u = Me^{Jt}M^{-1}u_0$, and that the S and Λ in the old formula $Se^{\Lambda t}S^{-1}u_0$ have been replaced by M and J.

EXERCISE B.1   Find the Jordan forms (in three steps!) of

$$
A = \begin{bmatrix} 1 & 1 \\ 1 & 1 \end{bmatrix} \quad \text{and} \quad B = \begin{bmatrix} 0 & 1 & 2 \\ 0 & 0 & 0 \\ 0 & 0 & 0 \end{bmatrix}.
$$

EXERCISE B.2   Show that the special solution u_2 in Eq. (8) does satisfy $du/dt = Au$, exactly because of the string $Ax_1 = 8x_1$, $Ax_2 = 8x_2 + x_1$.

EXERCISE B.3   For the matrix B above, use $Me^{Jt}M^{-1}$ to compute the exponential e^{Bt}, and compare it with the power series $I + Bt + ((Bt)^2/2!) + \cdots$.

REFERENCES

Abstract Linear Algebra

F. R. Gantmacher, "Theory of Matrices." Chelsea, New York, 1959.

P. R. Halmos, "Finite-Dimensional Vector Spaces." Van Nostrand-Reinhold, Princeton, New Jersey, 1958.

K. Hoffman and R. Kunze, "Linear Algebra," 2nd ed. Prentice-Hall, Englewood Cliffs, New Jersey, 1971.

T. Muir, "Determinants." Dover, New York, 1960.

Applied Linear Algebra

A. Ben-Israel and T. N. E. Greville, "Generalized Inverses: Theory and Applications." Wiley, New York, 1974.

R. Bellman, "Introduction to Matrix Analysis." McGraw-Hill, New York, 1960.

D. Gale, "The Theory of Linear Economic Models." McGraw-Hill, New York, 1960.

D. G. Luenberger, "Introduction to Linear and Nonlinear Programming." Addison-Wesley, Reading, Massachusetts, 1973.

B. Noble, "Applied Linear Algebra." Prentice-Hall, Englewood Cliffs, New Jersey, 1969.

Numerical Linear Algebra

G. Forsythe and C. Moler, "Computer Solution of Linear Algebraic Systems." Prentice-Hall, Englewood Cliffs, New Jersey, 1967.

C. L. Lawson and R. J. Hanson, "Solving Least Squares Problems." Prentice-Hall, Englewood Cliffs, New Jersey, 1974.

G. W. Stewart, "Introduction to Matrix Computations." Academic Press, New York, 1973.

R. S. Varga, "Matrix Iterative Analysis." Prentice-Hall, Englewood Cliffs, New Jersey, 1962.

J. M. Wilkinson, "Rounding Errors in Algebraic Processes." Prentice-Hall, Englewood Cliffs, New Jersey, 1963.

J. M. Wilkinson, "The Algebraic Eigenvalue Problem." Oxford Univ. Press, London and New York, 1965.

J. M. Wilkinson and C. Reinsch, Eds., "Handbook for Automatic Computation II, Linear Algebra." Springer, Berlin and New York, 1971.

D. M. Young, "Iterative Solution of Large Linear Systems." Academic Press, New York, 1971.

SOLUTIONS TO
EXERCISES

Chapter 1

1.2.1. $u = 3, v = 1, w \neq 0$, with pivots 2, 1, -5.

1.2.2. $u = 1, v = 2, w = 3, z = 4$.

1.2.3. A coefficient $+1$ would make the system singular.

1.2.4. Elimination computes the multiplier $l = c/a$, and then $d - lb$.

1.2.5. The given computer does 3,600,000 operations for \$1, so in theory it could solve a system of order $n = 10{,}800{,}000^{1/3} = 221$; in practice there would also be overhead costs. Multiplying the expense by 1000 will only multiply n by 10.

1.2.6. The second term $bc + ad$ is $(a + b)(c + d) - ac - bd$.

1.2.7. $u = 1, v = 3, w = 2$.

1.3.1. $Ax = \begin{bmatrix} 17 \\ 4 \\ 17 \end{bmatrix}$.

1.3.2. $Ax = \begin{bmatrix} -1 \\ 6 \\ -4 \\ -1 \end{bmatrix}$ and $[26]$.

1.3.3. $u_{\text{end}} = .8u_{\text{start}} + .1v_{\text{start}}$, $A = \begin{bmatrix} .8 & .1 \\ .2 & .9 \end{bmatrix}$
$v_{\text{end}} = .2u_{\text{start}} + .9v_{\text{start}}$

(i) $\begin{bmatrix} u_e \\ v_e \end{bmatrix} = A \begin{bmatrix} 30 \\ 200 \end{bmatrix} = \begin{bmatrix} 44 \\ 186 \end{bmatrix}$,

(ii) $\begin{bmatrix} 30 \\ 200 \end{bmatrix} = A \begin{bmatrix} u_s \\ v_s \end{bmatrix}$, so $\begin{bmatrix} u_s \\ v_s \end{bmatrix} = \begin{bmatrix} 10 \\ 220 \end{bmatrix}$.

(iii) $\begin{bmatrix} u \\ v \end{bmatrix} = A \begin{bmatrix} u \\ v \end{bmatrix}$, or $\begin{bmatrix} .2 & -.1 \\ -.2 & .1 \end{bmatrix} \begin{bmatrix} u \\ v \end{bmatrix} = 0$, or $v = 2u$.

1.3.4.   The corners of the parallelogram are $(2, 1)$, $(0, 3)$, $(2, 4)$, and $(0, 0)$.

1.3.5. $EA = \begin{bmatrix} 2 & 1 & 1 \\ 0 & -1 & -2 \\ 0 & 3 & 2 \end{bmatrix}$, the result of elimination steps 1 and 2.

1.3.6. $EA = \begin{bmatrix} 2 & 1 & 1 \\ 0 & -1 & -2 \\ 0 & 0 & -4 \end{bmatrix}$, the result of elimination steps 1, 2, 3.

1.3.7. $EA = [0]$; there are ln entries in the product, each requiring m multiplications.

1.3.8. $EA = \begin{bmatrix} 13 \\ -8 \\ 2 \end{bmatrix}$.

1.3.9. ABC is the result of elimination steps 1, 2.

1.3.10. $\begin{bmatrix} a & b \\ c & d \end{bmatrix} \begin{bmatrix} 1 & 0 \\ 0 & 0 \end{bmatrix} = \begin{bmatrix} 1 & 0 \\ 0 & 0 \end{bmatrix} \begin{bmatrix} a & b \\ c & d \end{bmatrix}$ forces $b = c = 0$,

then $\begin{bmatrix} a & 0 \\ 0 & d \end{bmatrix} \begin{bmatrix} 0 & 1 \\ 0 & 0 \end{bmatrix} = \begin{bmatrix} 0 & 1 \\ 0 & 0 \end{bmatrix} \begin{bmatrix} a & 0 \\ 0 & d \end{bmatrix}$ forces $a = d$.

1.3.11. $A = \begin{bmatrix} 0 & 1 \\ -1 & 0 \end{bmatrix}$, $B = \begin{bmatrix} 0 & 1 \\ 0 & 0 \end{bmatrix}$, $C = \begin{bmatrix} 0 & 1 \\ 1 & 0 \end{bmatrix}$, $D = A$, $E = F = \begin{bmatrix} 1 & -1 \\ 1 & -1 \end{bmatrix}$.

1.3.12.   True; false; true; false.

1.3.13.   The weights are the entries in the first row of A.

1.4.2. $A = \begin{bmatrix} 1 & 0 \\ 4 & 1 \end{bmatrix} \begin{bmatrix} 2 & 1 \\ 0 & 3 \end{bmatrix} = LU$; $A = L$, $U = I$.

1.4.3. $LU = \begin{bmatrix} 1 & 0 & 0 \\ 0 & 1 & 0 \\ 3 & 0 & 1 \end{bmatrix} \begin{bmatrix} 2 & 3 & 3 \\ 0 & 5 & 7 \\ 0 & 0 & -1 \end{bmatrix}, c = \begin{bmatrix} 2 \\ 2 \\ -1 \end{bmatrix}.$

1.4.4. $\begin{bmatrix} a & b \\ c & d \end{bmatrix} = \begin{bmatrix} 1 & 0 \\ c/a & 1 \end{bmatrix} \begin{bmatrix} a & 0 \\ 0 & (ad - bc)/a \end{bmatrix} \begin{bmatrix} 1 & b/a \\ 0 & 1 \end{bmatrix}.$

1.4.5. $L = \begin{bmatrix} 1 & 0 & 0 \\ -\frac{1}{2} & 1 & 0 \\ 0 & -\frac{2}{3} & 1 \end{bmatrix}, \quad D = \begin{bmatrix} 2 & & \\ & \frac{3}{2} & \\ & & \frac{4}{3} \end{bmatrix}, \quad U = \begin{bmatrix} 1 & -\frac{1}{2} & 0 \\ 0 & 1 & -\frac{2}{3} \\ 0 & 0 & 1 \end{bmatrix},$

$c = \begin{bmatrix} 6 \\ 3 \\ -4 \end{bmatrix}, \quad x = \begin{bmatrix} 3 \\ 0 \\ -3 \end{bmatrix}$

1.4.6.   The ratio of $(150)^3/3$ to $(150)^2$ is 50.

1.4.7. $c = \begin{bmatrix} 2 \\ -1 \\ 3 \end{bmatrix}, \quad x = \begin{bmatrix} 4 \\ 2 \\ 3 \end{bmatrix}.$

1.5.1. $u = 2, v = -3, w = 4$; exchange rows 2 and 3.

1.5.2.   The second equation is $0u + 0v = b_2 - 3b_1$, so there is a solution only if $b_2 = 3b_1$.

1.5.3. $P = \begin{bmatrix} 0 & 1 \\ 1 & 0 \end{bmatrix}, \quad L = I, \quad D = \begin{bmatrix} 2 & 0 \\ 0 & 1 \end{bmatrix}, \quad U = \begin{bmatrix} 1 & \frac{3}{2} \\ 0 & 1 \end{bmatrix}.$

1.5.4. $P = \begin{bmatrix} 1 & 0 & 0 \\ 0 & 0 & 1 \\ 0 & 1 & 0 \end{bmatrix}.$

1.5.5.   Both singular; no solution to the first, $u = v = w = 7$ (or anything else) in the second.

1.5.6. $C^{-1}BA^{-1}.$

1.5.7. $A^{-1} = A$, $B^{-1} =$ transpose of $B = \begin{bmatrix} \cos\theta & \sin\theta \\ -\sin\theta & \cos\theta \end{bmatrix}.$

1.5.9.   8 times row 3 is added to row 1; E^{-1} has -8 in place of 8.

1.5.10. $A^{-1} = \dfrac{1}{ad - bc} \begin{bmatrix} d & -b \\ -c & a \end{bmatrix}, \quad A^{-1} = \dfrac{1}{4} \begin{bmatrix} 3 & 2 & 1 \\ 2 & 4 & 2 \\ 1 & 2 & 3 \end{bmatrix}.$

1.5.11. $A^{-1} = \begin{bmatrix} 9 & -36 & 30 \\ -36 & 192 & -180 \\ 30 & -180 & 180 \end{bmatrix}$.

1.5.12.   The pivots are .001, 1000; after exchange they are 1, -1.

1.5.13.   If the pivot is large, then it is multiplied by less than one in eliminating each entry below it. An extreme case, with all multipliers equal to 1 or -1, is

$$A = \begin{bmatrix} \frac{1}{2} & \frac{1}{2} & 1 \\ -\frac{1}{2} & 0 & 1 \\ -\frac{1}{2} & -1 & 1 \end{bmatrix}.$$

1.6.1. $L = \begin{bmatrix} 1 & & & \\ -1 & \cdot & & \\ & \cdot & \cdot & \\ & & \cdot & \cdot \\ & & & -1 & 1 \end{bmatrix}$, $D = I$, $U = L^{\mathrm{T}}$.

1.6.2.   The lower 2 by 2 matrix becomes the symmetric

$$\begin{bmatrix} d - b^2/a & e - bc/a \\ e - bc/a & f - c^2/a \end{bmatrix}.$$

1.6.3. $A = \begin{bmatrix} 1 & -1 & & & \\ -1 & 2 & -1 & & \\ & -1 & 2 & -1 & \\ & & -1 & 2 & -1 \\ & & & -1 & 1 \end{bmatrix}$.

1.6.4. $(u_1, u_2, u_3) = (\pi^2/8, 0, -\pi^2/8)$ instead of the true values $(1, 0, -1)$.

Chapter 2

2.1.1.   The points with integer coordinates; the cross formed by the axes.

2.1.2.   Yes, no, no, yes, yes, yes.

2.1.3.   If $Ax = 0$, $Ay = 0$, then $A(cx + dy) = 0$. If $b \neq 0$, the solutions do not include $x = 0$ and cannot form a subspace.

2.2.1. $\begin{bmatrix} 1 & 1 \\ 0 & 0 \end{bmatrix} + \begin{bmatrix} -1 & 0 \\ 0 & 1 \end{bmatrix} = \begin{bmatrix} 0 & 1 \\ 0 & 1 \end{bmatrix}$, not in echelon form.

2.2.2. $x + y + z = 1$, $x + y + z = 0$.

2.2.3. $LU = \begin{bmatrix} 1 & 0 & 0 \\ 0 & 1 & 0 \\ 1 & 0 & 1 \end{bmatrix} \begin{bmatrix} 1 & 2 & 0 & 1 \\ 0 & 1 & 1 & 0 \\ 0 & 0 & 0 & 0 \end{bmatrix}.$

The first two columns of U have pivots, so $r = 2$. The other two variables w and y are free, and the general solution is

$$x = \begin{bmatrix} 2w - y \\ -w \\ w \\ y \end{bmatrix} = w\begin{bmatrix} 2 \\ -1 \\ 1 \\ 0 \end{bmatrix} + y\begin{bmatrix} -1 \\ 0 \\ 0 \\ 1 \end{bmatrix}.$$

2.2.4.   The second variable is basic (nonzero pivot) and the variables u, w, and y are free. The rank is one, and the general solution is $x = (u, -4w, w, y)^{\mathrm{T}}$. With $b \neq 0$, the system is consistent only if $b_2 - 2b_1 = 0$, and then

$$x = \begin{bmatrix} u \\ b_1 - 4w \\ w \\ y \end{bmatrix} = u\begin{bmatrix} 1 \\ 0 \\ 0 \\ 0 \end{bmatrix} + w\begin{bmatrix} 0 \\ -4 \\ 1 \\ 0 \end{bmatrix} + y\begin{bmatrix} 0 \\ 0 \\ 0 \\ 1 \end{bmatrix} + \begin{bmatrix} 0 \\ b_1 \\ 0 \\ 0 \end{bmatrix}.$$

2.2.5. u is basic, v is free, $r = 1$, $x = v\begin{bmatrix} -2 \\ 1 \end{bmatrix}$. In general the system needs $b_1 = 0$, $b_3 = 4b_2$, $b_4 = 0$, and then $x = v\begin{bmatrix} -2 \\ 1 \end{bmatrix} + \begin{bmatrix} b_2 \\ 0 \end{bmatrix}$.

2.2.6.   Elimination gives $u + 2v + 2w = 1$, $w = 2$, so v is free:

$$\begin{bmatrix} u \\ v \\ w \end{bmatrix} = \begin{bmatrix} -2v - 3 \\ v \\ 2 \end{bmatrix} = v\begin{bmatrix} -2 \\ 1 \\ 0 \end{bmatrix} + \begin{bmatrix} -3 \\ 0 \\ 2 \end{bmatrix}.$$

2.2.7.   The third equation requires $0 = b_3 - 2b_1 - 3b_2$; rank two.

2.2.8. $c = 7$.

2.3.1. $v_1 - v_2 + v_3 - v_4 = 0$.

2.3.2.   The rows are independent.

2.3.4.   If $c_1(v_1 + v_2) + c_2(v_1 + v_3) + c_3(v_2 + v_3) = 0$, then $(c_1 + c_2)v_1 + (c_1 + c_3)v_2 + (c_2 + c_3)v_3 = 0$. By the independence of the v's, this forces $c_1 + c_2 = 0$, $c_1 + c_3 = 0$, $c_2 + c_3 = 0$, and then all the c's are zero. Therefore the w's are also independent.

2.3.5.   The column space is the line of all multiples of $\begin{bmatrix} 1 \\ 3 \end{bmatrix}$; the row space is also a line, but going through $(1, 2)$.

2.3.6.   Columns 2 and 3 are a basis for the column space, with the last column equal to $7 \,(\mathrm{col}\,2) - (\mathrm{col}\,3)$. Rows 1 and 2 are a basis for the row space.

2.3.7.   One basis is $\begin{bmatrix} 1 & 0 \\ 0 & 0 \end{bmatrix}$, $\begin{bmatrix} 0 & 1 \\ 0 & 0 \end{bmatrix}$, $\begin{bmatrix} 0 & 0 \\ 1 & 0 \end{bmatrix}$, $\begin{bmatrix} 0 & 0 \\ 0 & 1 \end{bmatrix}$. The echelon matrices span the upper tri-angular matrices.

2.3.8. $v_1 = \begin{bmatrix} 1 \\ 1 \\ 0 \end{bmatrix}$ and $v_2 = \begin{bmatrix} 0 \\ 0 \\ 1 \end{bmatrix}$, or $\begin{bmatrix} 1 \\ 1 \\ 1 \end{bmatrix}$ and $\begin{bmatrix} 0 \\ 0 \\ 1 \end{bmatrix}$.

2.3.9.   Let $v_1 = (1, 0, 0, 0), \ldots, v_4 = (0, 0, 0, 1)$ be the coordinate vectors. If W is the line through $(1, 2, 3, 4)$, none of the v's are in W.

2.3.10.   Dimension zero, with $x = y = z = 0$; dimension one, with $x = (1, 0, 0) = -y$, $z = 0$; dimension two, with $x = (1, 0, 0)$, $y = (0, 1, 0)$, $z = (-1, -1, 0)$; since they are dependent they cannot span the whole space.

2.3.12.   (i)   If it were not a basis, we could add more independent vectors, which would exceed the given dimension k. Contradiction.   (ii)   If it were not a basis, we could delete some vectors, leaving less than the given dimension k. Contradiction.

2.3.13.   Dimension 6.

2.3.14.   If v_1, v_2, v_3 are a basis for V, and w_1, w_2, w_3 are a basis for W, then these six vectors cannot be independent and some combination is zero: $\sum c_i v_i + \sum d_i w_i = 0$, or $\sum c_i v_i = -\sum d_i w_i$ is a vector in both subspaces.

2.4.1.   False; their dimensions are equal.

2.4.2.   Column 2 is a basis for $\mathcal{R}(A)$, and row 1 is a basis for $\mathcal{R}(A^T)$. The vector $(-2, 1)^T$ spans the left nullspace, and the vectors $(1, 0, 0, 0)^T$, $(0, -4, 1, 0)^T$, $(0, 0, 0, 1)^T$ are a basis for $\mathcal{N}(A)$.

2.4.3.   Columns 1 and 2 are a basis for $\mathcal{R}(A)$, and rows 1 and 2 for $\mathcal{R}(A^T)$. The vector $(1, 0, -1)^T$ spans the left nullspace, and $(-1, 0, 0, 1)^T$, $(2, -1, 1, 0)^T$ are a basis for $\mathcal{N}(A)$.

2.4.4.   The column space has all fourth components zero; the row space has all first components zero. The nullspace is spanned by $(1, 0, 0, 0)^T$ and the left null-space by $(0, 0, 0, 1)^T$.

2.4.5.   If $AB = 0$, then A takes every column of B into zero; therefore it does the same to any combination of the columns, and all such combinations must be in $\mathcal{N}(A)$.

2.4.6.   If the dimension is not increased by including the new column b, then b must be a combination of the columns of A and therefore $Ax = b$ is solvable.

2.4.7. $d = bc/a$; $A = \begin{bmatrix} a \\ c \end{bmatrix} \begin{bmatrix} 1 & b/a \end{bmatrix}$.

2.4.8.　$A = \begin{bmatrix} 2 \\ 4 \\ 0 \end{bmatrix} [1 \ -1], \quad B = \begin{bmatrix} 1 \\ 3 \end{bmatrix} [1 \ 1 \ 2],$

$AB = u(v^{\mathrm{T}}w)z^{\mathrm{T}} = \begin{bmatrix} 2 \\ 4 \\ 0 \end{bmatrix} (-2)[1 \ 1 \ 2] = \begin{bmatrix} -4 & -4 & -8 \\ -8 & -8 & -16 \\ 0 & 0 & 0 \end{bmatrix}.$

2.4.10.　$y = e^x$ is one-to-one but never takes the value zero; $z = \tan x$ takes all values, but more than once.

2.4.11.　If existence holds for A, then $r = m$, and the m columns of A^{T} must be independent. If uniqueness holds for A, then $r = n$, and the columns of A^{T} must span all of $\mathbf{R}^n$.

2.4.12.　A has no left-inverse; its right-inverses are all of the form

$$C = \begin{bmatrix} 1 - u \\ u \\ v \end{bmatrix}.$$

2.5.1.　$\| x \|^2 = 21, \| y \|^2 = 18, x^{\mathrm{T}}y = 0.$

2.5.2.　$(1, 0)$ and $(1, 1)$ are independent but not orthogonal; $(0, 0)$ and $(1, 1)$ are orthogonal but not independent.

2.5.3.　If $(x_2/x_1)(y_2/y_1) = -1$, then $x_2 y_2 = -x_1 y_1$, or $x_1 y_1 + x_2 y_2 = 0.$

2.5.4.　If $i \neq j$, the i, j entry of $BB^{-1} = I$ is zero.

2.5.5.　$v_1^{\mathrm{T}}v_3 = 0, v_2^{\mathrm{T}}v_3 = 0.$

2.5.6.　Solving $u + v + w = 0, u - v = 0$, any multiple of $(1, 1, -2)$ is orthogonal to the two given vectors. An orthonormal system is $(1, 1, 1)/\sqrt{3}, (1, -1, 0)/\sqrt{2}, (1, 1, -2)/\sqrt{6}$.

2.5.7.　$w_2 = (0, 0, 0, 1); v_3 = (0, 0, 5, -4).$

2.5.8.　Any vector in $V \cap W$ must be orthogonal to itself.

2.5.9.　Solving $u + v + 2w = 0, u + 2v + 3w = 0$, the orthogonal complement is the line spanned by $(1, 1, -1).$

2.5.10.　The system is $x + y - z = 0.$

2.5.11.　Take y to be the left nullspace component of b.

2.5.12.　The vector $(2, 2, -1)$ is a basis for the nullspace; $(3, 3, 3) = (1, 1, 4) + (2, 2, -1) = x_{\mathrm{r}} + x_{\mathrm{n}}.$

2.5.13.　x is orthogonal to $V + W$ if and only if it is orthogonal to every vector in V and in W; it is in the intersection of $V^{\perp}$ and $W^{\perp}$.

2.5.15. The voltage drop equals 5 from 2 to 3, from 3 to 1, from 4 to 1, and from 2 to 5. There is no drop from 3 to 4, which have $p_3 = p_4 = 5$. There is a jump of 10 from 1 to 2, giving $p_2 = 10$.

2.5.16. The current from 1 to 3 is $I = 3$, with half flowing back through 2, half through 4, and none from 2 to 4.

2.6.1. $V \cap W =$ space of bidiagonal matrices, with zero entries except on the main diagonal and first subdiagonal; dimension 7. $V + W =$ matrices with zero entries below the first subdiagonal; dimension 13. Originally dim $V = 10$ and dim $W = 10$.

2.6.2. If $x = v + w$ and also $x = v' + w'$, then $v + w = v' + w'$, or $v - v' = w' - w$. This last vector is in both V and W, and must be zero.

2.6.3. One possibility: W is spanned by $(1, 0, 0, 0)$ and $(0, 0, 1, 0)$. Altogether, the four vectors are independent.

2.6.4. The vectors v_1, v_2, and w_1 are a basis for $V + W$; $V \cap W$ is one dimensional, and is spanned by the vector $v_1 - v_2$ $(= w_1 - w_2)$.

2.6.6. $A = \begin{bmatrix} 0 & 1 \\ 0 & 0 \end{bmatrix}$, $B = \begin{bmatrix} 0 & 0 \\ 1 & 0 \end{bmatrix}$, $AB = \begin{bmatrix} 1 & 0 \\ 0 & 0 \end{bmatrix}$.

2.6.7. Let $A = [0 \ \ 0]$, $B = \begin{bmatrix} 0 \\ 0 \end{bmatrix}$, $AB = [0]$, $\nu(AB) = 1 < \nu(A) = 2$.

2.6.8. $A = \begin{bmatrix} 1 & 0 \\ 2 & 1 \end{bmatrix} \begin{bmatrix} 0 & 1 & 4 & 0 \\ 0 & 0 & 0 & 0 \end{bmatrix} = \begin{bmatrix} 1 \\ 2 \end{bmatrix} [0 \ \ 1 \ \ 4 \ \ 0] = \bar{L}\bar{U}$.

2.6.9. $A = P^{-1}LU = \begin{bmatrix} 0 & 1 & 0 & 0 \\ 1 & 0 & 0 & 0 \\ 0 & 0 & 1 & 0 \\ 0 & 0 & 0 & 1 \end{bmatrix} \begin{bmatrix} 1 & 0 & 0 & 0 \\ 0 & 1 & 0 & 0 \\ 4 & 0 & 1 & 0 \\ 0 & 0 & 0 & 1 \end{bmatrix} \begin{bmatrix} 1 & 2 \\ 0 & 0 \\ 0 & 0 \\ 0 & 0 \end{bmatrix} = \begin{bmatrix} 0 \\ 1 \\ 4 \\ 0 \end{bmatrix} [1 \ \ 2] = \bar{L}\bar{U}$.

2.6.10. $A = \begin{bmatrix} 1 & 0 & 0 \\ 2 & 1 & 0 \\ -1 & 2 & 1 \end{bmatrix} \begin{bmatrix} 1 & 3 & 3 & 2 \\ 0 & 0 & 3 & 1 \\ 0 & 0 & 0 & 0 \end{bmatrix} = \begin{bmatrix} 1 \\ 2 \\ -1 \end{bmatrix} [1 \ \ 3 \ \ 3 \ \ 2]$
$+ \begin{bmatrix} 0 \\ 1 \\ 2 \end{bmatrix} [0 \ \ 0 \ \ 3 \ \ 1]$.

2.6.11. The rank is one; any submatrix of this size is invertible.

2.6.12. rank $(AB) \leq$ rank$(B) \leq n < m$, but AB is m by m.

2.6.13.   If $v_1, \ldots, v_r$ is a basis for the column space of A, and $w_1, \ldots, w_s$ is the same for B, then any column of $A + B$ is a combination of the v's and w's; they span the column space of $A + B$, so its dimension cannot exceed $r + s$.

2.6.14.   Suppose y is in $\Re(B) \cap \Re(A)$, so $y = Bx$ for some x and $Ay = 0$. Then $ABx = 0$, and x is in $\Re(AB)$; but this equals $\Re(B)$, so $Bx = 0$. Thus the only vector in $\Re(B) \cap \Re(A)$ is $y = Bx = 0$.

2.6.16.   (i)   Any $B = \begin{bmatrix} 0 & 0 \\ c & d \end{bmatrix}$ gives $AB = 0$, so there is a two-dimensional nullspace.
(ii)   If $B = \begin{bmatrix} a & b \\ c & d \end{bmatrix}$, then $AB = \begin{bmatrix} a & b \\ 2a & 2b \end{bmatrix}$ which is a two-dimensional subspace (the range).

Chapter 3

3.1.1.   (a) $a^{\mathrm{T}}b = 2\sqrt{xy} \le \|a\| \|b\| = x + y$. (b) $\|x + y\|^2 \le (\|x\| + \|y\|)^2$ is the same as $x^{\mathrm{T}}x + 2x^{\mathrm{T}}y + y^{\mathrm{T}}y \le x^{\mathrm{T}}x + 2\|x\| \|y\| + y^{\mathrm{T}}y$, which is the Schwarz inequality $x^{\mathrm{T}}y \le \|x\| \|y\|$.

3.1.2. $Op^2 = (a^{\mathrm{T}}b/a^{\mathrm{T}}a)^2 a^{\mathrm{T}}a$; added to Eq. (5) this leaves $b^{\mathrm{T}}b$, the square of the hypotenuse.

3.1.3. $p = (a^{\mathrm{T}}b/a^{\mathrm{T}}a)a = \frac{10}{3}(1, 1, 1)$; the other projection is $(b^{\mathrm{T}}a/b^{\mathrm{T}}b)b$.

3.1.4.   It is an equality in both cases; the angle is 0 or π, the cosine is 1 or -1, and the error bp is zero because b coincides with its projection p.

3.1.5. θ is the angle whose cosine is $1/\sqrt{n}$.

3.1.6.   For each j, $|a_j|^2 + |b_j|^2 - 2|a_j| |b_j|$ equals $(|a_j| - |b_j|)^2$, and cannot be negative.

3.1.7.   The transpose of A^{-1} was shown to be $(A^{\mathrm{T}})^{-1}$, which is the same as A^{-1} itself if $A = A^{\mathrm{T}}$.

3.1.8. $A = \begin{bmatrix} 0 & 1 \\ 1 & 0 \end{bmatrix} = A^{\mathrm{T}}, B = \begin{bmatrix} 1 & 0 \\ 0 & 0 \end{bmatrix} = B^{\mathrm{T}}, AB = \begin{bmatrix} 0 & 0 \\ 1 & 0 \end{bmatrix} \ne (AB)^{\mathrm{T}}$.

3.1.9.   Let A have more rows than columns, say $A = \begin{bmatrix} 1 \\ 0 \end{bmatrix}$.

3.1.10. Typical rays to the vertices are $v = (-\frac{1}{2}, -\frac{1}{2}, -\frac{1}{2})$ and $w = (\frac{1}{2}, \frac{1}{2}, -\frac{1}{2})$, with $\cos\theta = v^{\mathrm{T}}w/\|v\| \|w\| = -\frac{1}{3}$.

3.2.1. $\bar{x} = 151$.

3.2.2. $\bar{x} = 2$.

3.2.3. $\bar{u} = \bar{v} = \frac{1}{3}$; $\bar{x} = 3$.

3.2.4. $E^2 = (u - 1)^2 + (v - 3)^2 + (u + v - 4)^2$; $dE^2/du = 2(u - 1) + 2(u + v - 4)$, $dE^2/dv = 2(v - 3) + 2(u + v - 4)$, $\bar{u} = 1$, $\bar{v} = 3$; $p = A\begin{bmatrix} 1 \\ 3 \end{bmatrix} = b$ because b is itself in the column space.

3.2.5. $P = \dfrac{1}{3} \begin{bmatrix} 2 & 1 & 1 \\ 1 & 2 & -1 \\ 1 & -1 & 2 \end{bmatrix}$, $Pb = \begin{bmatrix} 1 \\ 1 \\ 0 \end{bmatrix}$, $(I - P)b = \begin{bmatrix} -1 \\ 1 \\ 1 \end{bmatrix}$, $\bar{x} = \begin{bmatrix} 1 \\ 0 \end{bmatrix}$.

3.2.6.   (a) $(I - P)^2 = I - 2P + P^2 = I - P$; $(I - P)^T = I - P^T = I - P$.
(b) $(P_1 + P_2)^2 = P_1^2 + P_1P_2 + P_2P_1 + P_2^2 = P_1 + P_2$; $(P_1 + P_2)^T = P_1^T + P_2^T = P_1 + P_2$.

3.2.7. $(2P - I)^2 = 4P^2 - 4P + I = I$.

3.2.8. $P^2 = uu^T uu^T = uu^T = P$ since $u^T u = 1$; uu^T is symmetric.

3.2.9. $P = \begin{bmatrix} 0 & 0 \\ 0 & 1 \end{bmatrix}$; then $P\begin{bmatrix} x \\ y \end{bmatrix} = \begin{bmatrix} 0 \\ y \end{bmatrix}$.

3.2.11. $y = -\frac{3}{10} - \frac{12}{5}t$.

3.2.12. $\begin{bmatrix} 1 & -1 & 1 \\ 1 & 0 & 0 \\ 1 & 1 & 1 \\ 1 & 2 & 4 \end{bmatrix} \begin{bmatrix} C \\ D \\ E \end{bmatrix} = \begin{bmatrix} 2 \\ 0 \\ -3 \\ -5 \end{bmatrix}$.

3.3.1. $\begin{bmatrix} 1 & -2 \\ 1 & -1 \\ 1 & 1 \\ 1 & 2 \end{bmatrix} \begin{bmatrix} C \\ D \end{bmatrix} = \begin{bmatrix} -4 \\ -3 \\ -1 \\ 0 \end{bmatrix}$.

$c = 2C$, $d = \sqrt{10}\,D$; $y = -2 + t$ is optimal and actually goes through all four points; $E^2 = 0$ and $b = p$ because b is in the column space.

3.3.2.   The separate projections are $2a_1$ and $2a_2$; $p = 2a_1 + 2a_2$.

3.3.3.   This projection is $-a_3$; the sum $2a_1 + 2a_2 - a_3 = (0, 3, 0)$ is b itself; the projection operator onto the whole space is the identity.

3.3.4. $(Q_1Q_2)^T(Q_1Q_2) = Q_2^T Q_1^T Q_1 Q_2 = Q_2^T I Q_2 = I$.

3.3.5. $Q^T Q = (I - 2uu^T)(I - 2uu^T) = I - 4uu^T + 4uu^T uu^T = I$ since $u^T u = 1$.

3.3.6.   The third column is $\pm(1, -2, 1)/\sqrt{6}$.

3.3.7. $\| v \|^2 = (x_1q_1 + \cdots + x_nq_n)^T(x_1q_1 + \cdots + x_nq_n)$
$= x_1^2 q_1^T q_1 + \cdots + x_n^2 q_n^T q_n = x_1^2 + \cdots + x_n^2$.

3.3.8. Q is orthogonal.

3.3.9. $q_1 = a_1$, $q_2 = a_2 - a_1$, $q_3 = a_3 - a_2$,

$$A = \begin{bmatrix} 0 & 0 & 1 \\ 0 & 1 & 1 \\ 1 & 1 & 1 \end{bmatrix} = \begin{bmatrix} 0 & 0 & 1 \\ 0 & 1 & 0 \\ 1 & 0 & 0 \end{bmatrix} \begin{bmatrix} 1 & 1 & 1 \\ 0 & 1 & 1 \\ 0 & 0 & 1 \end{bmatrix} = QR.$$

3.3.10. $\begin{bmatrix} 3 & 0 \\ 4 & 5 \end{bmatrix} = \begin{bmatrix} \frac{3}{5} & -\frac{4}{5} \\ \frac{4}{5} & \frac{3}{5} \end{bmatrix} \begin{bmatrix} 5 & 4 \\ 0 & 3 \end{bmatrix}$.

3.3.11. $\begin{bmatrix} 1 & 1 \\ 2 & 3 \\ 2 & 1 \end{bmatrix} = \begin{bmatrix} \frac{1}{3} & 0 \\ \frac{2}{3} & 1/\sqrt{2} \\ \frac{2}{3} & -1/\sqrt{2} \end{bmatrix} \begin{bmatrix} 3 & 3 \\ 0 & \sqrt{2} \end{bmatrix}$;

   A and Q are m by n, R is n by n.

3.3.12. $\bar{x} = R^{-1}Q^{\mathrm{T}}b = \begin{bmatrix} \frac{5}{9} \\ 0 \end{bmatrix}$.

3.3.13. Q has the same column space as A, so $P = Q(Q^{\mathrm{T}}Q)^{-1}Q^{\mathrm{T}} = QQ^{\mathrm{T}}$.

3.3.14. $v_3 = \left(c - \dfrac{v_1^{\mathrm{T}}c}{v_1^{\mathrm{T}}v_1} v_1 \right) - \dfrac{v_2^{\mathrm{T}}c}{v_2^{\mathrm{T}}v_2} v_2$ because $v_2^{\mathrm{T}}v_1 = 0$.

3.3.15. $\| v \|^2 = \frac{1}{2} + \frac{1}{4} + \cdots = 1$, $\| e^x \|^2 = \int e^{2x}\, dx = (e^2 - 1)/2$, $\int e^x e^{-x}\, dx = 1$.

3.3.16. $b_1 = \int y \sin x / \int \sin^2 x$; if $y = \cos x$, its Fourier sine coefficient b_1 is zero.

3.3.17. $a_0 = 1$, $a_1 = 0$, $b_1 = 2/\pi$.

3.3.18. x^3 is already orthogonal to the even functions 1 and $x^2 - \frac{1}{3}$; $\int_{-1}^{1} x^3 x\, dx = \frac{2}{5}$, so the next Legendre polynomial is $x^3 - \frac{2}{5}x$.

3.3.19. The horizontal line $y = \frac{1}{3}$.

3.4.1. If $A = 0$, then the row space is 0, $\bar{x} = 0$, and $A^+ = 0$.

3.4.2. $A = \begin{bmatrix} 1 \\ 1 \\ 1 \end{bmatrix} \begin{bmatrix} 1 & 1 & 1 \end{bmatrix} = \bar{L}\bar{U}$; $A^+ = \dfrac{1}{6}\begin{bmatrix} 1 & 1 \\ 1 & 1 \\ 1 & 1 \end{bmatrix}$,

   $\bar{x} = A^+b = \dfrac{1}{6}\begin{bmatrix} 1 \\ 1 \\ 1 \end{bmatrix}$, $A\bar{x} = \begin{bmatrix} \frac{1}{2} \\ \frac{1}{2} \end{bmatrix} = p$.

3.4.3. $A^+ = A^{\mathrm{T}}$ since the least squares solution to $Ax = b$ is $\bar{x} = A^{\mathrm{T}}b$.

3.4.4. $\bar{u} = \bar{v} = \frac{3}{2}$.

3.5.1. From (59), $\bar{x}_W = (4b_1 + b_2)/(4 + 1) = \frac{17}{5}$.

3.5.2. c is W-perpendicular to b if $(c_1\ c_2)W^{\mathrm{T}}Wb = 0$, or $4c_1 + c_2 = 0$; $\| b \|_W = \| Wb \| = \sqrt{5}$.

3.5.3. $\bar{x}_W = \frac{8}{13}$, the reverse of Fig. 3.9.

3.5.4. Any orthogonal matrix.

Chapter 4

4.2.1. $\det 2A = 2^n \det A$; $\det -A = (-1)^n \det A$; $\det A^2 = (\det A)^2$.

4.2.2. Rules (5) and (1) imply (2).

4.2.3. $\det A = 20$, $\det A = 5$.

4.2.4. $\det A^{\mathrm{T}} = \det A$, $\det(-A) = (-1)^3 \det A$, and therefore $\det A = -\det A$ or $\det A = 0$.

4.2.5.   (a) 0   (b) 16   (c) 16   (d) $\frac{1}{16}$ (e) 16

4.2.7. $\det Q^{\mathrm{T}} \det Q = 1$, or $(\det Q)^2 = 1$; a unit cube.

4.3.1.   columns 2, 1, 4, 3; an even permutation, so $|A| = 1$.

4.3.4. $|A_4| = -3$, $|A_3| = 2$, $|A_2| = -1$, $|A_n| = (-1)^n(1 - n)$.

4.3.5. $|A| = 20$, $\quad B = \begin{bmatrix} 20 & -10 & -12 \\ 0 & 5 & 0 \\ 0 & 0 & 4 \end{bmatrix}$, $\quad A^{-1} = B/20$.

4.4.1. $A^{-1} = \begin{bmatrix} 1 & -1 & 0 \\ 0 & 1 & -1 \\ 0 & 0 & 1 \end{bmatrix}$.

4.4.2. $x = 3$, $y = -1$, $z = -2$.

4.4.3. $J = r^2 \cos \phi$ ($\phi = 0$ is the equator!).

4.4.4.   The triangle $A'B'C'$ has the same area as ABC; it is just moved to the origin.

4.4.5.   All the edges are three times as long.

4.4.6.   The pivots are 2, $\frac{8}{2}$, $\frac{9}{8}$; no, A is singular.

4.4.7.   Starting from $n = 2$, the permutations are odd, odd, even, even, odd, odd, and so on. The problem for $n + 1$ is like the problem for n, except the number $n + 1$ has to be moved from one end to the other, requiring n additional exchanges. Therefore the parity stays the same when n is even and reverses when n is odd.

4.4.8.   Even.

4.4.9.   Dependent since their determinant is zero.

Chapter 5

5.1.1. $\lambda = 2$, $\lambda = 3$, trace $= 5$, determinant $= 6$.

5.1.2. $u = c_1 \begin{bmatrix} 1 \\ -1 \end{bmatrix} e^{2t} + c_2 \begin{bmatrix} 1 \\ -2 \end{bmatrix} e^{3t} = 6 \begin{bmatrix} 1 \\ -1 \end{bmatrix} e^{2t} - 6 \begin{bmatrix} 1 \\ -2 \end{bmatrix} e^{3t}$.

5.1.3. $\lambda = -5$, $\lambda = -4$; both reduced by 7, with unchanged eigenvectors.

5.1.4. $c_1 x_1 + c_2 x_2 + c_3 x_3 = \begin{bmatrix} 1 \\ 3 \\ 1 \end{bmatrix}$, $\quad$ or $\quad c_1 = \frac{5}{3}$, $c_2 = 0$, $c_3 = -\frac{2}{3}$.

5.1.5. $\begin{bmatrix}1\\0\\0\end{bmatrix}, \begin{bmatrix}1\\-1\\0\end{bmatrix}, \begin{bmatrix}1\\-2\\1\end{bmatrix}; \ \lambda = 3, 1, 2; \quad \begin{bmatrix}1\\0\\0\end{bmatrix}, \begin{bmatrix}0\\1\\0\end{bmatrix}, \begin{bmatrix}0\\0\\1\end{bmatrix}.$

5.1.7.   If $Ax = \lambda x$, then $(A - 7I)x = (\lambda - 7)x$; if $Ax = \lambda x$, then $x = \lambda A^{-1}x$ or $A^{-1}x = \lambda^{-1}x$.

5.1.8.   Choose $\lambda = 0$.

5.1.11. $\det(A - \lambda I) = \det(A^{\mathrm{T}} - \lambda I)$.

5.1.12. $\lambda_1 = 5, \lambda_2 = -5, x_1 = \begin{bmatrix}2\\1\end{bmatrix}, x_2 = \begin{bmatrix}1\\-2\end{bmatrix}$.

5.1.13. $\lambda_1 = 2, \lambda_2 = 0, \lambda_3 = -2$.

5.2.1.   The columns of S are x_1, x_2, x_3 ; $S^{-1}A\,S = \mathrm{diag}(0, 1, 3)$.

5.2.2. $A = \begin{bmatrix}-5 & 18\\-3 & 10\end{bmatrix}$.

5.2.3. $\lambda = 0, 0, 3$; the third column of S is a multiple of $\begin{bmatrix}1\\1\\1\end{bmatrix}$, and the other columns are orthogonal to it.

5.2.4. $S^{\mathrm{T}}A^{\mathrm{T}}(S^{-1})^{\mathrm{T}} = \Lambda^{\mathrm{T}}$, so $(S^{-1})^{\mathrm{T}}$ is the eigenvector matrix for A^{T}.

5.2.5. $AB = S\Lambda_1 S^{-1}S\Lambda_2 S^{-1} = S\Lambda_1\Lambda_2 S^{-1} = S\Lambda_2\Lambda_1 S^{-1} = S\Lambda_2 S^{-1}S\Lambda_1 S^{-1} = BA$; any $B = \begin{bmatrix}\alpha & \beta\\\beta & \alpha\end{bmatrix}$.

5.2.6. $R = 45° \text{ rotation} = \dfrac{1}{\sqrt{2}}\begin{bmatrix}1 & -1 & 0\\1 & 1 & 0\\0 & 0 & \sqrt{2}\end{bmatrix}$.

5.2.7. $\begin{bmatrix}1 & 1\\-1 & -1\end{bmatrix}$ has only one eigenvector.

5.3.1. $u_0 = \begin{bmatrix}3\\1\end{bmatrix}, c_1 = \dfrac{3 - \lambda_2}{\lambda_1 - \lambda_2}, c_2 = \dfrac{-3 + \lambda_1}{\lambda_1 - \lambda_2}, F_k = c_1\lambda_1{}^k + c_2\lambda_2{}^k; F_{k+1}/F_k \to \lambda_1$.

5.3.2. $\begin{bmatrix}G_{k+2}\\G_{k+1}\end{bmatrix} = \begin{bmatrix}\frac{1}{2} & \frac{1}{2}\\1 & 0\end{bmatrix}\begin{bmatrix}G_{k+1}\\G_k\end{bmatrix}, \quad \lambda_1 = 1, \ \lambda_2 = -\tfrac{1}{2},$

$\begin{bmatrix}G_{k+1}\\G_k\end{bmatrix} = S\Lambda^k S^{-1}\begin{bmatrix}G_1\\G_0\end{bmatrix} = \begin{bmatrix}1 & 1\\1 & -2\end{bmatrix}\begin{bmatrix}1^k & \\ & (-\frac{1}{2})^k\end{bmatrix}\begin{bmatrix}\frac{2}{3} & \frac{1}{3}\\\frac{1}{3} & -\frac{1}{3}\end{bmatrix}\begin{bmatrix}\frac{1}{2}\\0\end{bmatrix},$

$G_k = \tfrac{1}{3}[1 - (-\tfrac{1}{2})^k] \to \tfrac{1}{3}$.

5.3.3. $\begin{bmatrix}3000\\3000\\3000\end{bmatrix}, \begin{bmatrix}18000\\1500\\1000\end{bmatrix}, \begin{bmatrix}6000\\9000\\500\end{bmatrix}, \begin{bmatrix}3000\\3000\\3000\end{bmatrix}, \begin{bmatrix}18000\\1500\\1000\end{bmatrix}, \begin{bmatrix}6000\\9000\\500\end{bmatrix}, \begin{bmatrix}3000\\3000\\3000\end{bmatrix}.$

5.3.4. $A = \begin{bmatrix} \frac{1}{2} & 0 & \frac{1}{2} \\ 0 & \frac{1}{2} & \frac{1}{2} \\ \frac{1}{2} & \frac{1}{2} & 0 \end{bmatrix} \begin{matrix} \text{Boston} \\ \text{L.A.} \\ \text{Chicago} \end{matrix}$, $u_\infty = \begin{bmatrix} \frac{1}{3} \\ \frac{1}{3} \\ \frac{1}{3} \end{bmatrix}$.

5.3.5. In the steady state everyone is dead: $d_\infty = 1$, $s_\infty = 0$, $w_\infty = 0$.

5.3.6. $A = \begin{bmatrix} \frac{1}{2} & \frac{1}{4} & 0 \\ \frac{1}{2} & \frac{1}{2} & \frac{1}{2} \\ 0 & \frac{1}{4} & \frac{1}{2} \end{bmatrix}$, $u_\infty = \begin{bmatrix} \frac{1}{4} \\ \frac{1}{2} \\ \frac{1}{4} \end{bmatrix}$.

5.3.7. $A^{\mathrm{T}}A - I = \begin{bmatrix} -1 & 0 \\ 0 & a_{12}^2 - \frac{3}{4} \end{bmatrix}$;

the eigenvalues are on the diagonal, and are both negative if $a_{12}^2 < \frac{3}{4}$.

5.3.8. It is unstable for $|\alpha| > \frac{1}{2}$, and stable for $|\alpha| < \frac{1}{2}$.

5.3.9. If A is increased, then more goods are consumed in the production, and the net expansion must be slower.

5.3.10. $A^2 = \begin{bmatrix} 0 & 0 & 1 \\ 0 & 0 & 0 \\ 0 & 0 & 0 \end{bmatrix}$, $A^3 = 0$, $(I - A)^{-1} = I + A + A^2 = \begin{bmatrix} 1 & 1 & 2 \\ 0 & 1 & 1 \\ 0 & 0 & 1 \end{bmatrix}$.

5.4.1. $e^{At} = Se^{\Lambda t}S^{-1} = \begin{bmatrix} 1 & 1 \\ 1 & -1 \end{bmatrix} \begin{bmatrix} e^{-t} & \\ & e^{-3t} \end{bmatrix} \begin{bmatrix} \frac{1}{2} & \frac{1}{2} \\ \frac{1}{2} & -\frac{1}{2} \end{bmatrix}$

$= \frac{1}{2} \begin{bmatrix} e^{-t} + e^{-3t} & e^{-t} - e^{-3t} \\ e^{-t} - e^{-3t} & e^{-t} + e^{-3t} \end{bmatrix} > 0$.

5.4.2. The eigenvalues are $\lambda = 1$, $\lambda = 3$; $u = \frac{1}{2}\begin{bmatrix} 1 \\ 1 \end{bmatrix}e^t + \frac{1}{2}\begin{bmatrix} 1 \\ -1 \end{bmatrix}e^{3t} \to \infty$.

5.4.3. $\lambda_1 = 0$, $\lambda_2 = -2$, $x_1 = \begin{bmatrix} 1 \\ 1 \end{bmatrix}$, $x_2 = \begin{bmatrix} 1 \\ -1 \end{bmatrix}$, $u(t) = c_1 x_1 + c_2 x_2 e^{-2t}$;

$u_0 = \begin{bmatrix} 3 \\ 1 \end{bmatrix}$ gives $u = \begin{bmatrix} 2 + e^{-2t} \\ 2 - e^{-2t} \end{bmatrix} \to \begin{bmatrix} 2 \\ 2 \end{bmatrix}$ as $t \to \infty$.

5.4.4. The change gives $dv/dt = \Lambda v$, $v = e^{\Lambda t}v_0 = e^{\Lambda t}S^{-1}u_0$, or $u = Se^{\Lambda t}S^{-1}u_0$ as before.

5.4.5. $e^{tA}e^{sA} = (Se^{t\Lambda}S^{-1})(Se^{s\Lambda}S^{-1}) = Se^{t\Lambda}e^{s\Lambda}S^{-1} = Se^{(t+s)\Lambda}S^{-1} = e^{(t+s)A}$.

5.4.6. $\dfrac{d}{dt}(x_1y_1 + \cdots) = \dfrac{dx_1}{dt}y_1 + x_1\dfrac{dy_1}{dt} + \cdots = \left(\dfrac{dx}{dt}\right)^{\mathrm{T}} y + x^{\mathrm{T}}\left(\dfrac{dy}{dt}\right)$.

5.4.7. A has imaginary eigenvalues and is neutrally stable.

5.4.8. A has $\lambda_1 = \lambda_2 = -1$, giving decay at $t = \infty$; $A^{\mathrm{T}} + A$ has $\lambda_1 = 1$, $\lambda_2 = -5$, so there may not be immediate decay at $t = 0$.

5.4.9. The system is unstable.

5.4.10. $u = \begin{bmatrix} 1 \\ 0 \\ -1 \end{bmatrix} \cos t + \begin{bmatrix} 1 \\ -2 \\ 1 \end{bmatrix} \dfrac{\sin\sqrt{3}\,t}{\sqrt{3}}.$

5.4.11. $u = \dfrac{1}{2}\begin{bmatrix} 1 \\ 1 \end{bmatrix} \cos\sqrt{6}\,t + \dfrac{1}{2}\begin{bmatrix} 1 \\ -1 \end{bmatrix} \cos 2t.$

5.4.12. $\rho^2 e^{\rho t} x = -\rho F e^{\rho t} x + A e^{\rho t} x$, or $(A - \rho F - \rho^2 I)x = 0$.

5.4.14. $u = \begin{bmatrix} 1 & 1 \\ 1 & -1 \end{bmatrix}\begin{bmatrix} \sin t & 0 \\ 0 & (\sin\sqrt{3}\,t)/\sqrt{3} \end{bmatrix}\begin{bmatrix} \frac{1}{2} & \frac{1}{2} \\ \frac{1}{2} & -\frac{1}{2} \end{bmatrix}\begin{bmatrix} 1 \\ 0 \end{bmatrix}$

$= \dfrac{1}{2}\begin{bmatrix} \sin t + (\sin\sqrt{3}\,t)/\sqrt{3} \\ \sin t - (\sin\sqrt{3}\,t)/\sqrt{3} \end{bmatrix}.$

5.5.1. The whole exercise is outside the unit circle.

5.5.2. It is real; it is also on the unit circle; it is also on the unit circle; it is on or inside the circle of radius two.

5.5.4. $\| x \| = 6 = \| y \|$, $x^H y = -12$.

5.5.5. $C = \begin{bmatrix} 1 & -i \\ -i & 0 \\ 0 & 1 \end{bmatrix}\begin{bmatrix} 1 & i & 0 \\ i & 0 & 1 \end{bmatrix} = \begin{bmatrix} 2 & i & -i \\ -i & 1 & 0 \\ i & 0 & 1 \end{bmatrix}$, $C = C^H$ because

$(A^H A)^H = A^H A^{HH} = A^H A$.

5.5.6. $\begin{bmatrix} 1 & i & 0 \\ i & 0 & 1 \end{bmatrix} \rightarrow \begin{bmatrix} 1 & i & 0 \\ 0 & 1 & 1 \end{bmatrix} = U$; $Ax = 0$ if x is a multiple of $\begin{bmatrix} i \\ -1 \\ 1 \end{bmatrix}$; this vector

is orthogonal not to the columns of A^T (rows of A) but to the columns of A^H.

5.5.7. $A = \begin{bmatrix} 0 & 1 \\ 0 & 0 \end{bmatrix}$, $x = \begin{bmatrix} 1 \\ i \end{bmatrix}$, $x^H A x = i$.

5.5.8. $\lambda_1 = 1$, $\lambda_2 = -1$, $x_1 = \begin{bmatrix} 1 \\ -i \end{bmatrix}$, $x_2 = \begin{bmatrix} 1 \\ i \end{bmatrix}$, $A = x_1 x_1^H - x_2 x_2^H$; the product $x_1^H x_2$ is zero because the eigenvectors are orthogonal (Property 3).

5.5.9. $U = \begin{bmatrix} 1/\sqrt{3} & 1/\sqrt{2} & 1/\sqrt{6} \\ 1/\sqrt{3} & 0 & -2/\sqrt{6} \\ 1/\sqrt{3} & -1/\sqrt{2} & 1/\sqrt{6} \end{bmatrix}$, $A = 0 + \begin{bmatrix} \frac{1}{2} & 0 & -\frac{1}{2} \\ 0 & 0 & 0 \\ -\frac{1}{2} & 0 & \frac{1}{2} \end{bmatrix} + 3\begin{bmatrix} \frac{1}{6} & -\frac{2}{6} & \frac{1}{6} \\ -\frac{2}{6} & \frac{4}{6} & -\frac{2}{6} \\ \frac{1}{6} & -\frac{2}{6} & \frac{1}{6} \end{bmatrix}.$

5.5.10. The determinant is the product of the eigenvalues.

5.5.12. 1 is not an eigenvalue because it is not imaginary.

5.5.13. $K = \begin{bmatrix} 0 & 1 \\ -1 & 0 \end{bmatrix}$, $x = \begin{bmatrix} 4 \\ 5 \end{bmatrix}$, $x^H K x = 0$ (it must be imaginary by 1′, and real because K and x are real, therefore zero).

5.5.14. $K = (Z - Z^H)/2 = -K^H$; then $Z = A + K$,

$$\begin{bmatrix} 3+i & 4+2i \\ 0 & 5 \end{bmatrix} = \begin{bmatrix} 3 & 2+i \\ 2-i & 5 \end{bmatrix} + \begin{bmatrix} i & i \\ i & 0 \end{bmatrix}.$$

5.5.15. $(UV)^H (UV) = V^H U^H UV = V^H I V = I$.

5.5.16. The determinant is the product of the eigenvalues, and the modulus of $\lambda_1 \cdots \lambda_n$ is $|\lambda_1| \cdots |\lambda_n| = 1$; we could have the 1 by 1 $U = [i]$;

$$U = \begin{bmatrix} e^{i\theta} & 0 \\ 0 & e^{i\psi} \end{bmatrix}$$

for any angles θ and ψ.

5.5.17. The last column is $(1, -2, 1)/\sqrt{6}$, multiplied by any number (real or complex) which has modulus one.

5.5.18. $\Lambda = \begin{bmatrix} 0 & \\ & 2i \end{bmatrix}$, $S = \begin{bmatrix} 1 & 1 \\ -1 & 1 \end{bmatrix}$, $e^{Kt} = Se^{\Lambda t}S^{-1} = \frac{1}{2}\begin{bmatrix} 1+e^{2it} & -1+e^{2it} \\ -1+e^{2it} & 1+e^{2it} \end{bmatrix}$;

at $t = 0$, $\dfrac{de^{Kt}}{dt} = \begin{bmatrix} i & i \\ i & i \end{bmatrix} = K$.

5.5.19. A has $+1$ or -1 in each diagonal entry; eight possibilities.

5.6.1. $C = (M')^{-1}BM' = (M')^{-1}M^{-1}AMM'$, or $C = (MM')^{-1}A(MM')$. The only matrix similar to I is $M^{-1}IM = I$.

5.6.2. The (3, 1) entry is $g\cos\theta + h\sin\theta$, which is zero if $\tan\theta = -g/h$.

5.6.3. Take M to equal A; then $M^{-1}(AB)M = BA$ is similar to AB. If the eigenvalues are the same, so is their sum the trace. If AB and BA have the same trace, then $AB - BA$ has trace zero and cannot equal I.

5.6.4. $\lambda_1 = -1$ with unit eigenvector $(1, 1)^T/\sqrt{2}$; this goes in the first column of U, and the second column $(-1, 1)^T/\sqrt{2}$ makes U unitary; then $T = U^{-1}AU = \begin{bmatrix} -1 & -7 \\ 0 & 2 \end{bmatrix}$. There are other possible T, with 7 instead of -7 and/or with the 2 and -1 reversed.

5.6.5. $T = \begin{bmatrix} 0 & \pm 2 \\ 0 & 0 \end{bmatrix}$.

5.6.6. $p(T) = \begin{bmatrix} 0 & * & * \\ 0 & * & * \\ 0 & 0 & * \end{bmatrix}\begin{bmatrix} * & * & * \\ 0 & 0 & * \\ 0 & 0 & * \end{bmatrix}\begin{bmatrix} * & * & * \\ 0 & * & * \\ 0 & 0 & 0 \end{bmatrix} = 0$, so $P(A) = 0$.

5.6.7. $U = \begin{bmatrix} 1/\sqrt{2} & 1/\sqrt{6} & 1/\sqrt{3} \\ 0 & -2/\sqrt{6} & 1/\sqrt{3} \\ -1/\sqrt{2} & 1/\sqrt{6} & 1/\sqrt{3} \end{bmatrix};$

the first two columns could also be $x_1 = (1, -1, 0)^T/\sqrt{2}$, $x_2 = (1, 1, -2)^T/\sqrt{6}$ (or many other possibilities); always

$$x_1 x_1^H + x_2 x_2^H = I - x_3 x_3^H = \frac{1}{3}\begin{bmatrix} 2 & -1 & -1 \\ -1 & 2 & -1 \\ -1 & -1 & 2 \end{bmatrix}.$$

5.6.8.   (i) $TT^H = U^{-1}AUU^H A^H (U^{-1})^H = I$; (ii) if T is triangular and unitary, then its diagonal entries (the eigenvalues) must have modulus one. Then if the columns are to be unit vectors, all off-diagonal entries must be zero.

5.6.9. $A = \begin{bmatrix} 1 & i \\ i & 1 \end{bmatrix}$.

5.6.11.   If $N = UAU^{-1}$, then $NN^H = (U\Lambda U^{-1})(U\Lambda U^{-1})^H = U\Lambda\Lambda^H U^H = U\Lambda^H\Lambda U^H = (U\Lambda U^{-1})^H (U\Lambda U^{-1}) = N^H N$.

5.6.12. $BM = MJ_2$ holds as long as $m_{11} = m_{32}$, $m_{21} = m_{31} = m_{33} = 0$.

5.6.13. $M^{-1}J_3M = 0$, so the last two inequalities are easy; $MJ_1 = J_2M$ forces the first column of M to be zero, so it cannot be invertible.

5.6.14. $J_1 = 4$ by 4 block; $J_2 = 3$ by 3 and 1 by 1; $J_3 = 2$ by 2 and 2 by 2; $J_4 = 2$ by 2, 1 by 1, 1 by 1; $J_5 =$ four 1 by 1 blocks $= 0$. Both J_2 and J_3 have two eigenvectors.

Chapter 6

6.1.1. $ac - b^2 = 2 - 4 = -2$; $f = (\sqrt{2}x + \sqrt{2}y)^2 - y^2$.

6.1.2.   No, no, no, yes (indefinite, positive semidefinite, indefinite, positive definite).

6.1.4.   Second-derivative matrices $\begin{bmatrix} 4 & -5 \\ -5 & 12 \end{bmatrix}$ (minimum), $\begin{bmatrix} -2 & 0 \\ 0 & -1 \end{bmatrix}$ (maximum).

6.2.1.   Semidefinite, semidefinite, positive definite: $x^TBx = (u + w)^2 + v^2$.

6.2.2. $\det A = 1 + 2\alpha^2\beta - \beta^2 - 2\alpha^2 < 0$ if $\alpha = -\beta = -\frac{2}{3}$.

6.2.3.   If A has positive eigenvalues λ_i, then the eigenvalues of A^2 are λ_i^2 and the eigenvalues of A^{-1} are $1/\lambda_i$, also all positive; use condition II.

6.2.4.   If $x^TAx > 0$ and $x^TBx > 0$, then $x^T(A + B)x > 0$; condition I.

6.2.5. $A = \begin{bmatrix} 1 & 0 \\ -\frac{1}{2} & 1 \end{bmatrix}\begin{bmatrix} 2 & \\ & \frac{3}{2} \end{bmatrix}\begin{bmatrix} 1 & -\frac{1}{2} \\ 0 & 1 \end{bmatrix} = \begin{bmatrix} 1/\sqrt{2} & 1/\sqrt{2} \\ 1/\sqrt{2} & -1/\sqrt{2} \end{bmatrix}\begin{bmatrix} 1 & \\ & 3 \end{bmatrix}\begin{bmatrix} 1/\sqrt{2} & 1/\sqrt{2} \\ 1/\sqrt{2} & -1/\sqrt{2} \end{bmatrix},$

so $W = \sqrt{D}L^T = \begin{bmatrix} \sqrt{2} & -1/\sqrt{2} \\ 0 & \sqrt{3}/\sqrt{2} \end{bmatrix}$ or $W = \sqrt{\Lambda}Q^T = \begin{bmatrix} 1/\sqrt{2} & 1/\sqrt{2} \\ \sqrt{3}/\sqrt{2} & -\sqrt{3}/\sqrt{2} \end{bmatrix}.$

6.2.6. $R = \dfrac{1}{2}\begin{bmatrix} 1+\sqrt{3} & 1-\sqrt{3} \\ 1-\sqrt{3} & 1+\sqrt{3} \end{bmatrix}$, $R^T = Q(\sqrt{\Lambda})^T Q^T = R$, $R^2 = Q\sqrt{\Lambda}\,Q^T Q\sqrt{\Lambda}\,Q^T =$

$Q\Lambda Q^T = A$ (because $Q^T Q = I$).

6.2.8. $x^H A x = x^H(\alpha + i\beta)x$, and if the real part of the left side is positive, then so is the real part $x^H \alpha x$; thus $\alpha > 0$. On the other hand, $A = \begin{bmatrix} 1 & 3 \\ 0 & 1 \end{bmatrix}$ has positive eigenvalues but $A^H + A$ is not positive definite.

6.2.9. A is negative definite if $x^T A x < 0$ for all nonzero x; all $\lambda_i < 0$; the determinants of the A_k alternate in sign (not det $A_k < 0!$); negative pivots; $A = -W^T W$.

6.2.10. $\lambda_1 = 1$, $\lambda_2 = 4$, semiaxes go through $(\pm 1, 0)$ and $(0, \pm\tfrac{1}{2})$.

6.2.11. $A = \begin{bmatrix} 3 & -\sqrt{2} \\ -\sqrt{2} & 2 \end{bmatrix}$ with eigenvalues 1 and 4.

6.2.12.   One zero eigenvalue pulls the ellipsoid into an infinite cylinder $\lambda_1 y_1^2 + \lambda_2 y_2^2 = 1$ along the third axis; two zero eigenvalues leave only the two planes $y_1 = \pm 1/\sqrt{\lambda_1}$; three zero eigenvalues leave $0 = 1$ (no graph).

6.3.1. $W = \sqrt{D}\,L^T = \dfrac{1}{\sqrt{2}}\begin{bmatrix} 2 & -1 & -1 \\ 0 & \sqrt{3} & -\sqrt{3} \\ 0 & 0 & 0 \end{bmatrix}$;

$x^T A x = \left(\sqrt{2}u - \dfrac{\sqrt{2}}{2}v - \dfrac{\sqrt{2}}{2}w\right)^2 + \left(\dfrac{\sqrt{3}}{\sqrt{2}}v - \dfrac{\sqrt{3}}{\sqrt{2}}w\right)^2$; the surface is a cylinder

(see   6.2.12).

6.3.3. $C^T A C = \begin{bmatrix} 4 & -2 \\ -2 & 1 \end{bmatrix}$ has a positive and a zero eigenvalue, as A has;

$C(t) = \begin{bmatrix} 2-t & 0 \\ 0 & -1 \end{bmatrix}$.

6.3.4.   The last pivot is negative for A, but all pivots are positive for $A + 2I$.

6.3.5. $\begin{bmatrix} \sqrt{3}-1 & 1 \end{bmatrix}\begin{bmatrix} 1 & 0 \\ 0 & 2 \end{bmatrix}\begin{bmatrix} 1+\sqrt{3} \\ -1 \end{bmatrix} = 0$; $u_0 = \begin{bmatrix} -1 \\ 1 \end{bmatrix}$, $a_1 = \tfrac{1}{2}$,

$a_2 = -\tfrac{1}{2}$; the smaller mass reaches as far as $\sqrt{3}$ although the larger mass never exceeds its initial displacement 1.

6.3.6. $\lambda_1 = 54$, $\lambda_2 = \tfrac{54}{5}$, $x_1 = \begin{bmatrix} 1 \\ -1 \end{bmatrix}$, $x_2 = \begin{bmatrix} 1 \\ 1 \end{bmatrix}$.

6.3.7. $A = \begin{bmatrix} 1 & 0 \\ 0 & -1 \end{bmatrix}$, $B = \begin{bmatrix} 0 & 1 \\ 1 & 0 \end{bmatrix}$, $|A - \lambda B| = -\lambda^2 - 1$, $\lambda = \pm i$.

6.4.1. $P = x_1^2 - x_1 x_2 + x_2^2 - x_2 x_3 + x_3^2 - 4x_1 - 4x_3$;

$\partial P/\partial x_1 = 2x_1 - x_2 - 4$, $\partial P/\partial x_2 = -x_1 + 2x_2 - x_3$,

$\partial P/\partial x_3 = -x_2 + 2x_3 - 4$.

6.4.2. $\partial P_1/\partial x = x + y = 0$, $\partial P_1/\partial y = x + 2y - 3 = 0$, $x = -3$, $y = 3$. P_2 has no minimum (let $y \to \infty$) and corresponds to the indefinite matrix $\left[\begin{smallmatrix} -1 & 0 \\ 0 & 0 \end{smallmatrix}\right]$.

6.4.3. Minimizing Q leads to the normal equations $A^T A x = A^T b$.

6.4.4. $R(x) = a_{11}$, so that $\lambda_1 = \min R(x) \le a_{11}$; $x = e_i$ gives $R(x) = a_{ii} \ge \lambda_1$.

6.4.5. The minimum value is $R(x) = 1$, with $x = (1, 1)$.

6.4.6. Since $x^T B x > 0$ for all nonzero x, $x^T(A + B)x$ will be larger than $x^T A x$.

6.4.7. If $(A + B)x = \theta_1 x$, then

$$\lambda_1 + \mu_1 \le \frac{x^T A x}{x^T x} + \frac{x^T B x}{x^T x} = \theta_1 .$$

6.4.8. The maximum is $\frac{3}{2}$.

6.4.9. Since $\lambda_1 = \lambda_2 = 1$, μ_1 must equal one; football (or rugby).

6.4.10. When two rows and columns are removed, $\lambda_{\min} \le$ original λ_3.

6.4.11. $\mu_2 = 3$, $2 < 3 < 2 + \sqrt{2}$.

6.4.12. The extreme S_j is spanned by the first j eigenvectors.

6.4.13. The maximum is over the subspace S_j orthogonal to the z's.

6.4.14. Take $x = (1, 0, \ldots, 0)^T$; then $\lambda_1 \le R(x) = a_{11}/b_{11}$.

6.5.1. $b_j = h$ (the area under V_4 in Fig. 6.6); $Ay = b$ becomes

$$\frac{-y_{j-1} + 2y_j - y_{j+1}}{h}$$

$$= \frac{-(x - h) + (x - h)^2 + 2x - 2x^2 - (x + h) + (x + h)^2}{2h} = h.$$

6.5.2. $P(y) - P(b) = \frac{1}{2}y^T y - y^T b + \frac{1}{2}b^T b = \frac{1}{2}\| y - b \|^2$; minimizing $P(y)$ over the trial functions will also minimize $P(y) - P(b)$, and therefore the distance to b.

6.5.3. $\Lambda_1 = \frac{54}{5} > \lambda_1 = \pi^2$.

6.5.4. The minimax principle says that λ_2 is the minimum value for $\lambda_{\max}$ on all two-dimensional subspaces; Λ_2 raises this minimum because its two-dimensional subspaces are restricted to lie within the trial space.

Chapter 7

7.2.1. If Q is orthogonal, then $\| Q \| = \max \| Qx \|/\| x \| = 1$ because Q preserves length: $\| Qx \| = \| x \|$ for every x. Also Q^{-1} is orthogonal and has norm one, so $c(Q) = 1$.

7.2.2. The triangle inequality for vectors gives $\| Ax + Bx \| \le \| Ax \| + \| Bx \|$, and when we divide by $\| x \|$ and maximize each term, the result is the triangle inequality for matrix norms.

7.2.3. $\| ABx \| \le \| A \| \, \| Bx \|$, by the definition of the norm of A, and then $\| Bx \| \le \| B \| \, \| x \|$. Dividing by $\| x \|$ and maximizing, $\| AB \| \le \| A \| \, \| B \|$. The same is true for the inverse, $\| B^{-1}A^{-1} \| \le \| B^{-1} \| \, \| A^{-1} \|$; then $c(AB) \le c(A)c(B)$ by multiplying these two inequalities.

7.2.4. $\| A^{-1} \| = 1$, $\| A \| = 3$, $c(A) = 3$; take $b = x_2 = \left[\begin{smallmatrix} 1 \\ -1 \end{smallmatrix} \right]$, $\delta b = x_1 = \left[\begin{smallmatrix} 1 \\ 1 \end{smallmatrix} \right]$.

7.2.5. In the definition $\| A \| = \max \| Ax \| / \| x \|$, choose the vector x to be the particular eigenvector in question; then $\| Ax \| = | \lambda | \, \| x \|$, and the ratio is $| \lambda |$.

7.2.6. $A^T A = \left[\begin{smallmatrix} 1 & 100 \\ 100 & 10001 \end{smallmatrix} \right]$, $\lambda^2 - 10002\lambda + 1 = 0$, $\lambda_{\max} = 5001 + (5001^2 - 1)^{1/2}$. The norm is the square root, and is the same as $\| A^{-1} \|$.

7.2.7. $A^T A$ and $A A^T$ have the same eigenvalues (even if A were singular, which is a limiting case of the exercise), and equality of the largest eigenvalues gives $\| A \| = \| A^T \|$.

7.2.8. Since $A = W^T W$ and $A^{-1} = W^{-1}(W^T)^{-1}$, we have $\| A \| = \| W \|^2$ and $\| A^{-1} \| = \|(W^T)^{-1} \|^2 = \| W^{-1} \|^2$. (From the previous exercise, the transpose has the same norm.) Therefore $c(A) = (c(W))^2$.

7.2.9.
$$\begin{bmatrix} .01 & 1 \\ 1 & 0 \end{bmatrix} = \begin{bmatrix} 1 & 0 \\ 100 & 1 \end{bmatrix} \begin{bmatrix} .01 & 1 \\ 0 & -100 \end{bmatrix};$$

$c(A)$ is close to 1, whereas $c(L)$ and $c(U)$ are close to 10^4.

7.2.10. If $x = \left[\begin{smallmatrix} 1 \\ -1 \end{smallmatrix} \right]$, then $Ax = \left[\begin{smallmatrix} -1 \\ 7 \end{smallmatrix} \right]$ and $\| Ax \|_\infty / \| x \|_\infty = 7$; this is the extreme case, and $\| A \|_\infty = 7$.

7.3.1. $u_0 = \begin{bmatrix} 1 \\ 0 \end{bmatrix}$, $u_1 = \begin{bmatrix} 2 \\ -1 \end{bmatrix}$, $u_2 = \begin{bmatrix} 5 \\ -4 \end{bmatrix}$, $u_3 = \begin{bmatrix} 14 \\ -13 \end{bmatrix}$; $u_\infty = \begin{bmatrix} 1 \\ -1 \end{bmatrix}$.

7.3.2. $u_0 = \begin{bmatrix} 3 \\ 4 \end{bmatrix}$, $u_1 = \frac{1}{3}\begin{bmatrix} 10 \\ 11 \end{bmatrix}$, $u_2 = \frac{1}{9}\begin{bmatrix} 31 \\ 32 \end{bmatrix}$, $u_3 = \frac{1}{27}\begin{bmatrix} 94 \\ 95 \end{bmatrix}$;

with the shift, $\alpha = \dfrac{26}{25}$ and $u_1 = \dfrac{25}{49}\begin{bmatrix} 24 & 25 \\ 25 & 24 \end{bmatrix}\begin{bmatrix} 3 \\ 4 \end{bmatrix} \approx \begin{bmatrix} 172 \\ 171 \end{bmatrix}$.

7.3.3. $Hx = x - (x - y)\dfrac{2(x - y)^T x}{(x - y)^T(x - y)} = x - (x - y) = y$.

To prove $Hy = x$, either reverse x and y or else take $H(Hx) = Hy$ and use $H^2 = I$.

7.3.4. $\sigma = 5$, $v = \begin{bmatrix} 8 \\ 4 \end{bmatrix}$, $H = \dfrac{1}{5}\begin{bmatrix} -3 & -4 \\ -4 & 3 \end{bmatrix}$.

7.3.5. $U = \begin{bmatrix} 1 & 0 & 0 \\ 0 & -\frac{3}{5} & -\frac{4}{5} \\ 0 & -\frac{4}{5} & \frac{3}{5} \end{bmatrix} = U^{-1}, \quad U^{-1}AU = \begin{bmatrix} 1 & -5 & 0 \\ -5 & \frac{9}{25} & \frac{12}{25} \\ 0 & \frac{12}{25} & \frac{16}{25} \end{bmatrix}.$

7.3.6. $\begin{bmatrix} 2 & -1 \\ -1 & 2 \end{bmatrix} = Q_0R_0 = \frac{1}{\sqrt{5}}\begin{bmatrix} 2 & 1 \\ -1 & 2 \end{bmatrix}\frac{1}{\sqrt{5}}\begin{bmatrix} 5 & -4 \\ 0 & 3 \end{bmatrix},$

$A_1 = R_0Q_0 = \frac{1}{5}\begin{bmatrix} 14 & -3 \\ -3 & 6 \end{bmatrix}.$

7.3.7. $\begin{bmatrix} \cos\theta & \sin\theta \\ \sin\theta & 0 \end{bmatrix} = QR = \begin{bmatrix} \cos\theta & -\sin\theta \\ \sin\theta & \cos\theta \end{bmatrix}\begin{bmatrix} 1 & \cos\theta\sin\theta \\ 0 & -\sin^2\theta \end{bmatrix},$

$RQ = \begin{bmatrix} \cos\theta(1+\sin^2\theta) & -\sin^3\theta \\ -\sin^3\theta & -\sin^2\theta\cos\theta \end{bmatrix}.$

7.3.8. A is orthogonal, so $Q = A$, $R = I$, and $RQ = A$ again.

7.3.9. Assume that $(Q_0 \cdots Q_{k-1})(R_{k-1} \cdots R_0)$ is the QR factorization of A^k, which is certainly true if $k = 1$. By construction $A_{k+1} = R_kQ_k$, or

$$R_k = A_{k+1}Q_k^T = Q_k^T \cdots Q_0^T A Q_0 \cdots Q_kQ_k^T.$$

Postmultiplying by $(R_{k-1} \cdots R_0)$, and using the assumption, we have $R_k \cdots R_0 = Q_k^T \cdots Q_0^T A^{k+1}$; after moving the Q's to the left side, this is the required result for A^{k+1}.

7.4.1. $D^{-1}(-L - U) = \begin{bmatrix} 0 & \frac{1}{2} & 0 \\ \frac{1}{2} & 0 & \frac{1}{2} \\ 0 & \frac{1}{2} & 0 \end{bmatrix}, \mu = 0, \pm\frac{1}{\sqrt{2}};$

$(D+L)^{-1}(-U) = \begin{bmatrix} 1 & \frac{1}{2} & 0 \\ 0 & \frac{1}{4} & \frac{1}{2} \\ 0 & \frac{1}{8} & \frac{1}{4} \end{bmatrix},$

eigenvalues $0, 0, \frac{1}{2}$; $\omega_{opt} = 4 - 2\sqrt{2}$, with $\lambda_{max} = 3 - 2\sqrt{2} \approx .2$.

7.4.2. J has entries $\frac{1}{2}$ along the diagonals adjacent to the main diagonal, and zeros elsewhere;

$Jx_1 = \frac{1}{2}(\sin 2\pi h, \sin 3\pi h + \sin \pi h, \ldots) = (\cos \pi h)x_1.$

7.4.3. $Jx_k = \frac{1}{2}(\sin 2k\pi h, \sin 3k\pi h + \sin k\pi h, \ldots) = (\cos k\pi h)x_k.$

7.4.4. The circle around a_{ii} cannot reach zero if its radius r_i is less than $|a_{ii}|$; therefore zero is not an eigenvalue, and a diagonally dominant matrix cannot be singular.

7.4.5. $J = -\begin{bmatrix} 0 & \frac{1}{3} & \frac{1}{3} \\ 0 & 0 & \frac{1}{4} \\ \frac{2}{5} & \frac{2}{5} & 0 \end{bmatrix}$;

the radii are $r_1 = \frac{2}{3}$, $r_2 = \frac{1}{4}$, $r_3 = \frac{4}{5}$, the circles have centers at zero, so all $|\lambda_i| < 1$.

Chapter 8

8.1.1. The corners are at $(0, 6)$, $(2, 2)$, $(6, 0)$; see Fig. 8.4.

8.1.2. $x + y$ is minimized at $(2, 2)$, with cost 4; $3x + y$ is minimized at $(0, 6)$, with cost 6; the minimum of $x - y$ is $-\infty$, with $x = 0$, $y \to \infty$.

8.1.3. There are two equations $x + 2y - u = 6$, $2x + y - v = 6$, and four inequalities $x \geq 0$, $y \geq 0$, $u \geq 0$, $v \geq 0$. The three original corners have $(x, y, u, v) = (0, 6, 6, 0)$, $(2, 2, 0, 0)$, $(6, 0, 0, 6)$.

8.1.4. $A = [1 \quad 2 \quad -1]$, $b = [4]$, $c = [2 \quad 3 \quad 0]^\mathrm{T}$.

8.1.5. The constraints give $3(2x + 5y) + 2(-3x + 8y) \leq 9 - 10$, or $31y \leq -1$, which contradicts $y \geq 0$.

8.1.6. Take x and y to be equal and very large.

8.1.7. $x \geq 0$, $y \geq 0$, $x + y \leq 0$ admits only the point $(0, 0)$.

8.1.8. The feasible set is an equilateral triangle lying on the plane $x + y + z = 1$, with corners at $(x, y, z) = (1, 0, 0)$, $(0, 1, 0)$, $(0, 0, 1)$; the last corner gives the maximum value 3.

8.2.1. $r = [1 \quad 1]$, so the corner is optimal.

8.2.2. $r = [3 \quad -1]$, so the second column of F enters the basis; that column is $u = \begin{bmatrix} 0 \\ -1 \end{bmatrix}$, and $B^{-1}u = \begin{bmatrix} -2 \\ -1 \end{bmatrix}$ is negative, so the edge is infinitely long and the minimal cost is $-\infty$.

8.2.3. At P, $r = [-5 \quad 3]$; at Q, $r = [\frac{5}{3} \quad -\frac{1}{3}]$; at R, $r \geq 0$.

8.2.5. (a) The pair $x = 0$, $w = b$ is nonnegative, it satisfies $Ax + w = b$, and it is basic because $x = 0$ contributes n zero components.
(b) The auxiliary problem minimizes w_1, subject to $x_1 \geq 0$, $x_2 \geq 0$, $w_1 \geq 0$, $x_1 - x_2 + w_1 = 3$. Its Phase I vector is $x_1 = x_2 = 0$, $w_1 = 3$; its optimal vector is $x_1^* = 3$, $x_2^* = w_1^* = 0$. The corner is at $x_1 = 3$, $x_2 = 0$, and the feasible set is a line going up from this point with slope 1.

8.2.6. The stopping test becomes $r \leq 0$; if this fails, and the ith component is the largest, then that column of F enters the basis; the rule 8C for the vector leaving the basis is the same.

8.2.7. $\begin{bmatrix} B & F & b \\ c_B^\mathrm{T} & c_F^\mathrm{T} & 0 \end{bmatrix} = \begin{bmatrix} -1 & 2 & 1 & 0 & 6 \\ 0 & 1 & 2 & -1 & 6 \\ 0 & -1 & 1 & 0 & 0 \end{bmatrix} \rightarrow \begin{bmatrix} 1 & 0 & 3 & -2 & 6 \\ 0 & 1 & 2 & -1 & 6 \\ 0 & 0 & 3 & -1 & 6 \end{bmatrix}$;

$r = [3 \quad -1]$, but the column above -1 is negative.

8.2.9. $BE = B[\cdots v \cdots] = [\cdots u \cdots]$ since $Bv = u$.

8.3.1. Maximize $4y_1 + 11y_2$, with $y_1 \geq 0$, $y_2 \geq 0$, $2y_1 + y_2 \leq 1$, $3y_2 \leq 1$; $x_1^* = 2$, $x_2^* = 3$, $y_1^* = \frac{1}{3}$, $y_2^* = \frac{1}{3}$, cost = 5.

8.3.2. Minimize $3x_1$, subject to $x_1 \geq 0$, $x_1 \geq 1$; $y_1^* = 0$, $y_2^* = 3$, $x_1^* = 1$, cost = 3.

8.3.3. The dual maximizes $y^T b$, with $Iy \geq c$; the vectors $y = c$ and $x = b$ are feasible, and give the same value $c^T b$, so by 8F they must be optimal.

8.3.4. $A = [-1]$, $b = [1]$, $c = [0]$ is unfeasible; the dual maximizes y, with $y \geq 0$ and $-1y \leq 0$, and is unbounded.

8.3.5. $b = \left[\begin{smallmatrix} 0 \\ 1 \end{smallmatrix}\right]$, $c = \left[\begin{smallmatrix} -1 \\ 0 \end{smallmatrix}\right]$.

8.3.6. If x is very large, then $Ax \geq b$ and $x \geq 0$; if $y = 0$, then $A^T y \leq c$ and $y \geq 0$. Thus both are feasible.

8.3.7. Since $c^T x = 3 = y^T b$, x and y are optimal by 8F.

8.3.8. $Ax = [1 \ 1 \ 3 \ 1]^T \geq b = [1 \ 1 \ 1 \ 1]^T$, with strict inequality in the third component; therefore the third component of y is forced to be zero. Similarly $A^T y = [1 \ 1 \ 1 \ 1]^T \leq c = [1 \ 1 \ 1 \ 3]^T$, and the strict inequality forces $x_4 = 0$.

8.3.9. $x^* = \left[\begin{smallmatrix} 1 \\ 0 \end{smallmatrix}\right] = y^*$, with $(y^*)^T b = 1 = c^T x^*$; the second components of $Ax^* \geq b$ and $A^T y^* \leq c$ are both strict inequalities, producing zero second components in y^* and x^*.

8.3.10. $Ax = b$ leads to $y^T Ax = y^T b$, whether or not $y \geq 0$. $A^T y \leq c$ (or $y^T A \leq c^T$) leads as before to $y^T Ax \leq c^T x$, but only because $x \geq 0$. Comparing, $y^T b \leq c^T x$.

8.3.11. The columns generate the cone between the positive x axis and the ray $x = y$. In the first case $x = (1, 2)^T$; $y = (1, 1)^T$ satisfies the alternative.

8.3.12. The columns of $\begin{bmatrix} 1 & 0 & 0 & -1 & 0 & 0 \\ 0 & 1 & 0 & 0 & -1 & 0 \\ 0 & 0 & 1 & 0 & 0 & -1 \end{bmatrix}$.

8.3.13. Take $y = \left[\begin{smallmatrix} 2 \\ -1 \end{smallmatrix}\right]$; then $A^T y = 0$, $y^T b \neq 0$.

8.3.14. Take $y = \left[\begin{smallmatrix} 1 \\ -1 \end{smallmatrix}\right]$; then $A^T y \geq 0$, $y^T b < 0$.

8.3.15. $A^T y \geq 0$ leads to $y^T A \geq 0$, and thus $y^T Ax \geq 0$; on the other hand $Ax \geq b$ leads to $y^T Ax \leq y^T b < 0$.

8.4.1. $-10x_1 + 70(1 - x_1) = 10x_1 - 10(1 - x_1)$, or $x_1 = \frac{4}{5}$, $x_2 = \frac{1}{5}$; $-10y_1 + 10(1 - y_1) = 70y_1 - 10(1 - y_1)$, or $y_1 = \frac{1}{5}$, $y_2 = \frac{4}{5}$; average payoff 6.

8.4.2. If Y sticks to y_3, he cannot lose more than \$1; if he includes y_1 or y_2, X can win more than \$1; therefore $y^* = (0, 0, 1)$.

8.4.3. X can guarantee to win at least 3 by choosing always his second strategy; Y can guarantee to lose at most 3 by choosing the first column. Since this is an equilibrium, $x^* = (0, 1)$ and $y^* = (1, 0)^T$.

8.4.4. The entry $a_{21} = 3$ was the smallest in its row and largest in its column; if X moves from this row he wins less; and if Y moves, he loses more.

8.4.5.   If $x = (\frac{6}{11}, \frac{3}{11}, \frac{2}{11})$, then X will win $\frac{6}{11}$ against any strategy of Y; if $y = (\frac{6}{11}, \frac{3}{11}, \frac{2}{11})$, then Y loses $\frac{6}{11}$ against any strategy of X; this equilibrium solves the game.

8.4.6.   In (18), $\min x^*Ay \leq x^*Ay^*$ because the minimum over all y is not larger than the value for the particular y^*; similarly for $x^*Ay^* \leq \max xAy^*$. If equality holds in (18), so that $\min x^*Ay = x^*Ay^*$, then for all y this is less or equal to x^*Ay; that is the second half of (17), and the first half follows from $\max xAy^* = x^*Ay^*$.

8.4.7. $x^*A = [\frac{1}{2} \ \frac{1}{2}]$ and $x^*Ay = \frac{1}{2}y_1 + \frac{1}{2}y_2$, which equals $\frac{1}{2}$ for all strategies of Y; $Ay^* = [\frac{1}{2} \ \frac{1}{2} \ -1 \ -1]^T$ and $xAy^* = \frac{1}{2}x_1 + \frac{1}{2}x_2 - x_3 - x_4$, which cannot exceed $\frac{1}{2}$; in between is $x^*Ay^* = \frac{1}{2}$.

Appendix A

A.1. $[A] = \begin{bmatrix} 0 & 0 & 2 & 0 & 0 \\ 0 & 0 & 0 & 6 & 0 \\ 0 & 0 & 0 & 0 & 12 \end{bmatrix}$.

A.2. $[A][A]^T = I$, but $[A]^T[A]$ has a zero in the 1,1 entry. The first component of v is annihilated by the left shift, just like the constant in differentiation.

A.3.   Transposing leaves v_1 and v_4, and reverses v_2 and v_3 :

$[A] = \begin{bmatrix} 1 & 0 & 0 & 0 \\ 0 & 0 & 1 & 0 \\ 0 & 1 & 0 & 0 \\ 0 & 0 & 0 & 1 \end{bmatrix}$.

$[A]^2 = I$ because the transpose of the transpose is the original matrix.

A.4. $[A] = \begin{bmatrix} 1 & 0 & 0 & 0 \\ 0 & 1 & 0 & 0 \\ 2 & 0 & 0 & 0 \\ 0 & 2 & 0 & 0 \end{bmatrix}$.

A.5.   Any solution is a polynomial of degree $n - 1$, so $S_n = P_{n-1}$, with basis $1, t, \ldots, t^{n-1}$.

A.6. $[A] = [\begin{smallmatrix} 1 & 0 \\ 0 & -1 \end{smallmatrix}]$, $[B] = [\begin{smallmatrix} -1 & 0 \\ 0 & -1 \end{smallmatrix}]$, $[BA] = [\begin{smallmatrix} -1 & 0 \\ 0 & 1 \end{smallmatrix}]$ = mirror image in the y axis.

A.7. $[A] = [\begin{smallmatrix} 0 & 0 \\ 0 & 1 \end{smallmatrix}]$; all eigenvalues of a projection are 0 or 1.

A.8. $[A] = \begin{bmatrix} 1 & 2 \\ 3 & 0 \end{bmatrix} \begin{bmatrix} 4 & 0 \\ 0 & 0 \end{bmatrix} \begin{bmatrix} 1 & 2 \\ 3 & 0 \end{bmatrix}^{-1} = \begin{bmatrix} 0 & \frac{4}{3} \\ 0 & 4 \end{bmatrix}$.

Appendix B

B.1. $J = \begin{bmatrix} 2 & 0 \\ 0 & 0 \end{bmatrix}$, $J = \begin{bmatrix} 0 & 1 & 0 \\ 0 & 0 & 0 \\ 0 & 0 & 0 \end{bmatrix}$.

B.2. $\dfrac{du_2}{dt} = 8e^{8t}(tx_1 + x_2) + e^{8t}x_1$,

$Au_2 = e^{8t}(tAx_1 + Ax_2) = e^{8t}(8tx_1 + 8x_2 + x_1)$.

B.3. $e^{Bt} = \begin{bmatrix} 1 & t & 2t \\ 0 & 1 & 0 \\ 0 & 0 & 0 \end{bmatrix} = I + Bt$ since $B^2 = 0$.

INDEX

8
D 9
E 0
F 1
G 2
H 3
I 4
J 5